Quantum Probability

and Infinite Dimensional Analysis

Proceedings of the 29th Conference

QP–PQ: Quantum Probability and White Noise Analysis*

QP–PQ: Quantum Probability and White Noise Analysis

Vol. 26: Quantum Bio-Informatics III
From Quantum Information to Bio-Informatics
eds. L. Accardi, W. Freudenberg and M. Ohya

Vol. 25: Quantum Probability and Infinite Dimensional Analysis
Proceedings of the 29th Conference
eds. H. Ouerdiane and A. Barhoumi

Vol. 24: Quantum Bio-Informatics II
From Quantum Information to Bio-informatics
eds. L. Accardi, W. Freudenberg and M. Ohya

Vol. 23: Quantum Probability and Related Topics
eds. J. C. García, R. Quezada and S. B. Sontz

Vol. 22: Infinite Dimensional Stochastic Analysis
eds. A. N. Sengupta and P. Sundar

Vol. 21: Quantum Bio-Informatics
From Quantum Information to Bio-Informatics
eds. L. Accardi, W. Freudenberg and M. Ohya

Vol. 20: Quantum Probability and Infinite Dimensional Analysis
eds. L. Accardi, W. Freudenberg and M. Schürmann

Vol. 19: Quantum Information and Computing
eds. L. Accardi, M. Ohya and N. Watanabe

Vol. 18: Quantum Probability and Infinite-Dimensional Analysis
From Foundations to Applications
eds. M. Schürmann and U. Franz

Vol. 17: Fundamental Aspects of Quantum Physics
eds. L. Accardi and S. Tasaki

Vol. 16: Non-Commutativity, Infinite-Dimensionality, and Probability at the Crossroads
eds. N. Obata, T. Matsui and A. Hora

Vol. 15: Quantum Probability and Infinite-Dimensional Analysis
ed. W. Freudenberg

Vol. 14: Quantum Interacting Particle Systems
eds. L. Accardi and F. Fagnola

Vol. 13: Foundations of Probability and Physics
ed. A. Khrennikov

*For the complete list of the published titles in this series, please visit:
www.worldscibooks.com/series/qqpwna_series.shtml

QP–PQ
Quantum Probability and White Noise Analysis
Volume XXV

Quantum Probability

and Infinite Dimensional Analysis

Proceedings of the 29th Conference

Hammamet, Tunisia 13 – 18 October 2008

Editors

H Ouerdiane
University of Tunis El Manar, Tunisia

A Barhoumi
University of Sousse, Tunisia

World Scientific

NEW JERSEY • LONDON • SINGAPORE • BEIJING • SHANGHAI • HONG KONG • TAIPEI • CHENNAI

Published by

World Scientific Publishing Co. Pte. Ltd.

5 Toh Tuck Link, Singapore 596224

USA office: 27 Warren Street, Suite 401-402, Hackensack, NJ 07601

UK office: 57 Shelton Street, Covent Garden, London WC2H 9HE

Library of Congress Cataloging-in-Publication Data

Conference on Quantum Probability and Infinite Dimensional Analysis (29th : 2008 : Hammamet, Tunisia)

Quantum probability and infinite dimensional analysis : proceedings of the 29th conference, Hammamet, Tunisia 13-18 October 2008 / edited by H. Ouerdiane & A. Barhoumi.

p. cm. -- (QP-PQ, quantum probability and white noise analysis ; v. 25)

Includes bibliographical references and index.

ISBN-13: 978-981-4295-42-0 (hardcover : alk. paper)

ISBN-10: 981-4295-42-6 (hardcover : alk. paper)

1. Stochastic analysis--Congresses. 2. Probabilities--Congresses. 3. Mathematical physics--Congresses. I. Ouerdiane, Habib. II. Barhoumi, A. (Abdessatar) III. Title.

QA274.2.C66 2008

519.2'2--dc22

2009048003

British Library Cataloguing-in-Publication Data

A catalogue record for this book is available from the British Library.

Printed in Singapore by World Scientific Printers

PREFACE

This volume constitutes the proceedings of the 29th Conference on Quantum Probability and Related Topics held in Hammamet, Tunisia, October 13-18, 2008. This was the first time this important international event took place in the Africa. This Proceedings volume contains twenty refereed research articles. Almost 120 mathematicians and physical mathematicians participated in this conference from all over the world. This volume will be of interest to professional researchers and graduate students who will gain a perspective on current activity and background in the topics covered.

The goal of the conference was to communicate new results in the fields of quantum probability and infinite dimensional analysis, to bring together scientists with different backgrounds who work in related fields, and to stimulate new collaborations. The fact that contributions to this volume range from quantum probability, white noise and stochastic analysis, orthogonal polynomials and interacting Fock spaces, free probability and random matrices, quantum information, quantum statistics, control and filtering, Lévy processes, mathematical models in biology and foundation of quantum mechanics shows that research in quantum probability is very active and strongly involved in modern mathematical developments and applications.

The organizers of the conference were extremely happy to see many eminent mathematicians having contributed to the success of the conference and cultivated new ideas. We would like to thank all participants for their hard work and for helping to create a very stimulating atmosphere. To our great pleasure, important papers presented at the conference are published in this volume. As such, we are grateful to the respective authors and to the anonymous referees for their efficient help in evaluating each paper in this collection. We also express our deep appreciation to World Scientific Publishing, in particular to Ms E. H. Chionh as their acting representative for a remarkably pleasant cooperation.

We acknowledge gratefully the general support of the Tunisian Ministry of Higher Education Scientific Research and Technology, the Association of Infinite Dimensional Analysis and Quantum Probability (AQPIDA), the

Tunisian Mathematical Society, the University of Tunis El Manar and other Tunisian Research Institutes.

Finally we hope that this conference has contributed not only to promote scientific activities but also to mutual international understanding.

The editors:
Abdessatar Barhoumi
(Sousse University, Tunisia)
Habib Ouerdiane
(Tunis El-Manar University, Tunisia)

Hammamet, Tunisia
November 2009

CONTENTS

ON THE CENTRAL EXTENSIONS OF THE HEISENBERG ALGEBRA

Luigi ACCARDI

Centro Vito Volterra, Università di Roma Tor Vergata
via Columbia 2, 00133 Roma, Italy
E-mail:accardi@volterra.mat.uniroma2.it
URL:http://volterra.mat.uniroma2.it

Andreas BOUKAS

Department of Mathematics and Natural Sciences, American College of Greece
Aghia Paraskevi, Athens 15342, Greece
E-mail:andreasboukas@acgmail.gr

We describe the nontrivial central extensions $CE(Heis)$ of the Heisenberg algebra and their representation as sub–algebras of the Schroedinger algebra. We also present the characteristic and moment generating functions of the random variable corresponding to the self-adjoint sum of the generators of $CE(Heis)$.

Keywords: Heisenberg algebra; Schroedinger algebra; Central extension of a Lie algebra; Fock space.

1. Central extensions of Lie algebras

In the applications of Lie algebras to physical systems the symmetries of the system are frequently described at the level of classical mechanics by some Lie algebra L, and in the quantum theoretic description by L plus some extra, constant, not arbitrary terms which are interpreted as the eigenvalues of some new operators K^i which have constant eigenvalue on any irreducible module of L (by Schur's lemma the K^i must commute with all elements of L). The new generators K^i extend L to a new Lie algebra $\hat{L}$.

In general, given a Lie algebra L with basis $\{T^a \; ; \; a = 1, 2, ..., d\}$, by attaching additional generators $\{K^i \; ; \; i = 1, 2, .., l\}$ such that

$$[K^i, K^j] = [T^a, K^j] = 0 \tag{1}$$

we obtain an l–dimensional central extension $\hat{L}$ of L with Lie brackets

$$[T^a, T^b] = \sum_{c=1}^{d} f_c^{ab}\, T^c + \sum_{i=1}^{l} g_i^{ab}\, K^i \tag{2}$$

where f_c^{ab} are the structure constants of L in the basis $\{T^a \ ; \ a = 1, 2, ..., d\}$. If through a constant redefinition of the generators $\{T^a \ ; \ a = 1, 2, ..., d\}$ (i.e. if $\hat{L}$ is the direct sum of L and an Abelian algebra) the commutation relations of $\hat{L}$ reduce to those of L then the central extension is trivial.

A basis independent (or cocycle) definition of an one-dimensional (i.e. having only one central generator) central extension can be given as follows:

If L and $\hat{L}$ are two complex Lie algebras, we say that $\hat{L}$ is an one-dimensional central extension of L with central element E if

$$[l_1, l_2]_{\hat{L}} = [l_1, l_2]_L + \phi(l_1, l_2)\, E \ ; \ [l_1, E]_{\hat{L}} = 0 \tag{3}$$

for all $l_1, l_2 \in L$, where $[\cdot,\cdot]_{\hat{L}}$ and $[\cdot,\cdot]_L$ are the Lie brackets in $\hat{L}$ and L respectively, and $\phi : L \times L \mapsto C$ is a bilinear form (2-cocycle) on L satisfying the skew-symmetry condition

$$\phi(l_1, l_2) = -\phi(l_2, l_1) \tag{4}$$

and the Jacobi identity

$$\phi([l_1, l_2]_L, l_3) + \phi([l_2, l_3]_L, l_1) + \phi([l_3, l_1]_L, l_2) = 0 \tag{5}$$

A central extension is trivial if there exists a linear function $f : L \mapsto \mathbb{C}$ satisfying for all $l_1, l_2 \in L$

$$\phi(l_1, l_2) = f([l_1, l_2]_L) \tag{6}$$

For more information on central extensions we refer to Ref. 3. Detailed proofs of the material presented in sections 2–5 below will appear in Ref. 1.

2. Central extensions of the Heisenberg algebra

The Heisenberg $*$-Lie algebra $Heis$ is the 3-dimensional Lie algebra with generators $\{a^\dagger, a, h\}$, commutation relations

$$[a, a^\dagger]_{Heis} = h \;\; ; \;\; [a, h^\dagger]_{Heis} = [h, a]_{Heis} = 0 \tag{7}$$

and involution

$$(a^\dagger)^* = a \;\; ; \;\; (a)^* = a^\dagger \;\; ; \;\; (h)^* = h \tag{8}$$

All 2-cocycles ϕ corresponding to a central extension $CE(Heis)$ of $Heis$ are of the form

$$\phi(a, a^\dagger) = \lambda \tag{9}$$
$$\phi(h, a^\dagger) = z \tag{10}$$
$$\phi(a, h) = \bar{z} \tag{11}$$
$$\phi(h, h) = \phi(a^\dagger, a^\dagger) = \phi(a, a) = 0 \tag{12}$$

where $\lambda \in \mathbb{R}$ and $z \in \mathbb{C}$. To see that, let $l_i = a_i\, a^\dagger + b_i\, a + c_i\, h$ where $a_i, b_i, c_i \in \mathbb{C}$ for all $i \in \{1, 2, 3\}$, be three elements of $Heis$. Then

$$[l_1, l_2]_{Heis} = (b_1\, a_2 - a_1\, b_2)\, h \tag{13}$$
$$[l_2, l_3]_{Heis} = (b_2\, a_3 - a_2\, b_3)\, h \tag{14}$$
$$[l_3, l_1]_{Heis} = (b_3\, a_1 - a_3\, b_1)\, h \tag{15}$$

and

$$\phi([l_1, l_2]_{Heis}, l_3) = (b_1\, a_2\, a_3 - a_1\, b_2\, a_3)\, \phi(h, a^\dagger) + (b_1\, a_2\, b_3 - a_1\, b_2\, b_3)\, \phi(h, a) \tag{16}$$
$$\phi([l_2, l_3]_{Heis}, l_1) = (b_2\, a_3\, a_1 - a_2\, b_3\, a_1)\, \phi(h, a^\dagger) + (b_2\, a_3\, b_1 - a_2\, b_3\, b_1)\, \phi(h, a) \tag{17}$$
$$\phi([l_3, l_1]_{Heis}, l_2) = (b_3\, a_1\, a_2 - a_3\, b_1\, a_2)\, \phi(h, a^\dagger) + (b_3\, a_1\, b_2 - a_3\, b_1\, b_2)\, \phi(h, a) \tag{18}$$

and the Jacobi identity (5) for ϕ reduces to

$$0 \cdot \phi(h\,, a^\dagger\,) + 0 \cdot \phi(h\,, a\,) = 0 \tag{19}$$

which implies that $\phi(h\,, a^\dagger\,)$ and $\phi(h\,, a\,)$ are arbitrary complex numbers. Since it does not appear in (19), $\phi(a\,, a^\dagger\,)$ is also an arbitrary complex number. Therefore the, non–zero among generators, $CE(Heis)$ commutation relations have the form

$$[a\,, a^\dagger\,]_{CE(Heis)} = h + \phi(a\,, a^\dagger\,)\,E \tag{20}$$
$$[a\,, h\,]_{CE(Heis)} = \phi(a\,, h\,)\,E \tag{21}$$
$$[h\,, a^\dagger\,]_{CE(Heis)} = \phi(h\,, a^\dagger)\,E \tag{22}$$

where E is the, non–zero, central element. By skew-symmetry

$$\phi(a^\dagger\,, h\,) = -\phi(h\,, a^\dagger\,)\ ;\ \phi(a^\dagger\,, a\,) = -\phi(a\,, a^\dagger\,)\ ;\ \phi(a\,, h\,) = -\phi(h\,, a\,) \tag{23}$$

and

$$\phi(a\,, a\,) = \phi(a^\dagger\,, a^\dagger\,) = \phi(h\,, h\,) = 0 \tag{24}$$

By taking the adjoints of (20)-(22), assuming the involution conditions

$$(a^\dagger\,)^* = a \quad ; \quad (a)^* = a^\dagger \quad ; \quad (h\,)^* = h \ ; \ (E\,)^* = E \tag{25}$$

we find that

$$\phi(a\,, a^\dagger\,) = \overline{\phi(a\,, a^\dagger\,)} = \lambda \in \mathbb{R} \tag{26}$$

and

$$\phi(a\,, h) = \overline{\phi(h\,, a^\dagger)} = \bar{z} \tag{27}$$

where

$$z = \phi(h\,, a^\dagger) \in \mathbb{C} \tag{28}$$

If a central extension $CE(Heis)$ of $Heis$ is trivial then there exists a linear complex-valued function f defined on $Heis$ such that

$$f([a\,,a^\dagger\,]_{Heis}) = \lambda \tag{29}$$

$$f([a\,,h\,]_{Heis}) = \bar{z} \tag{30}$$

$$f([h\,,a^\dagger\,]_{Heis}) = z \tag{31}$$

Since $[h\,,a^\dagger\,]_{Heis} = 0$ and (for a linear f) $f(0) = 0$, by (31) we conclude that $z = 0$.

Conversely, suppose that $z = 0$. Define a linear complex-valued function f on $Heis$ by

$$f\left(z_1\,h\,+z_2\,a^\dagger\,+z_3\,a\right) = z_1\,\lambda \tag{32}$$

where λ is as above and $z_1, z_2, z_3 \in \mathbb{C}$. Then

$$f([a\,,a^\dagger\,]_{Heis}) = f(1\,h\,+0\,a^\dagger\,+0\,a\,) = \lambda = \phi(a\,,a^\dagger\,) \tag{33}$$

$$f([a\,,h\,]_{Heis}) = f(0\,h\,+0\,a^\dagger\,+0\,a\,) = 0 = \bar{z} = \phi(a\,,h) \tag{34}$$

$$f([h\,,a^\dagger\,]_{Heis}) = f(0\,h\,+0\,a^\dagger\,+0\,a\,) = 0 = z = \phi(h\,,a^\dagger\,) \tag{35}$$

which, by (6), implies that the central extension is trivial.

Thus, a central extension of $Heis$ is trivial if and only if $z = 0$.

The centrally extended Heisenberg commutation relations (20)-(22) now have the form

$$[a, a^\dagger]_{CE(Heis)} = h + \lambda\,E \;\;;\;\; [h, a^\dagger\,]_{CE(Heis)} = z\,E \;\;;\;\; [a, h]_{Heis} = \bar{z}\,E \tag{36}$$

Renaming $h + \lambda\,E$ to just h we obtain the equivalent (canonical) $CE(Heis)$ commutation relations

$$[a, a^\dagger]_{CE(Heis)} = h\;;\; [h, a^\dagger]_{CE(Heis)} = z\,E\;;\; [a, h]_{CE(Heis)} = \bar{z}\,E \tag{37}$$

For $z = 0$ we recover the Heisenberg commutation relations (7). Commutation relations (37) define a nilpotent (thus solvable) four–dimensional $*$–Lie algebra $CE(Heis)$ with generators $a, a^\dagger, h$ and E. Moreover, if we define p, q and H by

$$a^\dagger = p + i\,q \qquad ; \qquad a = p - i\,q \qquad ; \qquad H = -ih/2 \tag{38}$$

then p, q, E are self-adjoint, H is skew-adjoint, and $\{p, q, E, H\}$ are the generators of a real four-dimensional solvable $*$–Lie algebra with central element E and commutation relations

$$[p, q] = H \ ; \ [q, H] = c\,E \ ; \ [H, p] = b\,E \tag{39}$$

where b, c are (not simultaneously zero) real numbers given by

$$c = \frac{Re\,z}{2} \ , \ b = \frac{Im\,z}{2} \tag{40}$$

Conversely, if p, q, H, E are the generators (with p, q, E self-adjoint and H skew-adjoint) of a real four-dimensional solvable $*$–Lie algebra with central element E and commutation relations (39) with $b, c \in \mathbb{R}$ not simultaneously zero, then, defining z by (40), the operators $a, a^\dagger, h$ defined by (38) and E are the generators of the nontrivial central extension $CE(Heis)$ of the Heisenberg algebra defined by (37) and (25).

The real four–dimensional solvable Lie algebra generated by $\{p, q, E, H\}$ can be identified to the Lie algebra η_4 (one of the fifteen classified real four–dimensional solvable Lie algebras, see for example Ref. 4) with generators e_1, e_2, e_3, e_4 and (non-zero) commutation relations among generators

$$[e_4, e_1] = e_2 \qquad ; \qquad [e_4, e_2] = e_3 \tag{41}$$

This algebra has been studied by Feinsilver and Schott in Ref. 2.

3. Representations of $CE(Heis)$

The non–trivial central extensions of $CE(Heis)$ (corresponding to $z \neq 0$) can be realized as proper sub–algebras of the Schroedinger algebra, i.e. the six–dimensional $*$–Lie algebra generated by b, $b^\dagger$, b^2, ${b^\dagger}^2$, $b^\dagger\, b$ and 1 where $b^\dagger$, b and 1 are the generators of a Boson Heisenberg algebra with

$$[b, b^\dagger] = 1 \qquad ; \qquad (b^\dagger)^* = b \tag{42}$$

More precisely,

(i) If $z \in \mathbb{C}$ with $Re\,z \neq 0$, then for arbitrary $\rho, r \in \mathbb{R}$ with $r \neq 0$, letting

$$a = \left(\frac{4\,\rho\,Im\,z - r^2}{4\,Re\,z} + i\,\rho\right)(b - b^\dagger)^2 - \frac{i\,\bar{z}}{2\,r}\,(b + b^\dagger) \tag{43}$$

$$a^\dagger = \left(\frac{4\,\rho\,Im\,z - r^2}{4\,Re\,z} - i\,\rho\right)(b - b^\dagger)^2 + \frac{i\,z}{2\,r}\,(b + b^\dagger) \tag{44}$$

and

$$h = i\,r\,(b^\dagger - b) \tag{45}$$

we find that the quadruple $\{a^+, a, h, E = 1\}$ satisfies commutation relations (37) and duality relations (25) of $CE(Heis)$.

(ii) If $z \in \mathbb{C}$ with $Re\,z = 0$, then for arbitrary $\rho, r \in \mathbb{R}$ with $r \neq 0$, letting

$$a = \left(\rho - \frac{i\,Im\,z}{16\,r^2}\right)(b - b^\dagger)^2 + r\,(b + b^\dagger) \tag{46}$$

$$a^\dagger = \left(\rho + \frac{i\,Im\,z}{16\,r^2}\right)(b - b^\dagger)^2 + r\,(b + b^\dagger) \tag{47}$$

and

$$h = \frac{i\,Im\,z}{2\,r}\,(b^\dagger - b) \tag{48}$$

we find that the quadruple $\{a^+, a, h, E = 1\}$ satisfies commutation relations (37) and duality relations (25) of $CE(Heis)$.

Using the fact that for non-negative integers n, k

$$b^{\dagger^n}\, b^k\, y(\xi) = \xi^k\, \frac{\partial^n}{\partial\, \epsilon^n}|_{\epsilon=0}\, y(\xi + \epsilon) \tag{49}$$

where, for $\xi \in \mathbb{C}$, $y(\xi) = e^{\xi\, b}$ we may represent $CE(Heis)$ on the Heisenberg Fock space $\mathcal{F}$ defined as the Hilbert space completion of the linear span of the exponential vectors $\{y(\xi)\,;\, \xi \in \mathbb{C}\}$ with respect to the inner product

$$\langle y(\xi), y(\mu)\rangle = e^{\bar{\xi}\,\mu} \tag{50}$$

We have that:

(i) If $z \in \mathbb{C}$ with $Rez \neq 0$ then

$$a\,y(\xi)=\left(\left(\tfrac{4\,\rho\,Im\,z-r^2}{4\,Re\,z}+i\,\rho\right)(\xi^2-1)-\tfrac{i\,\bar{z}}{2\,r}\,\xi\right)\,y(\xi) \tag{51}$$
$$+\left(\left(\tfrac{4\,\rho\,Im\,z-r^2}{4\,Re\,z}+i\,\rho\right)\,\tfrac{\partial^2}{\partial\,\epsilon^2}|_{\epsilon=0}-\left(\left(\tfrac{4\,\rho\,Im\,z-r^2}{4\,Re\,z}+i\,\rho\right)2\,\xi+\tfrac{i\,\bar{z}}{2\,r}\right)\,\tfrac{\partial}{\partial\,\epsilon}|_{\epsilon=0}\right)\,y(\xi+\epsilon)$$

$$a^\dagger\,y(\xi)=\left(\left(\tfrac{4\,\rho\,Im\,z-r^2}{4\,Re\,z}-i\,\rho\right)(\xi^2-1)+\tfrac{i\,z}{2\,r}\,\xi\right)\,y(\xi) \tag{52}$$
$$+\left(\left(\tfrac{4\,\rho\,Im\,z-r^2}{4\,Re\,z}-i\,\rho\right)\,\tfrac{\partial^2}{\partial\,\epsilon^2}|_{\epsilon=0}-\left(\left(\tfrac{4\,\rho\,Im\,z-r^2}{4\,Re\,z}-i\,\rho\right)2\,\xi-\tfrac{i\,z}{2\,r}\right)\,\tfrac{\partial}{\partial\,\epsilon}|_{\epsilon=0}\right)\,y(\xi+\epsilon)$$

$$h\,y(\xi)=i\,r\,\left(\frac{\partial}{\partial\,\epsilon}|_{\epsilon=0}\,y(\xi+\epsilon)-\xi\,y(\xi)\right) \tag{53}$$

and

$$E\,y(\xi)=y(\xi) \tag{54}$$

(ii) If $z\in\mathbb{C}$ with $Re\,z=0$ then

$$a\,y(\xi)=\left(\left(\rho-\tfrac{i\,Im\,z}{16\,r^2}\right)(\xi^2-1)+r\,\xi\right)\,y(\xi) \tag{55}$$
$$+\left(\left(\rho-\tfrac{i\,Im\,z}{16\,r^2}\right)\,\tfrac{\partial^2}{\partial\,\epsilon^2}|_{\epsilon=0}+\left(r-\left(\rho-\tfrac{i\,Im\,z}{16\,r^2}\right)2\,\xi\right)\,\tfrac{\partial}{\partial\,\epsilon}|_{\epsilon=0}\right)\,y(\xi+\epsilon)$$

$$a^\dagger\,y(\xi)=\left(\left(\rho+\tfrac{i\,Im\,z}{16\,r^2}\right)(\xi^2-1)+r\,\xi\right)\,y(\xi) \tag{56}$$
$$+\left(\left(\rho+\tfrac{i\,Im\,z}{16\,r^2}\right)\,\tfrac{\partial^2}{\partial\,\epsilon^2}|_{\epsilon=0}+\left(r-\left(\rho+\tfrac{i\,Im\,z}{16\,r^2}\right)2\,\xi\right)\,\tfrac{\partial}{\partial\,\epsilon}|_{\epsilon=0}\right)\,y(\xi+\epsilon)$$

$$h\,y(\xi)=\frac{i\,Im\,z}{2\,r}\,\left(\frac{\partial}{\partial\,\epsilon}|_{\epsilon=0}\,y(\xi+\epsilon)-\xi\,y(\xi)\right) \tag{57}$$

and

$$E\,y(\xi)=y(\xi) \tag{58}$$

4. Random variables associated with $CE(Heis)$

Self-adjoint operators X on the Heisenberg Fock space $\mathcal{F}$ correspond to classical random variables with moment generating function $\langle \Phi, e^{s\,X}\,\Phi\rangle$ and characteristic function $\langle \Phi, e^{i\,s\,X}\,\Phi\rangle$, where $s \in \mathbb{R}$ and Φ is the Heisenberg Fock space cyclic vacuum vector such that $b\,\Phi = 0$.

Using the splitting (or disentanglement) formula

$$e^{s\,(L\,b^2+L\,b^{\dagger\,2}-2\,L\,b^\dagger\,b-L+M\,b+N\,b^\dagger)}\,\Phi = e^{w_1(s)\,b^{\dagger\,2}}\,e^{w_2(s)\,b^\dagger}\,e^{w_3(s)}\,\Phi \tag{59}$$

where $L \in \mathbb{R}$, $M, N \in \mathbb{C}$, $s \in \mathbb{R}$,

$$w_1(s) = \frac{L\,s}{2\,L\,s+1} \tag{60}$$

$$w_2(s) = \frac{L\,(M+N)\,s^2+N\,s}{2\,L\,s+1} \tag{61}$$

and

$$w_3(s) = \frac{(M+N)^2\,(L^2\,s^4+2\,L\,s^3)+3\,M\,N\,s^2}{6\,(2\,L\,s+1)} - \frac{\ln(2\,L\,s+1)}{2} \tag{62}$$

we find that the moment generating function MGF_X of the self-adjoint operator

$$X = a + a^\dagger + h \tag{63}$$

where $a, a^\dagger, h$ are three of the generators of $CE(Heis)$, is

$$MGF_X(s) = \langle \Phi, e^{s\,(a+a^\dagger+h)}\,\Phi\rangle = (2\,L\,s+1)^{-1/2}\,e^{\frac{(M+N)^2\,(L^2\,s^4+2\,L\,s^3)+3\,M\,N\,s^2}{6\,(2\,L\,s+1)}} \tag{64}$$

where $s \in \mathbb{R}$ is such that $2\,L\,s+1>0$.

Similarly, the characteristic function of X is

$$CF_X(s) = \langle \Phi, e^{i\,s\,(a+a^\dagger+h)}\,\Phi\rangle = (2\,i\,L\,s+1)^{-1/2}\,e^{\frac{(M+N)^2\,(L^2\,s^4-2\,i\,L\,s^3)-3\,M\,N\,s^2}{6\,(2\,i\,L\,s+1)}} \tag{65}$$

In both MGF_X and CF_X, in the notation of section 3:

(i) if $Re\,z \neq 0$ then

$$L = \frac{4\,\rho\,Im\,z - r^2}{2\,Re\,z} \tag{66}$$

$$M = -\left(\frac{Im\,z}{r} + i\,r\right) \tag{67}$$

$$N = -\left(\frac{Im\,z}{r} - i\,r\right) \tag{68}$$

(ii) if $Re\,z = 0$ then

$$L = 2\,\rho \tag{69}$$

$$M = 2\,r - i\,\frac{Im\,z}{2\,r} \tag{70}$$

$$N = 2\,r + i\,\frac{Im\,z}{2\,r} \tag{71}$$

Notice that, if $L = 0$ (corresponding to $\rho\,Im\,z > 0$ and $r^2 = 4\,\rho\,Im\,z$ in the case when $Re\,z \neq 0$ and to $\rho = 0$ in the case when $Re\,z = 0$) then

$$MGF_X(s) = e^{\frac{M\,N\,s^2}{2}} = \begin{cases} e^{\left(\frac{(Im\,z)^2}{2\,r^2} + \frac{r^2}{2}\right)\,s^2} & \text{if } Re\,z \neq 0 \\ e^{\left(2\,r^2 + \frac{(Im\,z)^2}{8\,r^2}\right)\,s^2} & \text{if } Re\,z = 0 \end{cases} \tag{72}$$

which means that X is a Gaussian random variable.

For $L \neq 0$ the term $(2\,L\,s + 1)^{-1/2}$ is the moment generating function of a gamma random variable.

We may also represent $CE(Heis)$ in terms of two independent CCR copies as follows:

For $j, k \in \{1, 2\}$ let $[q_j, p_k] = \frac{i}{2}\,\delta_{j,k}$ and $[q_j, q_k] = [p_j, p_k] = 0$ with $p_j^* = p_j$, $q_j^* = q_j$ and $i^2 = -1$.

(i) If $z \in \mathbb{C}$ with $Re\,z \neq 0$ and $Im\,z \neq 0$ then

$$a = i\,Re\,z\,q_1 + \frac{1}{Re\,z}\,p_1^2 - Im\,z\,p_2 - \frac{i}{Im\,z}\,q_2^2 \tag{73}$$

$$a^\dagger = -i\,Re\,z\,q_1 + \frac{1}{Re\,z}\,p_1^2 - Im\,z\,p_2 + \frac{i}{Im\,z}\,q_2^2 \tag{74}$$

$$h = -2\,(p_1 + q_2) \tag{75}$$

and $E = 1$ satisfy the commutation relations (37) and the duality relations (25) of $CE(Heis)$.

(ii) If $z \in \mathbb{C}$ with $Re\, z = 0$ and $Im\, z \neq 0$ then for arbitrary $r \in \mathbb{R}$ and $c \in \mathbb{C}$

$$a = c\, p_1^2 - Im\, z\, p_2 + \left(r - \frac{i}{Im\, z}\right) q_2^2 \tag{76}$$

$$a^\dagger = \bar{c}\, p_1^2 - Im\, z\, p_2 + \left(r + \frac{i}{Im\, z}\right) q_2^2 \tag{77}$$

$$h = -2\, q_2 \tag{78}$$

and $E = 1$ satisfy the commutation relations (37) and the duality relations (25) of $CE(Heis)$.

(iii) If $z \in \mathbb{C}$ with $Re\, z \neq 0$ and $Im\, z = 0$ then for arbitrary $r \in \mathbb{R}$ and $c \in \mathbb{C}$

$$a = i\, Re\, z\, q_1 + \left(\frac{1}{Re\, z} + i\, r\right) p_1^2 + c\, q_2^2 \tag{79}$$

$$a^\dagger = -i\, Re\, z\, q_1 + \left(\frac{1}{Re\, z} - i\, r\right) p_1^2 + \bar{c}\, q_2^2 \tag{80}$$

$$h = -2\, p_1 \tag{81}$$

and $E = 1$ satisfy the commutation relations (37) and the duality relations (25) of $CE(Heis)$.

We may take

$$q_1 = \frac{b_1 + b_1^\dagger}{2} \;\; ; \;\; p_1 = \frac{i\,(b_1^\dagger - b_1)}{2} \tag{82}$$

and

$$q_2 = \frac{b_2 + b_2^\dagger}{2} \;\; ; \;\; p_2 = \frac{i\,(b_2^\dagger - b_2)}{2} \tag{83}$$

where

$$[b_1, b_1^\dagger] = [b_2, b_2^\dagger] = 1 \tag{84}$$

and

$$[b_1^\dagger, b_2^\dagger] = [b_1, b_2] = [b_1, b_2^\dagger] = [b_1^\dagger, b_2] = 0 \tag{85}$$

.

In that case, MGF_X would be the product of the moment generating functions of two independent random variables defined in terms of the generators of two mutually commuting Schroedinger algebras.

5. The centrally extended Heisenberg group

For $u, v, w, y \in \mathbb{C}$ define

$$g(u, v, w, y) = e^{u\,a^\dagger} e^{v\,h} e^{w\,a} e^{y\,E} \tag{86}$$

The family of operators of the form (86) is a group with group law given by

$$g(\alpha, \beta, \gamma, \delta)\, g(A, B, C, D) = \tag{87}$$

$$= g(\alpha + A, \beta + B + \gamma\, A, \gamma + C, \left(\frac{\gamma\, A^2}{2} + \beta\, A\right)\, z + \left(\frac{\gamma^2\, A}{2} + \gamma\, B\right)\, \bar{z} + \delta + D)$$

Restricting to $u, v, w \in \mathbb{R}$ and $y \in \mathbb{C}$ we obtain the centrally extended Heisenberg group $\mathbb{R}^3 \times \mathbb{C}$ endowed with the composition law:

$$(\alpha, \beta, \gamma, \delta)\, (A, B, C, D) = \tag{88}$$

$$\left(\alpha + A, \beta + B + \gamma\, A, \gamma + C, \left(\frac{\gamma\, A^2}{2} + \beta\, A\right)\, z + \left(\frac{\gamma^2\, A}{2} + \gamma\, B\right)\, \bar{z} + \delta + D\right)$$

References

1. Accardi, L., Boukas, A.: *Central extensions of the Heisenberg algebra*, submitted (2008)
2. Feinsilver, P. J., Schott, R.: *Differential relations and recurrence formulas for representations of Lie groups*, Stud. Appl. Math., **96** no. 4 (1996), 387–406.
3. Fuchs, J., Schweigert C. : *Symmetries, Lie Algebras and Representations (A graduate course for physicists)*, Cambridge Monographs on Mathematical Physics, Cambridge University Press, 1997.
4. Ovando, G.: *Four dimensional symplectic Lie algebras*, Beitrage Algebra Geom. **47** no. 2 (2006), 419–434.

REPRESENTATIONS OF THE LÉVY-MEIXNER OSCILLATOR ALGEBRA AND THE OVERCOMPLETENESS OF THE ASSOCIATED SEQUENCES OF COHERENT STATES

Abdessatar BARHOUMI

Department of Mathematics
Higher School of Sci. and Tech. of Hammam-Sousse
University of Sousse, Sousse, Tunisia
E-mail: abdessatar.barhoumi@ipein.rnu

Habib OUERDIANE

Department of Mathematics
Faculty of Sciences of Tunis
University of Tunis El-Manar, Tunis, Tunisia
E-mail: habib.ouerdiane@fst.rnu.tn

Anis RIAHI

Department of Mathematics
Higher Institute of Appl. Sci. and Tech. of Gabes
University of Gabes, Gabes, Tunisia
E-mail: a1riahi@yahoo.fr

The main purpose of this paper is to investigate a generalized oscillator algebra, naturally associated to the Lévy-Meixner polynomials and a class of nonlinear coherent vector. We derive their overcompleteness relation, in so doing, the partition of the unity in terms of the eigenstates of the sequences of coherent vectors is established. An example of complex hypercontractivity property for an Hamiltonian is developed to illustrate our theory.

Keywords: Coherent state, Hypercontractivity, Lévy-Meixner polynomials, Overcompleteness relation.

1. Introduction

The original coherent states based on the Heisenberg-Weyl group has been extended for a number of Lie groups with square integrable representations,[3] and they have many applications in quantum mechanics.[19] In particular, they are used as bases of coherent states path integrals[19] or dynamical wavepackets for describing the quantum systems in semiclassical approximations.[27] This framework has been given in a general and elegant mathematical form through the work of Perelomov.[32]

Many definitions of coherent states exist. In this paper, we consider the so-called Klauder-Gazeau type.[18] The coherent state is defined as the eigenstates of the annihilation operator A for each individual oscillator mode of the electromagnetic field; namely, a system of coherent states is defined to be a set $\Omega(z), z \in \mathcal{X}$ of quantum states in some interacting Fock space Γ,[2,33] parameterized by some set $\mathcal{X}$, such that:
(*i*) $A\Omega(z) = z\Omega(z), \forall z \in \mathcal{X}$, (*ii*) the map $z \longmapsto \Omega(z)$ is smooth, and (*iii*) the system is overcomplete; i.e.

$$\int_{\mathcal{X}} |\Omega(z)\rangle\langle\Omega(z)|\nu(dz) = I. \tag{1}$$

Physicists usually call property (1) completeness relation.

Let us stress that coherent states have two important properties. First, they are not orthogonal to each other with respect to the positive measure in (1). Second, they provide a resolution of the identity, i.e., they form an overcomplete set of states in the interacting Fock space. In fact it is well known that they form a highly overcomplete set in the sense that there are much smaller subsets of coherent states which are also overcomplete. Using them one can express an arbitrary state as a line integral of coherent states.[34]

Our approach aims at generalizing the pioneering work of Bargmann[10] for the usual harmonic oscillator. It is well-known that the classical Segal-Bargmann transform in Gaussian analysis yields a unitary map of L^2 space of the Gaussian measure on $\mathbb{R}$ onto the space of L^2 holomorphic functions of the Gaussian measure on $\mathbb{C}$, see Refs. 10, 17, 21. Later on, based on the work by Accardi-Bożejko,[1] Asai[5] has extended the Segal-Bargmann transform to non-Gaussian cases. The crucial point is the introduction of a coherent vector as a kernel function in such a way that a transformed function, which is a holomorphic function on a certain domain, becomes a power series expression. Along this line, Asai-Kubo-Kuo[8] have considered the case of the Poisson measure compared with the case of the Gaussian measure. More recently, Asai[6,7] has constructed a Hilbert space of analytic L^2 functions with respect to a more general family and give examples including Laguerre, Meixner and Meixner-Pollaczek polynomials.

The present paper is organized as follows. In Section 2, we consider the Hilbert space of square integrable functions $L^2(\mathbb{R}, \mu)$ in which the normalized Lévy-Meixner polynomial system constitutes an orthonormal basis. Using the Poisson kernel, we define the generalized Fourier transform for this system of polynomials, which allows introducing the position and momentum operators. Then, considering the given L^2-space as a realization of the Fock space, the creation and annihilation operators can be standardly constructed. Together with the standard number operator in this Fock space, they satisfy commutation relations that generalize the Heisenberg relations and generate a Lie algebra that we naturally call the Lévy-Meixner oscillator algebra $\mathcal{A}$. In Section 3, we explicit an equivalent irreducible unitary representation of $\mathcal{A}$ on the basis of an adapted one-mode interacting Fock space. Sections 4 and 5 are devoted to the true subject of this paper.

We shall first define a class of nonlinear coherent vectors and we study the associated Bargmann representations. Secondly, we shall derive their overcompleteness relation. Finally, in Section 5, to illustrate our main results, we give a specific example of complex hypercontractivity property for a Meixner class Hamiltonian.

2. The Lévy-Meixner oscillator algebra

Let μ be an infinity divisible distribution such that its Laplace transform is given by

$$\mathcal{L}_\mu(x) = \int_{\mathbb{R}} e^{xt}\mu(dt) = e^{f(x)}, \tag{2}$$

where f is analytic in some neighborhood of zero with $f(0) = 0$. Suppose that the generating function of the orthogonal polynomials $\big(P_n\big)_{n=0}^{\infty}$ with respect to μ has the following form :

$$G_\mu(x,t) = \sum_{n=0}^{\infty} \frac{t^n}{n!} P_n(x) := \exp\Big\{x\rho(t) - f(\rho(t))\Big\}, \tag{3}$$

where ρ is analytic in some neighborhood of zero with $\rho(0) = 0$ and $\rho'(0) = 1$ and f is a Laplace exponent satisfying Eq. (2). By a multiplicative renormalization procedure (as in Ref. 9), $G_\mu(x,t)$ is a generating function of orthogonal polynomials if and only if there exists a function Ψ with $\Psi(0) = 0$ which is analytic in some neighborhood of zero, such that

$$f(\rho(t) + \rho(s)) - f(\rho(t)) - f(\rho(s)) = \Psi(st), \tag{4}$$

we call the function Ψ interaction exponent of μ. Moreover, there exists two constants $\beta, \gamma \in \mathbb{R}_+$ such that the family of polynomials $(P_n)_{n\geq 0}$ satisfies the recurrence formula (see Ref. 25)

$$x\,P_n(x) = P_{n+1}(x) + \upsilon_n P_n(x) + \omega_n P_{n-1}(x), \tag{5}$$

with

$$\upsilon_n = c_1 + \gamma n\,, \quad \omega_n = n(c_2 + \beta(n-1)), \tag{6}$$

where $c_1 = f'(0)$ and $c_2 = f''(0)$. The family $\{P_n\}_{n\in\mathbb{N}}$ is called the Lévy-Meixner polynomials. The measure μ satisfying condition (4) will be referred to us the *Lévy-Meixner distribution.* In general, the function Ψ only has two forms:

1. $\Psi(t) = c_2 t$, for $\beta = 0$;

2. $\Psi(t) = -\frac{c_2}{\beta}\ln(1-\beta t)$ for $\beta > 0$,

where for 1. includes the distributions of Gaussian and Poisson, and 2. is in the case of distributions of Meixner's type including the gamma, Pascal and Meixner ones. For more details see Ref. 25.

Now we will suppose that $\beta > 0$, then from the Favard theorem,[16] one can easily obtain

$$\|P_n\|^2_{L^2(\mathbb{R},\mu)} = \beta^n n!\Big(\frac{c_2}{\beta}\Big)_n. \tag{7}$$

Denote $Q_0(x) = 1$ and

$$Q_n(x) = \left[\beta^n n!\Big(\frac{c_2}{\beta}\Big)_n\right]^{-1/2} P_n(x). \tag{8}$$

Then $\{Q_n(x)\}_{n=0}^\infty$ is a complete orthonormal system in $L^2(\mu)$ and the recurrence relation (5) becomes

$$xQ_n = \Lambda_n Q_{n+1} + \upsilon_n Q_n + \Lambda_{n-1}Q_{n-1}, \quad n \geq 1 \tag{9}$$

with

$$\Lambda_n = [\omega_{n+1}]^{1/2}, \quad n \geq 0. \tag{10}$$

The relation (9) indicates a manner by which the position operator $\mathbb{X}$, on $L^2(\mu)$, acts on the elements of the basis $\{Q_n\}_{n=0}^\infty$:

$$(\mathbb{X}Q_n)(x) := xQ_n(x) = \Lambda_n Q_{n+1} + \upsilon_n Q_n + \Lambda_{n-1}Q_{n-1}, \quad n \geq 0. \tag{11}$$

Now we want to define the momentum operator $\mathbb{P}$ on $L^2(\mu)$. For this end, we do with the classical picture using the Poisson kernel. The Poisson kernel κ on $L^2(\mu)\otimes L^2(\mu)$ is defined by

$$\kappa_z(x,y) := \sum_{n=0}^\infty z^n Q_n(x)Q_n(y), \quad z \in \mathbb{C}.$$

We dcfine the integral operator $\mathfrak{K}_z$ on $L^2(\mu)$ by

$$(\mathfrak{K}_z\phi)(y) := \int_{\mathbb{R}} \kappa_z(x,y)\phi(x)\mu(dx).$$

It is noteworthy that $\mathfrak{K}_i$ is a unitary operator on $L^2(\mu)$ and

$$[\mathfrak{K}_i]^{-1} = [\mathfrak{K}_i]^* = \mathfrak{K}_{-i}. \tag{12}$$

The unitary operators $\mathfrak{K}_i$ and $\mathfrak{K}_{-i}$ are called the generalized Fourier transform and inverse Fourier transform, respectively. We shall denote $\mathfrak{K}_i$ simply by $\mathfrak{K}$.

We define the momentum operator $\mathbb{P}$ on $L^2(\mu)$ by

$$\mathbb{P} = \mathfrak{K}^{-1}\mathbb{X}\mathfrak{K}.$$

The energy operator is then defined by

$$\mathbb{H} = (\mathbb{X} - \upsilon_N)^2 + (\mathbb{P} - \upsilon_N)^2, \tag{13}$$

where υ_N is given by

$$\upsilon_N Q_n = \upsilon_n Q_n.$$

Proposition 2.1. *The operators $\mathbb{X}, \mathbb{P}$ and $\mathbb{H}$ acts on the basis element of $L^2(\mu)$ by*

(i) $\mathbb{X}Q_n = \Lambda_n Q_{n+1} + \upsilon_n Q_n + \Lambda_{n-1}Q_{n-1}$
(ii) $\mathbb{P}Q_n = -i\Lambda_n Q_{n+1} + \upsilon_n Q_n + i\Lambda_{n-1}Q_{n-1}$
(iii)

$$\mathbb{H}Q_n = \varepsilon_n Q_n \quad \textit{with} \quad \varepsilon_n = 4\beta n^2 + c_2(4n+2). \tag{14}$$

Proof. The statement (**i**) follows from the definition of $\mathbb{X}$. We easily verify that

$$\mathfrak{K}(z)Q_n = z^n Q_n, \quad z \in \mathbb{C}, \quad n \geq 0.$$

Thus, by using (12), one calculate

$$\begin{aligned}
\mathbb{P}Q_n &= \mathfrak{K}^{-1}\left[\mathbb{X}(i^n Q_n)\right] \\
&= i^n \mathfrak{K}^{-1}\left[\Lambda_n Q_{n+1} + \upsilon_n Q_n + \Lambda_{n-1}Q_{n-1}\right] \\
&= i^n[(-i)^{n+1}\Lambda_n Q_{n+1} + (-i)^n \upsilon_n Q_n + (-i)^{n-1}\Lambda_{n-1}Q_{n-1}] \\
&= -i\Lambda_n Q_{n+1} + \upsilon_n Q_n + i\Lambda_{n-1}Q_{n-1}.
\end{aligned}$$

This prove (**ii**). The identity (**iii**) follows immediately from (**i**), (**ii**) and (13). □

It is noteworthy that relation (14) tells us that the basis vectors are eigenfunctions of the self-adjoint operator $\mathbb{H}$.

Definition 2.1. The creation and annihilation operators are defined as follows

$$\mathrm{a}^{\dagger} = \frac{1}{2}\Big(\mathbb{X} + i\mathbb{P} - (1+i)\upsilon_N\Big)$$

$$\mathrm{a} = \frac{1}{2}\Big(\mathbb{X} - i\mathbb{P} - (1-i)\upsilon_N\Big).$$

Proposition 2.2.

$$(\mathrm{a}^{\dagger})^{*} = \mathrm{a}, \quad \mathrm{a}^{\dagger}Q_n = \Lambda_n Q_{n+1}, \quad \mathrm{a}Q_n = \Lambda_{n-1}Q_{n-1};$$

$$[\mathrm{a}, \mathrm{a}^{\dagger}] = 2\beta\mathbf{n} + c_2\mathbf{i}\,; \tag{15}$$

$$[\mathbf{n}, \mathrm{a}^{\dagger}] = \mathrm{a}^{\dagger}, \quad [\mathbf{n}, \mathrm{a}] = -\mathrm{a}\,, \tag{16}$$

where $\mathbf{n}$ *is the standard number operator acting on basis vectors by*

$$\mathbf{n}Q_n = nQ_n, \quad n \geq 0,$$

and $\mathbf{i}$ *is the identity operator on* $L^2(\mu)$.

Proof. A straightforward verification. □

Definition 2.2. The Lie algebra generated by the operators $\mathrm{a}^{\dagger}, \mathrm{a}, \mathbf{n}, \mathbf{i}$ with commutation relations (15) and (16) is called *the Lévy-Meixner oscillator* and will be denoted $\mathcal{A}$.

3. One-mode interacting Fock space representation of the Lévy-Meixner oscillator

In this Section we shall give a unitary equivalent irreducible Fock representation of the algebra $\mathcal{A}$. Let us consider the Hilbert space $\mathcal{K}$ to be the complex numbers which, in physical language, corresponds to a 1-particle space in zero space-time dimension. In this case, for each $n \in \mathbb{N}$, also $\mathcal{K}^{\otimes n}$ is 1-dimensional, so we identify it to the multiples of a number vector denoted by Φ^{+n}. The pre-scalar product on $\mathcal{K}^{\otimes n}$ can only have the form:

$$\langle z, w\rangle_{\otimes n} := \lambda_n \overline{z} w, \quad z, w \in \mathbb{C} \tag{17}$$

where the λ_n's are positive numbers.
According to our setting, we define the sequence $\lambda = \{\lambda_n\}_{n=0}^{\infty}$ by

$$\lambda_n := \omega_0\omega_1 \cdots \omega_n, \quad n \geq 0.$$

From (7), λ_n can be rewritten explicitly

$$\lambda_n = \beta^n n! \left(\frac{c_2}{\beta}\right)_n, \quad n \in \mathbb{N}. \tag{18}$$

Definition 3.1. The Lévy-Meixner one-mode interacting Fock space, denoted Γ_{LM}, is the Hilbert space given by taking quotient and completing the orthogonal direct sum

$$\bigoplus_{n=0}^{\infty} (\mathcal{H}^{\otimes n}, \langle\cdot,\cdot\rangle_{\otimes n, LM})$$

where $\langle\cdot,\cdot\rangle_{\otimes n, LM} = \langle\cdot,\cdot\rangle_{\otimes n}$ with the choice (18).

Denote $\Phi_0 = \Phi^{+(0)}$ the vacuum vector and for any $n \geq 1, \Phi_n = \Phi^{+(n)}$. For $\Phi = \sum_{n=0}^{\infty} a_n \Phi_n, \Psi = \sum_{n=0}^{\infty} b_n \Psi_n$ in Γ_{LM}, we have

$$\langle \Phi, \Psi \rangle_{\Gamma_{LM}} = \sum_{n=0}^{\infty} \lambda_n \overline{a_n} b_n. \tag{19}$$

The creation operator is the densely defined operator $A^\dagger$ on Γ_{LM} satisfying

$$A^\dagger : \Phi_n \longmapsto \Phi_{n+1}, \quad n \geq 0. \tag{20}$$

The annihilation A is given, according to the scalar product (19), as the adjoint of $A^\dagger$, by $A\Phi_0 = 0$ and

$$A\Phi_{n+1} = \frac{\lambda_{n+1}}{\lambda_n}\Phi_n, \quad n \geq 0. \tag{21}$$

Hence, for any $n \geq 0$ we have

$$AA^\dagger(\Phi_n) = \frac{\lambda_{n+1}}{\lambda_n}\Phi_n, \quad A^\dagger A(\Phi_n) = \frac{\lambda_n}{\lambda_{n-1}}\Phi_n \tag{22}$$

and the following relations arise

$$AA^\dagger = \frac{\lambda_{N+I}}{\lambda_N}, \quad A^\dagger A = \frac{\lambda_N}{\lambda_{N-I}}$$

where I is the identity operator on Γ_{LM} and N is the number operator satisfying $N\Phi_n = n\Phi_n, \quad n \geq 0$, and the right hand side of (22) is uniquely determined by the spectral theorem.

Proposition 3.1. *The Lie algebra $\mathcal{B}$ generated by $A^\dagger, A, N, I$ gives rise to a unitary equivalent irreducible representation of the Lévy-Meixner oscillator algebra $\mathcal{A}$.*

Proof. From the relations (18) and (22) we read, for any $n \geq 0$,

$$[A^\dagger, A]\Phi_n = \left[\frac{\lambda_{n+1}}{\lambda_n} - \frac{\lambda_n}{\lambda_{n-1}}\right]\Phi_n$$

$$= [\omega_{n+1} - \omega_n]\Phi_n$$

$$= [2\beta N + c_2 I]\Phi_n.$$

In a similar way we found

$$[N, A^\dagger]\Phi_n = A^\dagger\Phi_n, \quad [N, A]\Phi_n = -A\Phi_n.$$

The irreducibility property is obvious from the completeness of the family $\{\Phi_n\}_{n=0}^\infty$ in $\Gamma_{_{LM}}$. □

As it is expected, we want the Fock space representation of the Lévy-Meixner oscillator to be unitary equivalent to the one in the $L^2(\mu)$-space. This question is answered by Accardi-Bożejko isomorphism.[1]

Now we state the result of Accardi and Bożejko : there exists a unitary isomorphism $\mathfrak{U}_{_{LM}} : \Gamma_{_{LM}} \longrightarrow L^2(\mu)$ satisfying the following relations:

(1) $\mathfrak{U}_{_{LM}}\Phi_0 = 1$;
(2) $\mathfrak{U}_{_{LM}} A^\dagger \mathfrak{U}_{_{LM}}^{-1} Q_n = \Lambda_n Q_{n+1}$;
(3) $\mathfrak{U}_{_{LM}}\left(A + \upsilon_N + A^\dagger\right)\mathfrak{U}_{_{LM}}^{-1} Q_n = xQ_n$.

From the above relations $(1)-(3)$ we observe that Accardi-Bożejko isomorphism is uniquely determined by the correspondences

$$\Phi_0 \longmapsto Q_0, \quad \Phi_n \longmapsto (\Lambda_0\Lambda_1\cdots\Lambda_{n-1})\, Q_n, \quad n \geq 1. \tag{23}$$

Moreover, the unitary equivalence between the two representation of the Lévy-Meixner oscillator is obtained thought the formula

$$\mathfrak{U}_{_{LM}} A^\dagger \mathfrak{U}_{_{LM}}^{-1} = \mathrm{a}^\dagger, \quad \mathfrak{U}_{_{LM}} A \mathfrak{U}_{_{LM}}^{-1} = \mathrm{a}, \quad \mathfrak{U}_{_{LM}} N \mathfrak{U}_{_{LM}}^{-1} = \mathbf{n}.$$

4. Lévy-Meixner Coherent states

As indicated in the introduction, we consider in this paper, the Klauder-Gazeau type coherent states, $\Omega(z)$, which are considered to satisfy the following conditions:

(1) For any $z, \Omega(z)$ is an eigenvector of A, i.e., $A\Omega(z) = z\Omega(z)$.
(2) normalization, i.e., for any $z \in \mathbb{C}$, $\|\Omega(z)\|_{\Gamma_{LM}} = 1$;
(3) Continuity with respect to the complex number z;
(4) Overcompleteness, i.e., there exists a positive function $\mathcal{W}$ for which we have the resolution of identity

$$\int_{\mathbb{C}} |\Omega(z)\rangle\langle\Omega(z)|\mathcal{W}(|z|^2)dz = I,$$

where for vectors u, v, x, the one-rank projection $|u\rangle\langle v|$ is defined by

$$|u\rangle\langle v| := \langle v, x\rangle u.$$

4.1. *Bargmann space and holomorphic representation*

Let ν be an absolutely continuous measure on $\mathbb{C}$ with continuous Radom-Nikodym derivative

$$\mathcal{W}(|z|^2) = \frac{\nu(dz)}{dz},$$

where dz is the Lebesgue measure on $\mathbb{C}$. Our basis Hilbert space is the space $\mathcal{H}(\mathbb{C}) \cap L^2(\nu)$, i.e., the space of square integrable analytic functions. ($\mathcal{H}(\mathbb{C})$ is the Fréchet space of all analytic functions in $\mathbb{C}$). Clearly the space $\mathcal{H}(\mathbb{C}) \cap L^2(\nu)$ is a closed subspace of $L^2(\nu)$.

In the remainder of this subsection, we make the following assumption on ν

$$\langle z^n, z^m\rangle_{L^2(\nu)} = \lambda_n \delta_{n,m}, \quad \forall n, m = 0, 1, 2, \cdots \tag{24}$$

Later on, we shall delve a little more in detailed study of the existence of such positive measure ν with condition (24).

Under the assumption (24), we denote $\mathcal{H}L^2(\mathbb{C}, \nu)$ the Hilbert space $\mathcal{H}(\mathbb{C}) \cap L^2(\nu)$. $\mathcal{H}L^2(\mathbb{C}, \nu)$ is the (weighted) Bargmann space associated to the measure ν. It is easily seen that $\{\lambda_n^{-1/2} z^n\}_{n=0}^{\infty}$ is a complete orthonormal basis for $\mathcal{H}L^2(\mathbb{C}, \nu)$. Therefore, the reproducing kernel is

$$K(z, w) = \sum_{n=0}^{\infty} \frac{\overline{z}^n w^n}{\|z^n\|^2_{\mathcal{H}L^2(\nu)}} = \sum_{n=0}^{\infty} \frac{\overline{z}^n w^n}{\lambda_n} \tag{25}$$

with

$$L(|z|^2) := K(z, z) = \sum_{n=0}^{\infty} \frac{|z|^{2n}}{\lambda_n} < \infty, \quad \forall z \in \mathbb{C}. \tag{26}$$

Hence, the space $\mathcal{H}L^2(\mathbb{C},\nu)$ is given by

$$\mathcal{H}L^2(\mathbb{C},\nu)=\left\{F(z)=\sum_{n=0}^{\infty}a_nz^n, F \text{ analytic on } \mathbb{C} \text{ and } \sum_{n=0}^{\infty}\lambda_n|a_n|^2<\infty\right\}$$

The map

$$S:\Phi_n\in\Gamma_{LM}\longrightarrow z^n\in\mathcal{H}L^2(\mathbb{C},\nu)$$

can be uniquely extended to a unitary isomorphism. Moreover, the Lie algebra generated by $SA^\dagger S^{-1}, SAS^{-1}$ and SNS^{-1} gives rise to a unitary equivalent irreducible representation of the Lévy-Meixner oscillator algebra $\mathcal{A}$. This representation is called Bargmann (or holomorphic) representation of the Lévy-Meixner oscillator algebra.

4.2. *The coherent vectors* Ω_z

Theorem 4.1. *The family of Lévy-Meixner coherent vectors* $\Omega(z)$ *is given by*

$$\Omega(z)=\left[L(|z|^2)\right]^{-1/2}\sum_{n=0}^{\infty}\frac{z^n}{\lambda_n}\Phi_n,\qquad z\in\mathbb{C},$$

where L is defined in (26).

Proof. For $z\in\mathbb{C}$, let $\Omega(z)=\sum_{n=0}^{\infty}a_n\Phi_n$. By using (21), we have

$$\begin{aligned}A\Omega(z)&=\sum_{n=0}^{\infty}a_nA\Phi_n\\&=\sum_{n=0}^{\infty}a_n\frac{\lambda_n}{\lambda_{n-1}}\Phi_{n-1}\\&=\sum_{n=0}^{\infty}a_{n+1}\frac{\lambda_{n+1}}{\lambda_n}\Phi_n.\end{aligned}$$

From the equality $A\Omega(z)=z\Omega(z)$, we get

$$a_{n+1}=z\frac{\lambda_n}{\lambda_{n+1}}a_n,\qquad\forall n\geq 0.$$

Hence, since $\lambda_0=1$, we have

$$a_n=a_0\frac{z^n}{\lambda_n},\qquad(\lambda_0=1)$$

and therefore

$$\Omega(z) = a_0 \sum_{n=0}^{\infty} \frac{z^n}{\lambda_n} \Phi_n.$$

By the normalization property of $\Omega(z)$, we shall choose $a_0 \in \mathbb{C}$ in such a way that

$$1 = \langle \Omega(z), \Omega(z) \rangle_{\Gamma_{LM}} = |a_0|^2 \sum_{n=0}^{\infty} \frac{|z|^{2n}}{\lambda_n} = |a_0|^2 L(|z|^2).$$

which gives $|a_0| = [L(|z|^2)]^{-1/2}$. In conclusion, the eigenvectors of $\mathcal{A}$ are

$$\Omega(z) = \left[L(|z|^2)\right]^{-1/2} \sum_{n=0}^{\infty} \frac{z^n}{\lambda_n} \Phi_n, \quad z \in \mathbb{C},$$

as desired. □

Through Accardi-Bożejko isomorphism (23) one can define the Lévy-Meixner coherent vectors $\Omega(z;x)$ in $L^2(\mu)$:

$$\Omega(z;x) = \left[L(|z|^2)\right]^{-1/2} \sum_{n=0}^{\infty} \frac{z^n}{\sqrt{\lambda_n}} Q_n(x), \quad x \in \mathbb{R}, z \in \mathbb{C}. \tag{27}$$

In the following, for future use, we give representation of $\Omega(z;x)$ in terms of modified Bessel functions. Let us remember, that the modified Bessel function of the first kind and parameter $\gamma > -1$ is given by

$$I_\gamma(2\sqrt{x}) := x^{\frac{1}{2}\gamma} \sum_{n=0}^{\infty} \frac{x^n}{n!\Gamma(n+\gamma+1)}.$$

Theorem 4.2. *In $L^2(\mu)$, the Lévy-Meixner coherent vectors has the following integral representation*

$$\begin{aligned} \Omega(z;x) = &\frac{\sqrt{\Gamma(\frac{c_2}{\beta})}}{2i\pi} \left[\left(\frac{|z|^2}{\beta}\right)^{-\frac{\beta-c_2}{2\beta}} I_{\frac{c_2}{\beta}-1}\left(2\frac{|z|}{\sqrt{\beta}}\right) \right]^{-1/2} \\ &\times \int_{-\infty}^{0} u^{-\frac{c_2}{\beta}} \exp\left\{ u + x\rho\left(\frac{z}{\beta u}\right) - f \circ \rho\left(\frac{z}{\beta u}\right) \right\} du. \end{aligned} \tag{28}$$

Proof. It then follows, from (26),

$$
\begin{aligned}
L(|z|^2) &= \sum_{n=0}^{\infty} \frac{1}{(\frac{c_2}{\beta})_n} \frac{(\frac{|z|^2}{\beta})^n}{n!} \\
&= \Gamma(\tfrac{c_2}{\beta}) \sum_{n=0}^{\infty} \frac{1}{n!\Gamma(\frac{c_2}{\beta}+n)} \left(\frac{|z|^2}{\beta}\right)^n \\
&= \Gamma(\tfrac{c_2}{\beta}) \left(\frac{|z|^2}{\beta}\right)^{\frac{\beta-c_2}{2\beta}} I_{\frac{c_2}{\beta}-1}\left(\frac{2}{\sqrt{\beta}}|z|\right).
\end{aligned} \tag{29}
$$

On the other hand, Eq. (3) yields

$$
\sum_{n=0}^{\infty} \frac{z^n}{\lambda_n} P_n(x) = \Gamma(\tfrac{c_2}{\beta}) \sum_{n=0}^{\infty} \frac{z^n P_n(x)}{\beta^n n! \Gamma(\frac{c_2}{\beta}+n)}
$$

$$
\begin{aligned}
&= \frac{\Gamma(\frac{c_2}{\beta})}{2i\pi} \int_{-\infty}^{0} e^u u^{-\frac{c_2}{\beta}} \sum_{n=0}^{\infty} \frac{P_n(x)}{n!} \left(\frac{z}{\beta u}\right)^n du \\
&= \frac{\Gamma(\frac{c_2}{\beta}+1)}{2i\pi} \int_{-\infty}^{0} u^{-\frac{c_2}{\beta}} \exp\left\{u + x\rho\Big(\frac{z}{\beta u}\Big) - f\circ\rho\Big(\frac{z}{\beta u}\Big)\right\} du.
\end{aligned}
$$

This gives the statement. □

4.3. *Overcompletness of the Lévy-Meixner coherent vectors*

Now we shall investigate the assumption (24) in section 4.1. This is equivalent to problem of constructing a positive measure $\nu(dz) = \mathcal{W}(|z|^2)dz$ in the partition of unity

$$
\int_{\mathbb{C}} |\Omega_z\rangle\langle\Omega_z|\nu(dz) = \sum_{n=0}^{\infty} \left|\frac{\Phi_n}{\sqrt{\lambda_n}}\right\rangle\left\langle\frac{\Phi_n}{\sqrt{\lambda_n}}\right|. \tag{30}
$$

For $z \in \mathbb{C}$, write $z = re^{i\theta}$ with $r \in \mathbb{R}_+$ and $0 \le \theta \le 2\pi$, then

$$
\begin{aligned}
\sum_{n=0}^{\infty} \left|\frac{\Phi_n}{\sqrt{\lambda_n}}\right\rangle\left\langle\frac{\Phi_n}{\sqrt{\lambda_n}}\right| &= \int_{\mathbb{C}} |\mathcal{W}(|z|^2)\Omega_z\rangle\langle\Omega_z|\, dz \\
&= \sum_{n,m=0}^{\infty} \Big\{\frac{1}{\lambda_n\lambda_m} \int_0^{+\infty} \left(\frac{\mathcal{W}(r^2)}{L(r^2)}\right) r^{n+m} d(r^2) \\
&\times \int_0^{2\pi} e^{i\theta(n-m)} d\theta\Big\} |\Phi_n\rangle\langle\Phi_m| \\
&= \sum_{n=0}^{\infty} \left\{\frac{2\pi}{\lambda_n} \int_0^{+\infty} x^n \left(\frac{\mathcal{W}(x)}{L(x)}\right) dx\right\} \left|\frac{\Phi_n}{\sqrt{\lambda_n}}\right\rangle\left\langle\frac{\Phi_n}{\sqrt{\lambda_n}}\right|
\end{aligned}
$$

where we marked the change of variable $x = r^2$. Put $V(x) = \frac{\mathcal{W}(x)}{L(x)}$, one can deduce that the overcompleteness of the family of Lévy-Meixner coherent vectors is equivalent to the classical moment problem :

$$\int_0^{+\infty} x^n V(x)dx = \frac{1}{2\pi}\lambda_n, \quad n = 0, 1, 2, \cdots \tag{31}$$

To state theorem 4.3, we prepare the so-called modified Bessel function K_r given by

$$K_r(x) = \frac{\pi}{2\sin(\pi r)}\big(I_{-r}(x) - I_r(x)\big), \quad 0 < r < 1.$$

Theorem 4.3. *If* $1 < \frac{c_2}{\beta} < 2$, *the measure* ν *in partition of unity satisfied by the Lévy-Meixner coherent vectors* $\{\Omega_z\}_{z\in\mathbb{C}}$ *is given by*

$$d\nu(z) = \frac{1}{\beta\pi} I_{\frac{c_2}{\beta}-1}\Big(\frac{2}{\sqrt{\beta}}|z|\Big) K_{\frac{c_2}{\beta}-1}\Big(\frac{2}{\sqrt{\beta}}|z|\Big) dz.$$

Proof. In view of (18) we rewrite (31) as

$$\int_0^{+\infty} x^n V(x)dx = \frac{1}{2\pi}\beta^n n!\Big(\frac{c_2}{\beta}\Big)_n.$$

Since we have the formula

$$n!(\gamma)_n = \frac{2}{\Gamma(\gamma)}\int_0^{+\infty} x^{n+\frac{\gamma-1}{2}} K_{\gamma-1}(2\sqrt{x})dx$$

(see p. 193 in Ref. 23), then we get

$$\begin{aligned}\int_0^{+\infty} x^n V(x)dx &= \frac{\beta^n}{\pi\Gamma(\frac{c_2}{\beta})}\int_0^{+\infty} x^n x^{\frac{c_2-\beta}{2\beta}} K_{\frac{c_2}{\beta}-1}(2\sqrt{x})dx \\ &= \frac{1}{\beta\pi\Gamma(\frac{c_2}{\beta})}\int_0^{+\infty} x^n \Big(\frac{x}{\beta}\Big)^{\frac{c_2-\beta}{2\beta}} K_{\frac{c_2}{\beta}-1}\Big(2\sqrt{\frac{x}{\beta}}\Big)dx.\end{aligned}$$

Thus we deduce that V is uniquely given by

$$V(x) = \frac{1}{\beta\pi\Gamma(\frac{c_2}{\beta})}\Big(\frac{x}{\beta}\Big)^{\frac{c_2-\beta}{2\beta}} K_{\frac{c_2}{\beta}-1}\Big(2\sqrt{\frac{x}{\beta}}\Big).$$

Finally, we obtain

$$\begin{aligned}\mathcal{W}(|z|^2) &= V(|z|^2)L(|z|^2) \\ &= \frac{1}{\beta\pi} I_{\frac{c_2}{\beta}-1}\Big(\frac{2}{\sqrt{\beta}}|z|\Big) K_{\frac{c_2}{\beta}-1}\Big(\frac{2}{\sqrt{\beta}}|z|\Big).\end{aligned}$$

This gives the statement. □

Remark 4.1. If we choose $\beta = 1$ and $c_2 = \alpha \in]1, 2[$, then the measure $\nu_\alpha := \nu$ in partition of unity satisfied by the Gamma, Meixner and Pascal coherent vectors $\{\Omega_z\}_{z\in\mathbb{C}}$ is given by

$$\nu_\alpha(dz) = \frac{1}{\pi} I_{\alpha-1}(2|z|) K_{\alpha-1}\Big(2|z|\Big) dz.$$

5. Complex Hypercontractivity for the Meixner class

The quantum mechanical harmonic oscillator is essentially the Weyl representation of the Lie algebra associated to the Euclidean motion group. In Fock-Bargmann model, it can be described by the quadruple[10]

$$\{\mathcal{H}L^2(\mathbb{C}, \varrho), \partial, \partial^\dagger, H\}$$

where

$$\mathcal{H}L^2(\mathbb{C}, \varrho) =$$
$$\left\{ f : \mathbb{C} \to \mathbb{C}, \text{ holomorphic}, \ \|f\|_2 := \left(\int_{\mathbb{C}} |f(z)|^2 \varrho(dz) \right)^{1/2} < \infty \right\},$$

$$\partial f(z) = \frac{\partial}{\partial z} f(z), \quad \partial^\dagger f(z) = z f(z), \quad H f(z) = z \frac{\partial}{\partial z} f(z),$$

and $\varrho(dz) = \frac{1}{\pi} e^{-z\bar{z}} dz$ is the complex one-dimensional Gaussian measure. They satisfy the canonical commutation relations (CCR)

$$[\partial, \partial^\dagger] = I, \quad [\partial, I] = 0, \quad [\partial^\dagger, I] = 0 \tag{32}$$

and Wigner commutation relations (WCR)

$$[H, \partial] = -\partial, \quad [H, \partial^\dagger] = \partial^\dagger. \tag{33}$$

The Hamiltonian $H = \partial^\dagger \partial = z\frac{\partial}{\partial z}$ is diagonalized by the orthonormal basis $\{\frac{1}{\sqrt{n!}} z^n\}_{n=0}^\infty$ and has spectrum $\{0, 1, 2, \cdots\}$. It is remarkable that the semigroup $\{T_t := e^{-tH}, \ t \geq 0\}$ enjoys the following hypercontractivity property: for t satisfying $e^{-2t} \leq \frac{p}{q}$, T_t is a contraction from $\mathcal{H}L^p(\mathbb{C}, \varrho)$ to $\mathcal{H}L^q(\mathbb{C}, \varrho)$, where, for an integer $p \geq 1$, $\mathcal{H}L^p(\mathbb{C}, \varrho)$ is the Banach space defined by

$$\mathcal{H}L^p(\mathbb{C}, \varrho) =$$
$$\left\{ f : \mathbb{C} \to \mathbb{C}, \text{ holomorphic}, \ \|f\|_p := \left(\int_{\mathbb{C}} |f(z)|^p \varrho(dz) \right)^{1/p} < \infty \right\}.$$

This hypercontractivity plays an important role in the study of the Boson fields theory.[22]

For $p \geq 1$, let

$$\mathcal{H}L^p(\mathbb{C}, \nu_\alpha) =$$

$$\left\{ f : \mathbb{C} \to \mathbb{C}, \text{ holomorphic}, \ \|f\|_p := \left(\int_{\mathbb{C}} |f(z)|^p \nu_\alpha(dz) \right)^{1/p} < \infty \right\},$$

then $\mathcal{H}L^p(\mathbb{C}, \nu_\alpha)$ is a Banach space and $\mathcal{H}L^2(\mathbb{C}, \nu_\alpha)$ is a reproducing kernel Hilbert space with kernel given by (26). Recall that an orthonormal basis for $\mathcal{H}L^2(\mathbb{C}, \nu_\alpha)$ is

$$\left\{ \psi_{n,\alpha} := (\lambda_n)^{-1/2} \, z^n, \quad n \geq 0 \right\}.$$

The most important non compact non-Abelian group is the pseudounitary group SU(1,1) defined by

$$\mathrm{SU}(1,1) = \left\{ A = \begin{pmatrix} a & b \\ \overline{b} & \overline{a} \end{pmatrix} \ : \ |a|^2 - |b|^2 = 1 \right\}.$$

We recall that the derived Lie algebra of SU(1,1) is realized on $\mathcal{H}L^2(\mathbb{C}, \nu_\alpha)$ as

$$B^+ = z, \quad B^- = z\frac{\partial^2}{\partial z^2} + \alpha \frac{\partial}{\partial z}, \quad M = z\frac{\partial}{\partial z} + \frac{\alpha}{2}.$$

Here B^+, B^- and M are understood as creation, annihilation and neutral operators. It is noteworthy that B^+ and B^- are mutually adjoint and the following commutation relations hold :

$$[B^+, B^-] = 2M, \quad [M, B^+] = B^+, \quad [M, B^-] = -B^-.$$

Accordingly $\{\mathcal{H}L^2(\mathbb{C}\nu), B^-, B^+, M\}$ may be viewed as the Lévy-Meixner harmonic oscillator on $\mathbb{C}$. It is noteworthy that $\{B^-, B^+\}$ does not satisfy the CCR (32), while $\{B^-, B^+, M\}$ satisfy the WCR (33).

In the present context, the Hamiltonian is defined by

$$H_\alpha = B^+ B^- = \alpha z \frac{\partial}{\partial z} + z^2 \frac{\partial^2}{\partial z^2}.$$

By direct computation, we have

$$H_\alpha \psi_{n,\alpha} = \omega_n \psi_{n,\alpha}, \quad n = 0, 1, 2, \cdots$$

where $\omega_n = n(\alpha + n - 1)$ is the Jacobi parameter of the Meixner class defined in (6).

Let $T_\alpha = \{T_{\alpha,t} = e^{-tH_\alpha},\ t > 0\}$ be the semigroup generated by H_α. Motivated by the applications indicated in Ref. 22 we prove the following theorem.

Theorem 5.1. *If* $e^{-(\alpha-1)t} \leq 2^{-1/2}$, *then* $T_{\alpha,t}$ *is a contraction from* $\mathcal{H}L^2(\mathbb{C},\nu_\alpha)$ *to* $\mathcal{H}L^4(\mathbb{C},\nu_\alpha)$.

Proof. Recall that $\left\{\psi_{n,\alpha} = (\lambda_n)^{-1/2} z^n;\ n \geq 0\right\}$ is an orthonormal basis for $\mathcal{H}L^2(\mathbb{C},\nu_\alpha)$. Moreover, clearly $T_{\alpha,t}\psi_{n,\alpha} = e^{-n(n+\alpha-1)t}\psi_{n,\alpha}$. Then, for $f(z) = \sum_{n=0}^{\infty} \gamma_n \psi_{n,\alpha} \in \mathcal{H}L^2(\mathbb{C},\nu_\alpha)$, we have

$$(T_{\alpha,t}f)(z) = \sum_{n=0}^{\infty} \gamma_n e^{-n(n+\alpha-1)t}\psi_{n,\alpha}.$$

Therefore, one can estimate

$$\begin{aligned}
\|T_{\alpha,t}f\|_4^4 &= \int_{\mathbb{C}} |(T_{\alpha,t}f)(z)|^4 \,\nu(dz) = \int_{\mathbb{C}} \left((T_{\alpha,t}f)(z)\overline{(T_{\alpha,t}f)(z)}\right)^2 \nu_\alpha(dz) \\
&= \int_{\mathbb{C}} \left\{\sum_{n_1=0}^{\infty}\sum_{n_2=0}^{\infty} e^{-(n_1^2+n_2^2)t} e^{-(\alpha-1)t(n_1+n_2)} \gamma_{n_1}\overline{\gamma_{n_2}}\psi_{n_1,\alpha}\overline{\psi_{n_2,\alpha}}\right\}^2 \nu_\alpha(dz) \\
&= \int_{\mathbb{C}} \sum_{n_1,n_2,n_3,n_4=0}^{\infty} e^{-t(\sum_{k=1}^4 n_k^2)} e^{-(\alpha-1)t(\sum_{k=1}^4 n_k)} \gamma_{n_1}\gamma_{n_2}\overline{\gamma_{n_3}\gamma_{n_4}} \\
&\quad \times \psi_{n_1,\alpha}\psi_{n_2,\alpha}\overline{\psi_{n_3,\alpha}\psi_{n_4,\alpha}}\nu_\alpha(dz) \\
&= \int_{\mathbb{C}} \sum_{n_1,n_2,n_3,n_4=0}^{\infty} e^{-t(\sum_{k=1}^4 n_k^2)} e^{-(\alpha-1)t(\sum_{k=1}^4 n_k)} \gamma_{n_1}\gamma_{n_2}\overline{\gamma_{n_3}\gamma_{n_4}} \\
&\quad \times \left(\lambda_{n_1}\lambda_{n_2}\lambda_{n_3}\lambda_{n_4}\right)^{-1/2} z^{(n_1+n_2)}(\overline{z})^{(n_3+n_4)}\nu_\alpha(dz) \\
&\leq \sum_{n=0}^{\infty} e^{-2(\alpha-1)tn} \sum_{n_1+n_2=n}\ \sum_{n_3+n_4=n} \gamma_{n_1}\gamma_{n_2}\overline{\gamma_{n_3}\gamma_{n_4}} \\
&\quad \times \left(\lambda_{n_1}\lambda_{n_2}\lambda_{n_3}\lambda_{n_4}\right)^{-1/2} \lambda_n e^{-t(n_1^2+n_2^2)}
\end{aligned}$$

$$= \sum_{n=0}^{\infty} e^{-2(\alpha-1)tn} \left| \sum_{n_1+n_2=n} \gamma_{n_1}\gamma_{n_2} \left(\frac{\lambda_n e^{-t(n_1^2+n_2^2)}}{\lambda_{n_1}\lambda_{n_2}} \right)^{1/2} \right|^2$$

$$\leq \sum_{n=0}^{\infty} \left(\frac{1}{2}\right)^n \left\{ \sum_{n_1+n_2=n} \frac{n!}{n_1!\, n_2!} \left(\frac{e^{-t(n_1^2+n_2^2)}\Gamma(n+\alpha)\Gamma(\alpha)}{\Gamma(n_1+\alpha)\Gamma(n_2+\alpha)} \right) \right.$$

$$\left. \times \left(\sum_{n_1+n_2=n} |\gamma_{n_1}\gamma_{n_2}|^2 \right) \right\}$$

$$\leq \sum_{n=0}^{\infty} \left(\sum_{n_1+n_2=n} \frac{n!}{n_1!\, n_2!} \left(\frac{1}{2}\right)^{n_1} \left(\frac{1}{2}\right)^{n_2} \right) \left(\sum_{n_1+n_2=n} |\gamma_{n_1}\gamma_{n_2}|^2 \right)$$

$$= \|f\|_2^4.$$

This complete the proof. □

References

1. Accardi L., Bożejko M., *Interacting Fock space and Gaussianization of probability measures*, Infinite Dimensional Analysis, Quantum Probability and Related Topics 1 (1998), 663–670.
2. Accardi L. and Lu Y.G., *The Wigner Semi-circle law in Quantum Electrodynamics.*, Comm. Math. Phys. 180 (1996), 605-632.
3. Ali S.T., Antoine J.-P., Gazeau J.-P. and Müller U.-A., *Coherent states and their generalizations*, A mathematical overview. Rev. Math. Phys. 7 (1995), 1013-1104.
4. Antoine J.-P., Gazeau J.-P., Klauder J. R. and Monceau P., *Temporally stable coherent states for infinite well and Pöschl-Teller potentials*, J. Math. Phys. 42 (2001), 2349-2387.
5. N. Asai N., *Analytic characterization of one-mode interacting Fock space*, Infinite Dimensional Analysis, Quantum Probability and Related Topics. 4 (2001), 409-415.
6. N. Asai N., *Hilbert space of analytic functions associated to the modified Bessel function and related orthogonal polynomials*, Infinite Dimensional Analysis, Quantum Probability and Related Topics. 8 no. 3 (2005), 505-514.
7. N. Asai N., *Riesz potentials derived by one-mode interacting Fock space approach*, Colloquium Mathematicum. 109 (2007), 101-106.
8. Asai N., Kubo I. and Kuo H.-H., *Segal-Bargmann transforms of one-mode interacting Fock spaces associated with Gaussian and Poisson measures*, Proc. Amer. Math. Soc. 131, no. 3 (2003), 815-823.
9. Asai N., Kubo I. and Kuo H.-H., *Multiplicative renormalization and generating functions II*, Taiwanese J. Math. 8 (2004), 593-628.

10. Bargmann, V. : *On a Hilbert space of analytic functions and an associated integral transform I.* Commun. Pure Appl. Math. Vol. **14** (1961), 187-214.
11. Bargmann, V., *On a Hilbert space of analytic functions and an associated integral transform II*, Commun. Pure Appl. Math. Vol. 20 (1967), 1-101.
12. Berndt R. and Bocherer S., *Jacobi forms and discrete series representations of the Jacobi group*, Math. Z. 204 (1990), 13-44.
13. Berndt R. and Schmidt R., *Elements of the representation theory of the Jacobi group*, Progress in Mathematics 163, Birkhauser Verlag, Basel, 1998.
14. Burger M. and de la Harpe P., *Constructing irreducible representations of discrete groups*, Proc. Indian Acad. Sci. (Math. Sci.) 107 (1997), 223-235.
15. Carey A.I., *Group representations in reproducing kernel Hilbert spaces*, Report on Math. Physics 14 (1978), 247-259.
16. Chihara T. S., *An Introduction to orthogonal polynomials*, Gordon and Breach, 1978.
17. Dwyer T., *Partial differential equations in Fischer-Fock spaces for Hilbert-Schmidt holomorphy type*, Bull. Amer. Math. Soc. 77 (1971), 725-730.
18. Gazeau J.-P. and Klauder J.R, *Coherent states for systems with discrete and continuous spectrum*, J. Phys. A: Math. Gen. 32 (1999), 123-136.
19. Glauber R. J., *Coherent and incoherent states of the radiation field*, Phys. Rev. 131 (1963), 2766-2788.
20. Gradshtein I.S. and Ryzhik I.M., *Tables of integrals, series and products*, Acad. Press, San Diego, Calif. 2000.
21. Gross L. and Malliavin P., *Halls transform and the Segal-Bargmann map*, in: Ito Stochastic Calculus and Probability Theory, N. Ikeda et al. (eds.) Springer, (1996), 73-116.
22. Hoghn-Krohn R. and Simon B., *Hypercontractivitive semigroup and two dimensional self coupled Bose fields*, J. Funct. Anal. 9 (1972), 121-80.
23. Hitotsumatsu S., Moriguchi S. and Udagawa K., *Mathematical Formulas III, Special Functions*, (Iwanami, 1975).
24. Huang Z. Y. and Wu Y. *Interacting Fock expansion of Lévy white noise functionals*, Acta Appl. Math. 82(2004), pp. 333–352.
25. Huang Z. Y. and Wu Y. *Lévy white noise calculus based on Interaction exponents*, Acta Appl. Math. 88(2005), pp. 251–268.
26. Ismail M. E. H., *Classical and Quantum orthogonal polynomials in one variable*, Cambridge University Press. 2005.
27. Klauder J.R. and Skagerstam B.S., *Coherent states - Applications in Physics and Mathematical Physics*, World Scientific, Singapore, 1985.
28. Koekoek R. and Swarttouw R.F., *The Askey-scheme of hypergeometric orthogonal polynomials and its q-analogue*, Delft Univ. of Tech., Report No. 98-17 (1998). http://aw.twi.tudelft.nl/koekoek/askey/index.html.
29. Manko V.I., Marmo G., Porzio A., Solimeno S. and Zaccaria F., *Trapped ions in laser fields: a benchmark for deformed quantum oscillators*, Phys. Rev. A 62 (2000), 503-407.
30. Mueller C. E. and Wissler E B., *Hypercontractivity for the Heat Semigroup for ultraspherical polynomials on the n-sphere*, J. Funct. Anal. 48 (1982), 252-283.

31. Lisiecki L., *Coherent state representations*, A survey. Reports on Mathematical Physics, Vol. 35, no. 213 (1995), 327-358.
32. Perelomov A., *Generalized coherent states and their applications*, Texts and Monographs in Physics, Springer, Berlin, 1985.
33. Vitonofrio C., *Interacting Fock spaces: central limit theorems and quantum stochastic calculus*, Tesi di Dottorato, Universit'a Degli Studi Di Bari, 2006.
34. Vourdas A., *The growth of Bargmann functions and the completeness of sequences of coherent states*, J. Phys. A: Math. Gen. 30 (1997), 4867-4876.

SOME SYSTEMS OF DUALITIES IN WHITE NOISE ANALYSIS

Takeyuki Hida

Professor Emeritus, Nagoya University and Meijo University
Nagoya, Japan

The main aim of this report is to see characteristic properties of the space $(L^2)^-$ of generalized white noise functionals in terms of dualities between various pairings of subspaces of $(L^2)^-$. With this investigation we naturally come to the study of the subspaces of $(L^2)^-$ of degree n and the isomorphisms to the Sobolev spaces.

2000 Mathematics Subject Classification **60H40**

1. Introduction

First we review the Fock space of (ordinary) white noise functionals:

$$(L^2) = \oplus_{n=0}^{\infty} H_n,$$

where (L^2) is the complex Hilbert space involving square integrable functionals of white noise, i.e. $L^2(E^*, \mu)$, where the measure μ is the probability distribution of white noise $\dot{B}(t), t \in \mathbb{R}$, that is the white noise measure. It is given on the space E^* of generalized functions on $\mathbb{R}$.

The subspace H_n is the collection of homogeneous chaos in the sense of N. Wiener or that of multiple Wiener integrals in the sense of K. Itô, both are of degree n.

It is well-known that the space H_n is isomorphic to $\widehat{L}^2(\mathbb{R}^n)$, the symmetrized space of $L^2(\mathbb{R}^n)$ up to the constant $\sqrt{n!}$:

$$H_n \cong \widehat{L}^2(\mathbb{R}^n).$$

Such an isomorphism can be realized by the so-called S-transform defined as follows: for $\varphi(x) \in (L^2)$,

$$S(\varphi)(\xi) = C(\xi) \int \exp[\langle x, \xi \rangle] \varphi(x) d\mu(x),$$

where $C(\xi)$ is the characteristic functional of the white noise measure, the exact value is

$$C(\xi) = \exp[-\frac{1}{2}\|\xi\|^2].$$

See e.g. Refs. 3 and 5.

There can be a restriction of this isomorphism by introducing a stronger topology in such a way that

$$\widehat{K}^{(n+1)/2}(\mathbb{R}^n) \;\cong\; H_n^{(n)},$$

where we use the notation $\widehat{K}^m(\mathbb{R}^n)$ to denote the symmetric Sobolev space over $\mathbb{R}^n$ of degree m which can be positive or negative).

Then, we take the dual space of both sides of this isomorphism based on symmetric $\widehat{L}^2(R^n)$ and H_n, respectively. We can define $H_n^{(-n)}$ the space of generalized white noise functionals of degree n by the following isomorphism:

$$\widehat{K}^{-(n+1)/2}(\mathbb{R}^n) \;\cong\; H_n^{(-n)}$$

up to $\sqrt{n!}$.

Finally, with a suitable choice of a positive increasing sequence c_n, we have the *test functional space*

$$(L^2)^+ = \oplus c_n H_n^{(n)}$$

and its dual space

$$(L^2)^- = \oplus c_n^{-1} H_n^{(-n)},$$

which is called the space of *generalized white noise functionals.*

In this note we shall discuss various kind of dualities that exist among subspaces of $(L^2)^-$.

We have established in Ref. 5, Chapter 2, the structure of $H_1^{(-1)}$, where we have given the *identity* to the white noise $\dot{B}(t)$ (or its sample function $x(t)$ with $x \in E^*$). The space $H_1^{(-1)}$ is spanned by the $\dot{B}(t)$'s and each $\dot{B}(t)$ is taken to be the variables of generalized white noise functionals. This fact can be a basic notion in what we are going to discuss in this note.

2. Le passage du fini à l'infinit

2.1. *Finite dimensional approximations*

We often use finite dimensional approximation to Brownian motion $B(t)$ or to white noise $\dot{B}(t)$. There are, of course, many methods of approximations, and the choice of the method is crucial in our case, where we are interested in the stochastic analysis which is essentially infinite dimensional. We propose a method that comes from Lévy's method which uses successive interpolation so that uniformity in time t is taken into account. Indeed, the method is naive, but it meets exactly with our claim.

Actual method is as follows.

Construction of a Brownian motion (white noise).

We emphasize the significance of the Lévy construction of a Brownian motion which is done by successive interpolations. We follow the following steps. Prepare standard Gaussian i.i.d. (independent identically distributed) random variables $\{Y_n = Y_n(\omega), \omega \in \Omega, n \geq 1\}$ on a probability space $(\Omega, \mathbf{B}, P)$. Start with $\{X_1(t)\}$ given by

$$X_1(t) = tY_1. \tag{1}$$

The sequence of processes $\{X_n(t), t \in [0,1]\}$ is formed by induction. Let T_n be the set of binary numbers $k/2^{n-1}, k = 0, 1, 2, \cdots, 2^{n-1}$, and set $T_0 = \cup_{n\geq 1} T_n$. Assume that $X_j(t) = X_j(t,\omega), j \leq n$, are defined. Then, we set

$$X_{n+1}(t) =$$

$$\begin{cases} X_n(t), \quad t \in T_n, \\ \dfrac{X_n(t+2^{-n}) + X_n(t-2^{-n})}{2} + 2^{-\frac{n}{2}} Y_k, \quad t \in T_{n+1} - T_n, \\ \qquad\qquad k = k(t) = 2^{n-1} + \frac{2^n t+1}{2}, \\ (k+1-2^n t) X_{n+1}(k2^{-n}) + (2^n t - k) X_{n+1}((k+1)2^{-n}), \\ \qquad\qquad t \in [k2^{-n}, (k+1)2^{-n}], \end{cases} \tag{2}$$

where ω is omitted.

We now claim

Theorem 2.1. *i) The sequence $X_n(t), n \geq 1$, is consistent and that the* *uniform L^2-limit of the $X_n(t)$ exists. The limit is denoted by $\tilde{X}(t)$, which is a Brownian motion.*

ii) The time derivative $X_n'(t)$ converges to a white noise in $H_1^{(-1)}$.

Proof is easy and is omitted.

As was mentioned before, this method of approximation is significant in many ways. For instance, the approximation is *uniform* in t, and is consistent in the sense that it is getting finer and finer as $n \to \infty$ and the σ-field associated to $X_n(t)$ is strictly monotone increasing. Intuitively speaking, the system is like a martingale depending on n.

We can see that this property is realized in Ref. 9.

2.2. *Infinite dimensional rotation group*

Take a suitable nuclear space E and let $O(E)$ be the collection of linear isomorphisms of E which are orthogonal in $L^2(\mathbb{R}^1)$. It is topologized by the compact-open topology and call it *rotation group* of E, or if E is not specified, it is called *infinite dimensional rotation grloup.*

Let g^* be the adjoint of $g \in O(E)$, Each g^* is a μ measure preserving transformation acting on E^*.

Thus, our white noise analysis can be viewed as the *harmonic analysis* arising from the infinite dimensional rotation group.

The figures illustrate that the harmonic analysis can be, in some parts, approximated by finite dimensional analysis; on the other hand, there are places where essentially infinite dimensional and in fact, they can not be well approximated by finite dimensional subjects.

3. Quadratic functionals of white noise

We are ready to discuss nonlinear functions (actually functionals) of the $\dot{B}(t)$. We claim that, among others, the subspace $H_2^{(-2)}$ involving *quadratic generalized* white noise functionals is particularly important. We have the isomorphism

$$H_2^{(-2)} \cong \widehat{K}^{-3/2}(\mathbb{R}^2).$$

More explicitly, for $\varphi \in H_2^{(-2)}$ we find a function $F(u,v)$ in the space $\widehat{K}^{-3/2}(\mathbb{R}^2)$ to have a representation

$$\varphi(\dot{B}) = \int F(u,v) : \dot{B}(u)\dot{B}(v) : dudv.$$

We shall classify those quadratic functionals according to the analytic properties of the kernel.

We therefore start with a quadratic form in the elementary theory of linear algebra. A quadratic form $Q(x), x = (x_1, \cdots, x_n) \in R^n$, is expressed in the form:

$$Q(x) = \sum a_{j,k} x_j x_k.$$

It is significant to decompose $Q(x)$ into a sum of $Q_1(x)$ and $Q_2(x)$:

$$Q(x) = Q_1(x) + Q_2(x),$$

where

$$Q_1(x) = \sum a_j x_j^2, \text{ and } Q_2(x) = \sum_{j \neq k} a_{j,k} x_j x_k.$$

Taking the idea of passage to infinity, we can consider how to discriminate these two terms. Note that the x_j's are equally weighted variables since they are coordinates of finite or infinite dimensional vector space. We shall now make some quite elementary but important observations.

i) Suppose x_i's are mutually independent and subject to the standard Gaussian distribution $N(0,1)$. If both are infinite sum, in order that $Q_1(x)$ be convergent, the coefficients a_j's should be of trace class. But for $Q_2(x)$ the coefficients $a_{j,k}$ should be square summable. In short, the way of convergence is strictly different.

ii) As for the analytic property, any partial sum of $Q_2(x)$ is harmonic, while each term of $Q_1(x)$ is not.

iii) Start with a Brownian motion $B(t), t \in [0,1]$. Consider an approximation to white noise $\dot{B}(t), t \in [0,1]$ by taking $\frac{\Delta_j B(t)}{\Delta_j}$ for x_j. Let $|\Delta_j|$ tend to 0. Then, each term of Q_1 needs (additive) renormalization, but the trick is unnecessary for Q_2.

With the notes mentioned above we now come to the expression of generalized quadratic functionals of white noise, namely representation of quadratic functionals $\varphi(\dot{B}) \in H_2^{(-2)}$. It is expressed in the form

$$\varphi(\dot{B}) = \int\int F(u,v) : \dot{B}(u)\dot{B}(v) : dudv,$$

where $F \in \widehat{K}^{-3/2}(\mathbb{R}^2)$.

Applying the S-transform we have the U-functional of the form

$$U(\xi) = \int\int F(u,v)\xi(u)\xi(v)dudv,$$

which is a quadratic form of ξ.

We now remind the entire functionals of the second order due to P. Lévy (see Ref. 11 Part I, Chapter 3). He focuses his attention to the *normal* form, which is expressible as

$$U(\xi) = \int\int f(u,v)\xi(u)\xi(v)dudv + \int g(t)\xi(t)^2 dt.$$

We tacitly assume suitable conditions posed on f and g.

If we understand in our notation, the generalized function F, which is in the Sobolev space, is chosen with a restriction that a singularity is involved only *on the diagonal.* Namely, we may understand that $g(u)$ can be considered as $g(\frac{u+v}{2})\delta(u-v)$. Note that singularity does appear on the diagonal.

We are now in a position to remind the observations noted in i), ii) and iii) made just above. If we are permitted to say rather formally, the quadratic form $Q(x)$, which is divided into $Q_1(x)$ and $Q_2(x)$, goes to the Lévy's formula for normal functionals as the dimension of the vector x tends to infinity.

We understand that $Q_1(\xi) = \int g(t)\xi(t)^2 dt$ is in the domain of the Lévy Laplacian and the same for $Q_2(\xi) = \int\int f(u,v)\xi(u)\xi(v)dudv$, in addition, it is always harmonic.

There may arise a question. Why is a $H_2^{(-2)}$-functional having off-diagonal singularities of the kernel $F(u,v)$ not so important? There is an answer which is just simple; it is not in the domain of the Laplacian.

Remark. It is natural to ask what is the role of quadratic functional that has singularity is off diagonal. For example

$$\int g(u)\dot{B}(u)\dot{B}(u+1)du.$$

It is easy to see that the second order functional derivative does not exist, so that it is not in the domain of the Laplacian.

Taking the regularity of the functions in the Sobolev space into account, we take the order -3/2, which is important. We can now prove

Theorem 3.1. *If an $H_2^{(-3/2)}$-functional is in the domain of the Lévy Laplacian, then it is a normal functional in the sense of P. Lévy.*

For a higher chaos degree, say n, the Sobolev space order is taken to be $-\frac{n+1}{2}$. We shall see this choice in Section 5 for $n = 3$.

4. Duality in the space of quadratic generalized functionals

We can establish an identity of the renormalized square $:\dot{B}(t)^2:$ of white noise, as we did in the case of $\dot{B}(t)$ in $H_1^{(-1)}$ (see § 2.6 in Ref. 5).

Having done this, we can now introduce a subspace L_2^* spanned by quadratic normal functionals of the $\dot{B}(t)$'s. More precisely,

$$L_2^* = \left\{ \int g(u) :\dot{B}(u)^2: du; g \in K^{-1}(\mathbb{R}^1) \right\}.$$

It should be noted that the function g above may be regarded as the restriction of a function f in $\widehat{K}^{-3/2}(\mathbb{R}^2)$ down to the diagonal line of R^2. There the trace theorem for Sobolev space is applied.

Obviously the space can be made to be a subspace of $H_2^{(-2)}$ by viewing $g(u)$ to be $g(\frac{u+v}{2})\delta(u-v)$ as the integrand.

Our aim is to prove the following theorem

Theorem 4.1. *There exists a subspace L_2 of $H_2^{(2)}$ such that L_2^* is the dual space of L_2, where the topology of L_2 comes from that of $H_2^{(2)}$.*

Proof. Elementary computations can prove the theorem. But, in reality, there we can see some detailed structure of quadratic generalized white noise functionals. Step by step computations are now in order.

The Fourier transform of $g(\frac{u+v}{2})\delta(u-v)$ is

$$\frac{1}{2\pi}\int\int e^{i(\lambda_1 u+\lambda_2 v)}g(\frac{u+v}{2})\delta(u-v)dudv = \frac{1}{\sqrt{2\pi}}\widehat{g}(\lambda_1+\lambda_2),$$

where $\widehat{g}$ is the Fourier transform of g of one variable. By the definition of the Sobolev space of order $-3/2$ over R^2

$$\frac{1}{2\pi}\int\int \frac{|\widehat{g}(\lambda_1+\lambda_2)|^2}{(1+\lambda_1^2+\lambda_2^2)^{3/2}}d\lambda_1 d\lambda_2$$

is finite. This fact implies that $2^{-1/2}g(\frac{u}{\sqrt{2}})$ belongs to the Sobolev space $K^1(\mathbb{R}^1)$, in addition its norm is equal to the $K^{-3/2}(\mathbb{R}^2)$-norm of $g(\frac{u+v}{2})\delta(u-v)$ up to an universal constant.

Numerical values are as follows. Let $\|\cdot\|_{n,m}$ be the Sobolev norm of order m over $\mathbb{R}^n$. Then, actually we have already shown the following equality

$$\|g\|_{2,3/2}^2 = \frac{c}{2\pi}\|g'\|_{1,1}^2,$$

where $c = \int(1+x^2)^{-3/2}dx$ and $g'(u) = 2^{-1/2}g(\frac{u}{\sqrt{2}})$.

Finally, we come to the stage of determinations of the space L_2 and L_2^*. Remind (Ref. 5)

$$H_2^{(2)} = \left\{\varphi(\dot{B}) = \int\int f(u,v) : \dot{B}(u)\dot{B}(v) : dudv, f \in \widehat{K}^{3/2}(\mathbb{R}^2)\right\},$$

and introduce an equivalence relation $\sim$ in $H_2^{(2)}$ defined by

$$\int\int f_1(u,v) : \dot{B}(u)\dot{B}(v) : dudv \sim \int\int f_2(u,v) : \dot{B}(u)\dot{B}(v) : dudv$$

if and only if $f_1(u,u) = f_2(u,u)$ for all $u \in \mathbf{R}$.

Set

$$H_2^{(2)}/\sim \;\equiv\; L_2.$$

Note. Since $f_i, i = 1,2$ is in $K^{3/2}$, the relation to the diagonal $u = v$ is a continuous function. Hence the equivalence relation is defined without any ambiguity.

Then, what we have computed so far can prove that there is the dual pairing between L_2 and L_2^*, which proves the theorem.

This is somewhat a rephrasement, in a formal tone, of the Theorem 3. Suppose that $f \in \widehat{K}^{3/2}(\mathbb{R}^2)$ and that $g((u+v)/2)\delta(u-v) \in \widehat{K}^{-3/2}(\mathbb{R}^2)$ or $g \in K^1(\mathbb{R}^1)$. Then, formal computation shows

$$\left\langle \int g(u) : \dot{B}(u)^2 : du, \int\int f(u,v) : \dot{B}(u)\dot{B}(v) : dudv \right\rangle$$
$$= 2\int g(u)f(u,u)du.$$

This equality is derived from

$$E[(: \dot{B}(t)^2 :)^2] = 2\frac{1}{(dt)^2}.$$

Remark The relationship between $\int : \dot{B}(t)^2 : dt$ and the Lévy Laplacian has been discussed in Ref. 9 by Si Si.

5. White noise functionals of higher chaos degree

To fix the idea, we shall discuss dualities related to $H_3^{(-3)}$. Let φ be in $H_3^{(-3)}$. Its kernel function $F(u_1, u_2, u_3)$ is found in the Sobolev space $\widehat{K}^{-2}(\mathbb{R}^3)$. The S-transform $U(\xi) = (S\varphi)(\xi)$ can, therefore, be expressed in the form

$$U(\xi) = \int\int\int F(u_1, u_2, u_3)\xi(u_1)\xi(u_2)\xi(u_3)du^3.$$

Our method with the idea that *le passage du fini à l'infini* leads us to consider the class of *normal functionals*, namely we are interested in the following forms of degree three.

Type (2,1)

$$\int\int g(u,v)\xi(u)^2\xi(v)dudv.$$

To have a standard expression, we need to make the kernel g symmetric.

Type (3,0)

$$\int h(u)\xi(u)^3du.$$

We can define subspaces $L_{2,1}^*$ and $L_{3,0}^*$ of $H_3^{(-3)}$ spanned by generalized functionals of the types (2,1) and (3,0), respectively. Then, we have

Theorem 5.1. *There exist factor spaces $L_{2,1}$ and $L_{3,0}$ of subspaces of H_3 such that $(L_{2,1}, L_{2,1}^*)$ and $(L_{3,0}, L_{3,0}^*)$ are dual pairs, respectively.*

Proof is given by slight modifications of that of Theorem 3.

Now it is clear how to form dualities in the class of entire homogeneous functionals of each degree, by using singularities on the diagonals. The system of dual pairs is one of the characteristics of the space $(L^2)^-$ of generalized white noise functionals. There, the duality between $(L^2)^+$ and $(L^2)^-$ can be considered as the basic duality.

References

1. L. Accardi et al eds, Selected papers of Takeyuki Hida. World Sci. Pub. Co. 2001.
2. T. Hida, Analysis of Brownian functionals. Carleton Math. Lecture Notes no. 13, Carleton University, 1975.
3. T. Hida, Brownian motion. Springer-Verlag. 1980.
4. T. Hida and Si Si, An innovation approach to random fields. Application of white noise theory. World Scientific Pub. Co. 2004.
5. T. Hida and Si Si, Lectures on white noise functionals. World. Sci. Pub. Co. 2008.
6. Si Si and T. Hida, Some aspects of quadratic generalized white noise functionals. Proc. QBIC08 held at Tokyo Univ. of Science. 2009, to appear.
7. T. Hida, Si Si and T. Shimizu, The $\dot{B}(t)$'s as idealized elemental random variables. Volterra Center Notes N.614. 2008.
8. Si Si, Effective determination of Poisson noise. IDAQP 6 (2003), 609-617.
9. Si Si, An aspect of quadratic Hida distributions in the realization of a duality between Gaussian and Poisson noises. IDAQP 11 (2008) 109-118.
10. P. Lévy, Processus stochastiques et mouvement brownien. Gauthier-Villars. 1948. 2ème ed. with supplement 1965.
11. P. Lévy, Problèmes concrets d'analyse fonctionnelle. Gauthier-Villars. 1951.
12. J. Mikusiński, On the square of the Dirac delta-distribution. Bulletin de l'Academie Polonaise des Sciences. Ser. math, astro et Phys. 14 (1966), 511-513.
13. Si Si, Win Win Htay and L. Accardi, $\mathcal{T}$-transform of Hida distribution and factorizations. Volterra Center Notes, N.625, 2009.
14. Si Si, Introduction to Hida distributions. World Sci. Pub. Co. 2010, to appear.

QUANTUM WHITE NOISE DERIVATIVES AND ASSOCIATED DIFFERENTIAL EQUATIONS FOR WHITE NOISE OPERATORS

Un Cig Ji

Department of Mathematics
Research Institute of Mathematical Finance
Chungbuk National University
Cheongju 361-763, Korea
E-Mail: uncigji@chungbuk.ac.kr

Nobuaki Obata

Graduate School of Information Sciences
Tohoku University
Sendai 980-8579, Japan
E-Mail: obata@math.is.tohoku.ac.jp

We formulate a class of differential equations for white noise operators including the quantum white noise derivatives and obtain a general form of the solutions. As application we characterize intertwining operators appearing in the implementation problem for the canonical commutation relation.

Keywords: Fock space, implementation problem, quantum white noise, quantum white noise derivative, Wick product

1. Introduction

In the analysis of Boson Fock operators a fundamental role is played by the annihilation and creation operators since most important and interesting operators on Boson Fock space are their "polynomials" or "functions." This aspect has been widely accepted explicitly or implicitly [1,2,7,8,16,17], however, the systematic study has not actively developed mostly because of the cumbersome property of the annihilation and creation operators being formulated as (unbounded) operator-valued distributions. During the last fifteen years, we have developed the quantum white noise calculus to overcome this difficulty [9,19], where "every" operator on Boson Fock space is represented in a strict sense as a "function" of the annihilation and creation operators.

To be precise, let us consider the Boson Fock space $\Gamma(H)$ over the Hilbert space $H = L^2(T, dt)$, where (T, dt) is a certain measure space. We denote respectively by a_t and a_t^* the annihilation and creation operators at a point $t \in T$. In the quantum white noise calculus these operators are formulated as continuous operators for themselves. In this paper we consider the Hida–Kubo–Takenaka space:

$$(E) \subset \Gamma(H) \subset (E)^*,$$

which is constructed from a suitable Gelfand triple $E \subset H \subset E^*$ in the standard manner, see e.g., [14,15,19]. It is known that a_t is a continuous linear operator from (E) into itself and by duality a_t^* a continuous linear operator from $(E)^*$ into itself. Both together belong to the class of *white noise operators* $\mathcal{L}((E),(E)^*)$, i.e., continuous operators from (E) into $(E)^*$. The pair $\{a_t, a_t^*\}$ is called the *quantum white noise* on T.

It is a fundamental consequence of quantum white noise theory [9,19] that every white noise operator $\Xi \in \mathcal{L}((E),(E)^*)$ admits a Fock expansion:

$$\Xi = \sum_{l,m=0}^{\infty} \Xi_{l,m}(\kappa_{l,m}),$$

where $\Xi_{l,m}(\kappa_{l,m})$ is an integral kernel operator formally expressed as

$$\Xi_{l,m}(\kappa_{l,m}) = \int_{T^{l+m}} \kappa_{l,m}(s_1, \dots, s_l, t_1, \dots, t_m) \\ a_{s_1}^* \cdots a_{s_l}^* a_{t_1} \cdots a_{t_m} \, ds_1 \cdots ds_l dt_1 \cdots dt_m \, .$$

In this sense every white noise operator is a "function" of quantum white noise:

$$\Xi = \Xi(a_s \, , a_t^* \, ; s, t \in T),$$

and we are naturally interested in its derivatives:

$$D_t^+ \Xi = \frac{\delta \Xi}{\delta a_t^*}, \qquad D_t^- \Xi = \frac{\delta \Xi}{\delta a_t} \, . \tag{1}$$

These are called the *creation derivative* and *annihilation derivative*, respectively. This naive idea has developed into a new kind of differential calculus with applications to Hitsuda–Skorohod quantum stochastic integrals and quantum martingale representation theorem [10–13]. In this paper we introduce a new type of a differential equation for white noise operators involving quantum white noise derivatives and, as application, discuss the implementation problem for the canonical commutation relation.

This paper is organized as follows: In Section 2 we assemble some basic notions in quantum white noise calculus and recall the quantum white noise derivatives, where a precise meaning is given to (1). In Section 3 we prove that the quantum white noise derivatives are Wick derivations (Theorem 3.1). In Section 4 we study a new type of a differential equation for white noise operators Ξ of the form:

$$\mathcal{D}\Xi = G \diamond \Xi,$$

where $\mathcal{D}$ is a Wick derivation and G is a white noise operator, and give a general form of the solutions (Theorem 4.1). In Section 5 we discuss the implementation problem along our approach. Define transformed annihilation and creation operators by

$$b(\zeta) = a(S\zeta) + a^*(T\zeta), \quad b^*(\zeta) = a^*(S\zeta) + a(T\zeta).$$

It is an interesting problem to determine an operator U satisfying the intertwining property:

$$Ua(\zeta) = b(\zeta)U, \quad Ua^*(\zeta) = b^*(\zeta)U, \tag{2}$$

where ζ runs over E. We observe that (2) is equivalent to the differential equations:

$$\begin{aligned} D^+_{S\zeta}U &= [a(\zeta - S\zeta) - a^*(T\zeta)] \diamond U, \\ (D^-_\zeta - D^+_{T\zeta})U &= [a^*(S\zeta - \zeta) + a(T\zeta)] \diamond U, \end{aligned}$$

respectively. The solutions are obtained by applying the general results in Section 4 and, under some conditions on S and T, a white noise operator U satisfying (2) is characterized (Theorem 5.3).

2. Quantum white noise derivatives

2.1. *White noise operators*

Let T be a topological space equipped with a σ-finite Borel measure dt and $H = L^2(T)$ be the (complex) Hilbert space of L^2-functions on T. The Boson Fock space over H is defined by

$$\Gamma(H) = \left\{ \phi = (f_n)\,;\, f_n \in H^{\widehat{\otimes} n},\ \|\phi\|^2 = \sum_{n=0}^{\infty} n!|f_n|_0^2 < \infty \right\},$$

where $|f_n|_0$ is the usual L^2-norm of $H^{\widehat{\otimes} n} = L^2_{\mathrm{sym}}(T^n)$. Taking a suitable Gelfand triple $E \subset H \subset E^*$, we construct the Hida–Kubo–Takenaka space:

$$(E) \subset \Gamma(H) \subset (E)^*$$

in the standard manner, see e.g., [14,15,19]. The canonical $\mathbb{C}$-bilinear form on $(E)^* \times (E)$ takes the form:

$$\langle\langle \Phi, \phi \rangle\rangle = \sum_{n=0}^{\infty} n!\, \langle F_n, f_n \rangle, \quad \Phi = (F_n) \in (E)^*, \quad \phi = (f_n) \in (E).$$

A continuous operator from (E) into $(E)^*$ is called a *white noise operator*. The space of white noise operators is denoted by $\mathcal{L}((E), (E)^*)$ and is equipped with the bounded convergence topology. We note that $\mathcal{L}((E), (E))$, $\mathcal{L}((E)^*, (E)^*)$ and $\mathcal{L}(\Gamma(H), \Gamma(H))$ are subspaces of $\mathcal{L}((E), (E)^*)$. The *annihilation operator* at a point $t \in T$ is defined by

$$a_t : (0, \dots, 0, \xi^{\otimes n}, 0, \dots) \mapsto (0, \dots, 0, n\xi(t)\xi^{\otimes(n-1)}, 0, \dots), \quad \xi \in E.$$

It is shown that $a_t \in \mathcal{L}((E), (E))$. Its adjoint operator $a_t^* \in \mathcal{L}((E)^*, (E)^*)$ is called the *creation operator* at t. The pair $\{a_t, a_t^* \,;\, t \in T\}$ is called the *quantum white noise* on T. For more detailed account of quantum white noise calculus see e.g., [9,19].

2.2. *Quantum white noise derivatives*

For $f \in E^*$ we define white noise operators:

$$a(f) = \Xi_{0,1}(f) = \int_T f(t) a_t \, dt, \quad a^*(f) = \Xi_{1,0}(f) = \int_T f(t) a_t^* \, dt, \tag{3}$$

which are called respectively the *annihilation* and *creation operators* associated with f. We recall the following

Lemma 2.1. *For $\zeta \in E$, both $a(\zeta)$ and $a^*(\zeta)$ belong to $\mathcal{L}((E), (E)) \cap \mathcal{L}((E)^*, (E)^*)$.*

Thus, for any white noise operator $\Xi \in \mathcal{L}((E), (E)^*)$ and $\zeta \in E$ the commutators

$$[a(\zeta), \Xi] = a(\zeta)\Xi - \Xi a(\zeta), \qquad -[a^*(\zeta), \Xi] = \Xi a^*(\zeta) - a^*(\zeta)\Xi,$$

are well-defined white noise operators, i.e., belong to $\mathcal{L}((E), (E)^*)$. We define

$$D_\zeta^+ \Xi = [a(\zeta), \Xi], \qquad D_\zeta^- \Xi = -[a^*(\zeta), \Xi]. \tag{4}$$

These are called the *creation derivative* and *annihilation derivative* of Ξ, respectively. Both together are called the *quantum white noise derivatives*.

By definition we have

$$(D_\zeta^+ \Xi)^* = D_\zeta^-(\Xi^*), \qquad (D_\zeta^- \Xi)^* = D_\zeta^+(\Xi^*). \tag{5}$$

It is apparent that $D_\zeta^\pm$ becomes a linear map from $\mathcal{L}((E),(E)^*)$ into itself. Moreover, it is known [12] that $(\zeta,\Xi)\mapsto D_\zeta^\pm\Xi$ is a continuous bilinear map from $E\times\mathcal{L}((E),(E)^*)$ into $\mathcal{L}((E),(E)^*)$.

Remark 2.1. As in the case of annihilation and creation operators (3), it is natural to write

$$D_\zeta^+ = \int_T \zeta(t)D_t^+\,dt, \qquad D_\zeta^- = \int_T \zeta(t)D_t^-\,dt.$$

In fact, this expression is useful for computation. However, it is not straightforward to define $D_t^\pm$ for each point $t\in T$ because the compositions $a_t\Xi$ and Ξa_t^* are not well-defined in general and (4) makes no sense if $a(\zeta)$ is replaced with a_t. The pointwisely defined quantum white noise derivatives $D_t^\pm$ are discussed in [12,13].

We now show examples. For each $S\in\mathcal{L}(E,E^*)$, by the kernel theorem there exists a unique $\tau_S\in(E\otimes E)^*$ such that

$$\langle \tau_S,\,\eta\otimes\xi\rangle = \langle S\xi,\,\eta\rangle\,, \qquad \xi,\eta\in E.$$

The integral kernel operator

$$\Delta_{\mathrm{G}}(S) = \Xi_{0,2}(\tau_S) = \int_{T\times T}\tau_S(s,t)a_sa_t\,dsdt$$

is called the *generalized Gross Laplacian* associated with S, see [5]. Note that $\Delta_{\mathrm{G}}(S)\in\mathcal{L}((E),(E))$. The usual *Gross Laplacian* is $\Delta_{\mathrm{G}}=\Delta_{\mathrm{G}}(I)$. The adjoint of $\Delta_{\mathrm{G}}(S)$ is given by

$$\Delta_{\mathrm{G}}^*(S) = \Xi_{2,0}(\tau_S) = \int_{T\times T}\tau_S(s,t)a_s^*a_t^*\,dsdt$$

and belongs to $\mathcal{L}((E)^*,(E)^*)$. The quantum white noise derivatives are given by

$$D_\zeta^+\Delta_{\mathrm{G}}(S)=0, \qquad D_\zeta^-\Delta_{\mathrm{G}}(S)=a(S\zeta)+a(S^*\zeta), \tag{6}$$

$$D_\zeta^-\Delta_{\mathrm{G}}^*(S)=0, \qquad D_\zeta^+\Delta_{\mathrm{G}}^*(S)=a^*(S\zeta)+a^*(S^*\zeta). \tag{7}$$

The integral kernel operator

$$\Lambda(S) = \Xi_{1,1}(\tau_S) = \int_{T\times T}\tau_S(s,t)a_s^*a_t\,dsdt$$

is called the *conservation operator* associated with S. In general, $\Lambda(S)\in\mathcal{L}((E),(E)^*)$. Note that $N\equiv\Lambda(I)$ is the *number operator*. The quantum white noise derivatives are given by

$$D_\zeta^-\Lambda(S)=a^*(S\zeta), \qquad D_\zeta^+\Lambda(S)=a(S^*\zeta). \tag{8}$$

3. Wick derivations

We first recall the operator symbol. For a white noise operator $\Xi \in \mathcal{L}((E),(E)^*)$, we define its symbol by

$$\widehat{\Xi}(\xi,\eta) = \langle\langle \Xi\phi_\xi, \phi_\eta \rangle\rangle, \qquad \xi,\eta \in E,$$

where $\phi_\xi = (1,\xi,\cdots,\xi^{\otimes n}/n!,\cdots)$ is an *exponential vector*. Every white noise operator is uniquely determined by its symbol [4,18,19].

For $\Xi_1, \Xi_2 \in \mathcal{L}((E),(E)^*)$ we define the Wick (or normal-ordered) product $\Xi_1 \diamond \Xi_2$ by

$$(\Xi_1 \diamond \Xi_2)\widehat{\ }(\xi,\eta) = \widehat{\Xi}_1(\xi,\eta)\widehat{\Xi}_2(\xi,\eta)e^{-\langle\xi,\eta\rangle}, \qquad \xi,\eta \in E.$$

In particular, for any $\Xi \in \mathcal{L}((E),(E)^*)$ we have

$$a_t \diamond \Xi = \Xi \diamond a_t = \Xi a_t\,, \qquad a_t^* \diamond \Xi = \Xi \diamond a_t^* = a_t^*\Xi, \tag{9}$$

where the right-hand sides are usual composition of operators. Equipped with the Wick product, $\mathcal{L}((E),(E)^*)$ becomes a commutative algebra.

A continuous linear map $\mathcal{D} : \mathcal{L}((E),(E)^*) \to \mathcal{L}((E),(E)^*)$ is called a *Wick derivation* if

$$\mathcal{D}(\Xi_1 \diamond \Xi_2) = (\mathcal{D}\Xi_1) \diamond \Xi_2 + \Xi_1 \diamond (\mathcal{D}\Xi_2)$$

for all $\Xi_1, \Xi_2 \in \mathcal{L}((E),(E)^*)$.

Theorem 3.1. *The creation and annihilation derivatives $D_\zeta^\pm$ are Wick derivations for any $\zeta \in E$.*

Proof. By (5) it is sufficient to show that the creation derivative is a Wick derivation. In general, for $\Xi \in \mathcal{L}((E),(E)^*)$ we have

$$\begin{aligned}
(D_\zeta^+\Xi)\widehat{\ }(\xi,\eta) &= \langle\langle (a(\zeta)\Xi - \Xi a(\zeta))\phi_\xi, \phi_\eta \rangle\rangle \\
&= \langle\langle \Xi\phi_\xi, a^*(\zeta)\phi_\eta \rangle\rangle - \langle\langle \Xi a(\zeta)\phi_\xi, \phi_\eta \rangle\rangle \\
&= \frac{d}{dt}\bigg|_{t=0} \langle\langle \Xi\phi_\xi, \phi_{\eta+t\zeta} \rangle\rangle - \langle\xi,\zeta\rangle\langle\langle \Xi\phi_\xi, \phi_\eta \rangle\rangle \\
&= \frac{d}{dt}\bigg|_{t=0} \widehat{\Xi}(\xi,\eta+t\zeta) - \langle\xi,\zeta\rangle\widehat{\Xi}(\xi,\eta).
\end{aligned} \tag{10}$$

Then for $\Xi = \Xi_1 \diamond \Xi_2$ we have

$$\begin{aligned}(D_\zeta^+\Xi)\widehat{\ }(\xi,\eta) &= \frac{d}{dt}\bigg|_{t=0} \widehat{\Xi}_1(\xi, t\zeta+\eta)\widehat{\Xi}_2(\xi, t\zeta+\eta)e^{-\langle \xi,\, t\zeta+\eta\rangle} \\ &\qquad - \langle \xi, \zeta\rangle \widehat{\Xi}_1(\xi,\eta)\widehat{\Xi}_2(\xi,\eta)e^{-\langle \xi,\,\eta\rangle} \\ &= \left(\frac{d}{dt}\bigg|_{t=0} \widehat{\Xi}_1(\xi, t\zeta+\eta)\right)\widehat{\Xi}_2(\xi,\eta)e^{-\langle \xi,\,\eta\rangle} \\ &\qquad + \widehat{\Xi}_1(\xi,\eta)\left(\frac{d}{dt}\bigg|_{t=0} \widehat{\Xi}_2(\xi, t\zeta+\eta)\right)e^{-\langle \xi,\,\eta\rangle} \\ &\qquad - 2\langle \xi, \zeta\rangle \widehat{\Xi}_1(\xi,\eta)\widehat{\Xi}_2(\xi,\eta)e^{-\langle \xi,\,\eta\rangle}.\end{aligned}$$

Viewing (10) once again, we obtain

$$(D_\zeta^+\Xi)\widehat{\ }(\xi,\eta) = ((D_\zeta^-\Xi_1)\diamond\Xi_2)\widehat{\ }(\xi,\eta) + (\Xi_1\diamond(D_\zeta^-\Xi_2))\widehat{\ }(\xi,\eta),$$

which completes the proof. □

Remark 3.1. In general, a Wick derivation $\mathcal{D}$ is expressed in the form:

$$\mathcal{D} = \int_T F(t)\diamond D_t^+\,dt + \int_T G(t)\diamond D_t^-\,dt,$$

where $F, G \in E \otimes \mathcal{L}((E),(E)^*)$. This gives a quantum counterpart of the characterization of Wick derivations on the white noise functions [3]. The study in this line will appear elsewhere.

4. Differential equations associated with Wick derivations

In this section, we study a new type of differential equation for white noise operator of the form:

$$\mathcal{D}\Xi = G\diamond\Xi, \tag{11}$$

where $\mathcal{D}$ is a Wick derivation and $G \in \mathcal{L}((E),(E)^*)$ is a given white noise operator.

We need the Wick exponential. For $Y \in \mathcal{L}((E),(E)^*)$ we define

$$\operatorname{wexp} Y = \sum_{n=0}^{\infty} \frac{1}{n!} Y^{\diamond n},$$

whenever the series converges in $\mathcal{L}((E),(E)^*)$. For more details see e.g., [6].

Theorem 4.1. *Assume that there exists an operator $Y \in \mathcal{L}((E),(E)^*)$ such that $\mathcal{D}Y = G$ and $\operatorname{wexp} Y$ is defined in $\mathcal{L}((E),(E)^*)$. Then every solution to (11) is of the form:*

$$\Xi = (\operatorname{wexp} Y)\diamond F, \tag{12}$$

where $F \in \mathcal{L}((E),(E)^*)$ *satisfying* $\mathcal{D}F = 0$.

Proof. It is straightforward to see that (12) is a solution to (11). To prove the converse, let Ξ be an arbitrary solution to (11). Set

$$F = (\text{wexp}\,(-Y)) \diamond \Xi.$$

Obviously, $F \in \mathcal{L}((E),(E)^*)$ and $\Xi = (\text{wexp}\,Y) \diamond F$. We only need to show that $\mathcal{D}F = 0$. In fact,

$$\begin{aligned} \mathcal{D}F &= -\mathcal{D}Y \diamond (\text{wexp}\,(-Y)) \diamond \Xi + (\text{wexp}\,(-Y)) \diamond \mathcal{D}\Xi \\ &= -G \diamond (\text{wexp}\,(-Y)) \diamond \Xi + (\text{wexp}\,(-Y)) \diamond G \diamond \Xi = 0. \end{aligned}$$

This completes the proof. □

Example 4.1. Let us consider the (system of) differential equations:

$$D_\zeta^+ \Xi = 0, \qquad \zeta \in E. \tag{13}$$

The vanishing creation derivatives suggest that $\Xi = \Xi(a_s, a_t^*; s, t \in T)$ does not depend on the creation operators. In fact, as is verified easily by Fock expansion, the solutions to (13) are given by

$$\Xi = \sum_{m=0}^{\infty} \Xi_{0,m}(\kappa_{0,m}).$$

In a similar manner, the solutions to

$$D_\zeta^- \Xi = 0, \qquad \zeta \in E, \tag{14}$$

are given by

$$\Xi = \sum_{l=0}^{\infty} \Xi_{l,0}(\kappa_{l,0}).$$

Consequently, a white noise operator satisfying both (13) and (14) are the scalar operators. Thus, the irreducibility of the canonical commutation relation is reproduced.

Example 4.2. Let us consider the differential equation:

$$D_\zeta^- \Xi = 2a(\zeta) \diamond \Xi, \qquad \zeta \in E. \tag{15}$$

We need to find $Y \in \mathcal{L}((E),(E)^*)$ satisfying $D_\zeta^- Y = 2a(\zeta)$. In fact, $Y = \Delta_G$ is a solution, see (6). Moreover, it is easily verified that $\text{wexp}\,\Delta_G$ is defined in $\mathcal{L}((E),(E))$. Then, a general solution to (15) is of the form:

$$\Xi = (\text{wexp}\,\Delta_G) \diamond F, \tag{16}$$

where $D_\zeta^- F = 0$ for all $\zeta \in E$. Now we consider the differential equation:

$$\begin{cases} D_\zeta^+ \Xi = 0, \\ D_\zeta^- \Xi = 2a(\zeta) \diamond \Xi, \qquad \zeta \in E. \end{cases} \tag{17}$$

Since the solution is of the form (16), we need only to find additional conditions for F satisfying $D_\zeta^+ \Xi = 0$. Noting that $D_\zeta^+ \Delta_{\mathrm{G}} = 0$, we have

$$D_\zeta^+ \Xi = (\operatorname{wexp} \Delta_{\mathrm{G}}) \diamond D_\zeta^+ F = 0.$$

Hence $D_\zeta^+ F = 0$ for all $\zeta \in E$. Combining with the condition $D_\zeta^- F = 0$ for all $\zeta \in E$, we conclude that F is a scalar operator (see Example 4.1). Consequently, the solution to (17) is of the form:

$$\Xi = C \operatorname{wexp} \Delta_{\mathrm{G}}, \qquad C \in \mathbb{C}.$$

5. White noise operators implementing transformed annihilation and creation operators

In this section, let $S, T \in \mathcal{L}(E, E)$ and consider transformed annihilation and creation operators:

$$b(\zeta) = a(S\zeta) + a^*(T\zeta),$$
$$b^*(\zeta) = a^*(S\zeta) + a(T\zeta),$$

where $\zeta \in E$. Our implementation problem is to find a white noise operator $U \in \mathcal{L}((E), (E)^*)$ satisfying

$$Ua(\zeta) = b(\zeta)U, \tag{18}$$
$$Ua^*(\zeta) = b^*(\zeta)U. \tag{19}$$

We start with (18). By definition, we have

$$\begin{aligned} Ua(\zeta) = b(\zeta)U &= (a(S\zeta) + a^*(T\zeta))\, U \\ &= D_{S\zeta}^+ U + Ua(S\zeta) + a^*(T\zeta)U, \end{aligned}$$

and hence

$$D_{S\zeta}^+ U = Ua(\zeta) - Ua(S\zeta) - a^*(T\zeta)U.$$

Writing the right-hand side in terms of the Wick product (9), we come to the differential equation:

$$D_{S\zeta}^+ U = [a(\zeta - S\zeta) - a^*(T\zeta)] \diamond U, \tag{20}$$

which is equivalent to (18). Similarly, (19) is equivalent to

$$(D_\zeta^- - D_{T\zeta}^+) U = [a^*(S\zeta - \zeta) + a(T\zeta)] \diamond U. \tag{21}$$

Theorem 5.1. *Assume that S is invertible and that $T^*S = S^*T$. Then a white noise operator $U \in \mathcal{L}((E),(E)^*)$ satisfies the intertwining property:*

$$Ua(\zeta) = b(\zeta)U, \qquad \zeta \in E,$$

if and only if U is of the form

$$U = \mathrm{wexp}\left\{-\frac{1}{2}\Delta_{\mathrm{G}}^*(TS^{-1}) + \Lambda((S^{-1})^* - I)\right\} \diamond F, \tag{22}$$

where $F \in \mathcal{L}((E),(E)^*)$ is an arbitrary white noise operator satisfying $D_\zeta^+ F = 0$ for all $\zeta \in E$, see Example 4.1.

Proof. We only need to solve the differential equation (20). It follows from (7) and (8) that

$$D_{S\zeta}^+ \Lambda((S^{-1})^* - I) = a(\zeta - S\zeta), \quad D_{S\zeta}^+ \Delta_{\mathrm{G}}^*(TS^{-1}) = 2a^*(T\zeta).$$

Then by Theorem 4.1 a general form of the solutions to (20) is given by (22), where $F \in \mathcal{L}((E),(E)^*)$ is an arbitrary white noise operator satisfying $D_{S\zeta}^+ F = 0$ for all $\zeta \in E$. Since S is invertible, the last condition for F is equivalent to that $D_\zeta^+ F = 0$ for all $\zeta \in E$. □

Remark 5.1. We may obtain a conventional expression of U from the Wick product form in (22). Note that

$$\mathrm{wexp}\left\{-\frac{1}{2}\Delta_{\mathrm{G}}^*(TS^{-1})\right\} = e^{-\frac{1}{2}\Delta_{\mathrm{G}}^*(TS^{-1})},$$

$$\mathrm{wexp}\left\{\Lambda((S^{-1})^* - I)\right\} = \Gamma((S^{-1})^*),$$

where $\Gamma((S^{-1})^*)$ is the second quantization of $(S^{-1})^*$. Hence, (22) becomes

$$U = e^{-\frac{1}{2}\Delta_{\mathrm{G}}^*(TS^{-1})} \diamond \Gamma((S^{-1})^*) \diamond F.$$

Since F contains no creation operators, the last Wick products are reduced to the usual compositions of operators:

$$U = e^{-\frac{1}{2}\Delta_{\mathrm{G}}^*(TS^{-1})}\Gamma((S^{-1})^*)F.$$

Theorem 5.2. *Assume the following conditions:*

(i) *S is invertible;*
(ii) *$T^*S = S^*T$;*
(iii) *$S^*S - T^*T = I$;*
(iv) *$ST^* = TS^*$.*

Then a white noise operator $U \in \mathcal{L}((E),(E)^*)$ *satisfies the intertwining property:*

$$Ua^*(\zeta) = b^*(\zeta)U, \qquad \zeta \in E,$$

if and only if U *is of the form:*

$$U = \mathrm{wexp}\left\{-\frac{1}{2}\Delta_{\mathrm{G}}^*(TS^{-1}) + \Lambda((S^{-1})^* - I) + \frac{1}{2}\Delta_{\mathrm{G}}(S^{-1}T)\right\} \diamond G,$$

where $G \in \mathcal{L}((E),(E)^*)$ *is an arbitrary white noise operator satisfying* $(D_\zeta^- - D_{T\zeta}^+)G = 0$ *for all* $\zeta \in E$.

Proof. Our task is to solve the differential equation (21). First we need to find a solution to the differential equation:

$$(D_\zeta^- - D_{T\zeta}^+)Y = a^*(S\zeta - \zeta) + a(T\zeta). \tag{23}$$

As is easily verified,

$$Y = \Delta_{\mathrm{G}}^*(K) + \Lambda(L) + \Delta_{\mathrm{G}}(M), \qquad K = K^*, \quad M = M^*,$$

satisfies (23) if and only if

$$2M - L^*T = T, \qquad L - 2KT = S - I.$$

Thanks to the conditions (i)–(iv) we may choose

$$K = -\frac{1}{2}TS^{-1}, \quad L = (S^{-1})^* - I, \quad M = \frac{1}{2}S^{-1}T.$$

Then the assertion follows immediately from Theorem 4.1. □

Theorem 5.3. *Assumptions being the same as in Theorem 5.2, a white noise operator* $U \in \mathcal{L}((E),(E)^*)$ *satisfies the following intertwining properties:*

$$Ua(\zeta) = b(\zeta)U, \qquad Ua^*(\zeta) = b^*(\zeta)U, \qquad \zeta \in E,$$

if and only if U *is of the form:*

$$U = C\,\mathrm{wexp}\left\{-\frac{1}{2}\Delta_{\mathrm{G}}^*(TS^{-1}) + \Lambda((S^{-1})^* - I) + \frac{1}{2}\Delta_{\mathrm{G}}(S^{-1}T)\right\}, \tag{24}$$

or equivalently,

$$U = C\,e^{-\frac{1}{2}\Delta_{\mathrm{G}}^*(TS^{-1})}\Gamma((S^{-1})^*)\,e^{\frac{1}{2}\Delta_{\mathrm{G}}(S^{-1}T)}, \tag{25}$$

where $C \in \mathbb{C}$.

Proof. It follows from Theorems 5.1 and 5.2 that U is of the form

$$U = \mathrm{wexp}\left\{-\frac{1}{2}\Delta_{\mathrm{G}}^{*}(TS^{-1}) + \Lambda((S^{-1})^{*} - I)\right\} \diamond F$$

$$= \mathrm{wexp}\left\{-\frac{1}{2}\Delta_{\mathrm{G}}^{*}(TS^{-1}) + \Lambda((S^{-1})^{*} - I) + \frac{1}{2}\Delta_{\mathrm{G}}(S^{-1}T)\right\} \diamond G, \quad (26)$$

where $F, G \in \mathcal{L}((E),(E)^{*})$ satisfy

$$D_{\zeta}^{+}F = 0, \qquad (D_{\zeta}^{-} - D_{T\zeta}^{+})G = 0,$$

for all $\zeta \in E$. We see from (26) that

$$G = F \diamond \mathrm{wexp}\left\{-\frac{1}{2}\Delta_{\mathrm{G}}(S^{-1}T)\right\}.$$

Since the right hand side contains no creation operators, we have

$$D_{\zeta}^{+}G = 0, \qquad \zeta \in E. \quad (27)$$

Then,

$$0 = (D_{\zeta}^{-} - D_{T\zeta}^{+})G = D_{\zeta}^{-}G, \qquad \zeta \in E. \quad (28)$$

It follows from (27) and (28) that G is a scalar operator (Example 4.2). Consequently, (26) becomes (24). □

Remark 5.2. Condition (ii) in Theorem 5.2 is necessary and sufficient to have

$$[b(\zeta), b(\eta)] = [b^{*}(\zeta), b^{*}(\eta)] = 0, \qquad \zeta, \eta \in E.$$

While, condition (iii) therein is necessary and sufficient to have

$$[b(\zeta), b^{*}(\eta)] = \langle \zeta, \eta \rangle, \qquad \zeta, \eta \in E.$$

Remark 5.3. For $U \in \mathcal{L}(E, E^{*})$ we have $e^{\Delta_{\mathrm{G}}(U)} \in \mathcal{L}((E),(E))$ and for $V \in \mathcal{L}(E, E^{*})$ we have $\Gamma(V) \in \mathcal{L}((E),(E)^{*})$. Then their composition

$$\mathcal{G}_{U,V} = \Gamma(V)\, e^{\Delta_{\mathrm{G}}(U)}$$

becomes a white noise operator. This is called a *generalized Fourier–Gauss transform* and its adjoint operator $\mathcal{G}_{U,V}^{*}$ a *generalized Fourier–Mehler transform*, see [5]. The operator U in (25) is a composition of these transforms. The work in this line is in progress.

Acknowledgements

This work was supported by the Korea–Japan Basic Scientific Cooperation Program (2007–2009) "Noncommutative Stochastic Analysis and Its Applications to Network Science."

References

1. E. A. Berezin: "The Method of Second Quantization," Academic Press, 1966.
2. N. N. Bogoliubov, A. A. Logunov and I. T. Todorov: "Introduction to Axiomatic Quantum Field Theory," Benjamin, Massachusetts, 1975.
3. D. M. Chung and T. S. Chung: *Wick derivations on white noise functionals*, J. Korean Math. Soc. **33** (1996), 993–1008.
4. D. M. Chung, T. S. Chung and U. C. Ji: *A simple proof of analytic characterization theorem for operator symbols,* Bull. Korean Math. Soc. **34** (1997), 421–436.
5. D. M. Chung and U. C. Ji: *Transforms on white noise functionals with their applications to Cauchy problems*, Nagoya Math. J. **147** (1997), 1–23.
6. D. M. Chung, U. C. Ji and N. Obata: *Quantum stochastic analysis via white noise operators in weighted Fock space*, Rev. Math. Phys. **14**, 241–272 (2002)
7. J. Glimm and A. Jaffe: *Boson quantum field models,* in "Collected Papers, Vol.1," pp. 125–199, Birkhäuser, 1985.
8. R. Haag: *On quantum field theories,* Dan. Mat. Fys. Medd. **29** (1955), no. 12, 1–37.
9. U. C. Ji and N. Obata: *Quantum white noise calculus*, in "Non-Commutativity, Infinite-Dimensionality and Probability at the Crossroads (N. Obata, T. Matsui and A. Hora, Eds.)," pp. 143–191, World Scientific, 2002.
10. U. C. Ji and N. Obata: *Generalized white noise operators fields and quantum white noise derivatives*, Sem. Congr. Soc. Math. France **16** (2007), 17–33.
11. U. C. Ji and N. Obata: *Annihilation-derivative, creation-derivative and representation of quantum martingales*, Commun. Math. Phys. **286** (2009), 751–775.
12. U. C. Ji and N. Obata: *Quantum stochastic integral representations of Fock space operators*, to appear in Stochastics.
13. U. C. Ji and N. Obata: *Quantum stochastic gradients*, to appear in Interdiscip. Inform. Sci.
14. I. Kubo and S. Takenaka: *Calculus on Gaussian white noise I–IV*, Proc. Japan Acad. **56A** (1980), 376–380; 411–416; **57A** (1981), 433–437; **58A** (1982), 186–189.
15. H.-H. Kuo: "White Noise Distribution Theory," CRC Press, 1996.
16. P. Krée: *La théorie des distributions en dimension quelconque et l'intégration stochastique,* in "Stochastic Analysis and Related Topics (H. Korezlioglu and A. S. Ustunel eds.)," pp. 170–233, Lect. Notes in Math. Vol. 1316, Springer–Verlag, 1988.
17. P. A. Meyer: *Distributions, noyaux, symboles d'après Krée,* in "Séminaire de Probabilités XXII (J. Azéma et al. eds.)," pp. 467–476, Lect. Notes in Math. Vol. 1321, Springer–Verlag, 1988.
18. N. Obata: *An analytic characterization of symbols of operators on white noise functionals,* J. Math. Soc. Japan **45** (1993), 421–445.
19. N. Obata: "White Noise Calculus and Fock Space," Lect. Notes in Math. Vol. 1577, Springer, 1994.

THE GIBBS CONDITIONING PRINCIPLE FOR WHITE NOISE DISTRIBUTIONS: INTERACTING AND NON-INTERACTING CASES

Fernanda Cipriano

GFM e Dep. de Matemática FCT-UNL
Av. Prof. Gama Pinto 2, 1649-003, Lisboa, Portugal
E-mail:cipriano@cii.fc.ul.pt

Soumaya Gheryani* and Habib Ouerdiane

Department of Mathematics, Faculty of Sciences of Tunis
University of Tunis El Manar, Tunisia
**E-mail: soumaye_gheryani@yahoo.fr*
habib.ouerdiane@fst.rnu.tn

Let $Y = (Y_i)_{i \in \mathbb{N}}$ be a sequence of i. i. d. random variables taking values in the dual of a real nuclear Fréchet space. We denote by L_n^Y the empirical measure associated with these variables. In this paper we study the limit law of $(Y_1, ..., Y_k)$ under the constraint $E_U(L_n^Y) \in D$ where D is a measurable subset of $\mathbb{R}$, and E_U is the energy functional associated to a fixed positive test function U. For some particular choices of the functional E_U, we consider non interacting case , i. e., when the particles $Y_1, \cdots, Y_k$ do not affect each other and the case where interaction is present. In both cases, we show that the conditional law of Y_1 converges, as $n \to \infty$, to a Gibbs measure.

Keywords: Positive White Noise distributions; Large Deviation Principle; Gibbs measure.

1. Introduction

In recent years, large deviation techniques have been used extensively in connection with statistical mechanics and interacting particle systems. A particularly interesting application of large deviation techniques can be found in Ref. 3.

In this paper, let X be a real nuclear Fréchet space and X' its topological dual. Denote by N' the complexification of X'. For a Young function θ, we denote by $\mathcal{F}_\theta(N')$ the space of holomorphic functions on N' with θ-exponential growth of minimal type.[5] We define in Ref. 1 the space

$\mathcal{F}'_\theta(N')_{1,+}$ of positive distributions Ψ such that $\langle\langle \Psi, \mathbf{1} \rangle\rangle = 1$. Let us fix a positive distribution Φ in the space $\mathcal{F}'_\theta(N')_{1,+}$. Thus, we can associate, by Theorem 2.1, a probability measure μ_Φ on X' denoted by μ for simplicity. Consider a sequence $Y = (Y_i)_{i\in\mathbb{N}}$ of independent and identically distributed random variables defined on the probability space $(\Omega, \mathcal{A}, P)$ valued in X' with common law μ. Let L_n^Y denote the empirical measures associated with these variables, namely,

$$\begin{aligned} L_n^Y : (\Omega, \mathcal{A}, \mathrm{P}) &\to \mathcal{F}'_\theta(N')_{1,+} \\ \omega &\mapsto \frac{1}{n}\sum_{i=1}^{n} \delta_{Y_i(\omega)}, \end{aligned} \tag{1}$$

where for every measurable subset $A \subset X'$, $\delta_{Y_i(\omega)}(A) = 1$ if $Y_i(\omega) \in A$ and $\delta_{Y_i(\omega)}(A) = 0$ otherwise.
Let $(A_\delta)_{\delta>0}$ be a nested family of measurable sets in $\mathcal{F}'_\theta(N')_{1,+}$, i. e. $A_\delta \subseteq A_{\delta'}$, if $\delta < \delta'$ and $(F_\delta)_{\delta>0}$ a nested family of closed sets such that $A_\delta \subseteq F_\delta$. Then, it was proved in Ref. 1 that the set of minimizers of the entropy

$$\mathcal{M} = \{\Psi_0 \in F_0 : H(\Psi_0|\mu) = \inf_{\Psi\in F_0} H(\Psi|\mu)\} \tag{2}$$

is reduced to a unique measure Ψ_* when $A_0 = \cap_{\delta>0} A_\delta$ is convex, and therefore that

$$\lim_{\delta\to 0}\lim_{n\to\infty} \mathrm{P}((\mathrm{Y}_1, \cdots, \mathrm{Y}_\mathrm{k}) \in . \,|\mathrm{L}_\mathrm{n}^\mathrm{Y} \in \mathrm{A}_\delta) = \Psi_*^\mathrm{k}(.). \tag{3}$$

In the present paper, we investigate the case of non interacting particles $(Y_i)_{i\in\mathbb{N}}$ and also the case where the particles affect each other. Using basic techniques from the papers Refs. 1, 3 and 5, we show that in both cases the positive distribution Ψ_* correspond to a Gibbs measures. The paper is organized as follows: In section 2, we describe briefly the space of holomorphic functions of θ-exponential growth. Then, we give some basic results about large deviation theory and its application that will be needed during this work. In section 3, we study the case where there is no interaction between the particles $(Y_i)_{i\in\mathbb{N}}$. Let us consider the energy functional

$$\begin{aligned} E_U : \quad \mathcal{F}'_\theta(N')_{1,+} &\to [-1, \infty] \\ \Psi \quad &\mapsto E_U(\Psi) = \langle\langle \Psi, U \rangle\rangle - 1 \end{aligned}$$

for some test function $U \in \mathcal{F}_\theta(N')$. We prove in Theorem 3.1 that the set of minimizers $\mathcal{M}$ is reduced to the unique Gibbs measure γ_{β^*}, $\beta^* \in (0, \infty)$, defined by

$$\frac{d\gamma_{\beta^*}}{d\mu} = \frac{e^{-\beta^* U(x)}}{Z_{\beta^*}}, \tag{4}$$

where the normalizing constant $Z_{\beta^*} = \int_{X'} e^{-\beta^* U(x)} d\mu(x)$ is the partition function. In section 4, we consider the case where the interaction of the particles is present. Unlike the non-interacting case, here even the existence of the Gibbs measure needs to be proved (see Lemma 4.1). For that aim, we consider a test function $U : X' \times X' \to \mathbb{R}$ and the energy functional

$$\begin{aligned} E_U : \quad & \mathcal{F}'_\theta(N')_{1,+} \to [-1, \infty] \\ & \Psi \qquad\quad \mapsto E_U(\Psi) = \langle\langle \Psi \otimes \Psi, U \rangle\rangle - 1. \end{aligned}$$

Therefore, we define naturally the following interaction Hamiltonian

$$H_\beta(\Psi) = H(\Psi|\mu) + \frac{\beta}{2} \langle\langle \Psi \otimes \Psi, U \rangle\rangle, \quad \beta \in [0, \infty);$$

and we show, under some assumptions, the existence of a unique minimizer γ_{β^*} of the Hamiltonian H_β. In fact, that minimizer is the Gibbs measure γ_{β^*} given by

$$\frac{d\gamma_{\beta^*}}{d\mu} = \frac{e^{-\beta^* \langle\langle \gamma_{\beta^*}, U_x \rangle\rangle}}{Z_{\beta^*}}, \tag{5}$$

where U_x denotes the one variable test function $U_x(y) = U(x, y)$ and the normalizing constant $Z_{\beta^*} = \int_{X'} e^{-\beta^* \langle\langle \gamma_{\beta^*}, U_x \rangle\rangle} d\mu(x)$.

2. Background

2.1. *Definitions and results about the space of holomorphic functions with exponential growth*

In this section, elementary facts, notations and some results about the space of holomorphic functions with exponential growth and its dual space are recalled from the paper Ref. 5 (see also Refs. 8 and 7) for future use.

Let X be a real nuclear Fréchet space with topology given by an increasing family $\{|\cdot|_p; p \in \mathbb{N}_0\}$ of Hilbertian norms, $\mathbb{N}_0$ being the set of nonnegative integers. Then X is represented as

$$X = \bigcap_{p \in \mathbb{N}_0} X_p,$$

where X_p is the completion of X with respect to the norm $|\cdot|_p$. We use X_{-p} to denote the dual space of X_p. Then the dual space X' of X can be represented as

$$X' = \bigcup_{p \in \mathbb{N}_0} X_{-p}$$

and is equipped with the inductive limit topology.

Let $N = X + iX$ and $N_p = X_p + iX_p$, $p \in \mathbb{Z}$, be the complexifications of X and X_p, respectively. For $n \in \mathbb{N}_0$ we denote by $N^{\widehat{\otimes} n}$ the n-fold symmetric tensor product of N equipped with the π-topology and by $N_p^{\widehat{\otimes} n}$ the n-fold symmetric Hilbertian tensor product of N_p. We will preserve the notation $|\cdot|_p$ and $|\cdot|_{-p}$ for the norms on $N_p^{\widehat{\otimes} n}$ and $N_{-p}^{\widehat{\otimes} n}$, respectively.

Let θ be a Young function, i.e., it is a continuous, convex, and increasing function defined on $\mathbb{R}_+$ satisfying the condition $\lim_{x\to\infty} \theta(x)/x = \infty$. We define the conjugate function θ^* of θ by

$$\theta^*(x) = \sup_{t\geq 0} \big(tx - \theta(t)\big), \quad x \geq 0.$$

For a Young function θ, we denote by $\mathcal{F}_\theta(N')$ the space of holomorphic functions on N' with exponential growth of order θ and of minimal type. Similarly, let $\mathcal{G}_\theta(N)$ denotes the space of holomorphic functions on N with exponential growth of order θ and of arbitrary type. Moreover, for each $p \in \mathbb{Z}$ and $m > 0$, define $\mathrm{Exp}(N_p, \theta, m)$ to be the space of entire functions f on N_p satisfying the condition:

$$\|f\|_{\theta,p,m} = \sup_{x\in N_p} |f(x)| e^{-\theta(m|x|_p)} < \infty.$$

Then the spaces $\mathcal{F}_\theta(N')$ and $\mathcal{G}_\theta(N)$ can be represented as

$$\mathcal{F}_\theta(N') = \bigcap_{p\in\mathbb{N}_0,\, m>0} \mathrm{Exp}(N_{-p}, \theta, m),$$

$$\mathcal{G}_\theta(N) = \bigcup_{p\in\mathbb{N}_0,\, m>0} \mathrm{Exp}(N_p, \theta, m),$$

and are equipped with the projective limit topology and the inductive limit topology, respectively. The space $\mathcal{F}_\theta(N')$ is called the space of *test functions* on N'. We shall define its dual space $\mathcal{F}'_\theta(N')$ called the space of *distributions* on N'. This space is equipped with the strong topology $\sigma(\mathcal{F}'_\theta(N'), \mathcal{F}_\theta(N'))$ introduced by the smallest σ-field $\mathcal{E}_\mathcal{U}$ that makes all $p_\phi : \Phi \mapsto \langle\langle \Phi, \phi \rangle\rangle$ measurable.

For $p \in \mathbb{N}_0$ and $m > 0$, we define the Hilbert spaces

$$F_{\theta,m}(N_p) = \Big\{ \vec{\varphi} = (\varphi_n)_{n=0}^\infty \; ; \; \varphi_n \in N_p^{\widehat{\otimes} n}, \sum_{n\geq 0} \theta_n^{-2} m^{-n} |\varphi_n|_p^2 < \infty \Big\},$$

$$G_{\theta,m}(N_{-p}) = \Big\{ \vec{\Phi} = (\Phi_n)_{n=0}^\infty \; ; \; \Phi_n \in N_{-p}^{\widehat{\otimes} n}, \sum_{n\geq 0} (n!\theta_n)^2 m^n |\Phi_n|_{-p}^2 < \infty \Big\},$$

where $\theta_n = \inf_{r>0} e^{\theta(r)}/r^n$, $n \in \mathbb{N}_0$. Put

$$F_\theta(N) = \bigcap_{p\in\mathbb{N}_0, m>0} F_{\theta,m}(N_p),$$

$$G_\theta(N') = \bigcup_{p\in\mathbb{N}_0, m>0} G_{\theta,m}(N_{-p}).$$

The space $F_\theta(N)$ equipped with the projective limit topology is a nuclear Fréchet space.[5] The space $G_\theta(N')$ carries the dual topology of $F_\theta(N)$ with respect to the $\mathbb{C}$-bilinear pairing given by

$$\langle\langle \vec{\Phi}, \vec{\varphi} \rangle\rangle = \sum_{n\geq 0} n! \langle \Phi_n, \varphi_n \rangle, \tag{6}$$

where $\vec{\Phi} = (\Phi_n)_{n=0}^{\infty} \in G_\theta(N')$ and $\vec{\varphi} = (\varphi_n)_{n=0}^{\infty} \in F_\theta(N)$.

It was proved in Ref. 5 that the *Taylor map* defined by

$$T\colon\ \varphi \longmapsto \left(\frac{1}{n!}\varphi^{(n)}(0)\right)_{n=0}^{\infty}$$

is a topological isomorphism from $\mathcal{F}_\theta(N')$ onto $F_\theta(N)$.

The Taylor map T is also a topological isomorphism from $\mathcal{G}_{\theta^*}(N))$ onto $G_\theta(N'))$. The action of a distribution $\Phi \in \mathcal{F}'_\theta(N')$ on a test function $\varphi \in \mathcal{F}_\theta(N')$ can be expressed in terms of the Taylor map as follows:

$$\langle\langle \Phi, \varphi \rangle\rangle = \langle\langle \vec{\Phi}, \vec{\varphi} \rangle\rangle,$$

where $\vec{\Phi} = (T^*)^{-1}\Phi$ and $\vec{\varphi} = T\varphi$.

It is easy to see that for each $\xi \in N$, the exponential function

$$e_\xi(z) = e^{\langle z, \xi\rangle}, \quad z \in N',$$

is a test function in the space $\mathcal{F}_\theta(N')$ for any Young function θ. Thus we can define the *Laplace transform* of a distribution $\Phi \in \mathcal{F}'_\theta(N')$ by

$$\widehat{\Phi}(\xi) = \langle\langle \Phi, e_\xi \rangle\rangle, \quad \xi \in N. \tag{7}$$

From the paper Ref. 5, we have the duality Theorem which says that the Laplace transform is a topological isomorphism from $\mathcal{F}'_\theta(N')$ onto $\mathcal{G}_{\theta^*}(N)$. We denote by $\mathcal{F}_\theta(N')_+$ the cone of positive test functions, i.e.,

$$f \in \mathcal{F}_\theta(N')_+ \quad if \quad f(y + i0) \geq 0, \quad \forall y \in X'.$$

Then, the space $\mathcal{F}'_\theta(N')_+$ of positive distributions is defined as the space

$$\{\Phi \in \mathcal{F}'_\theta(N') : \langle\langle \Phi, f \rangle\rangle \geq 0, f \in \mathcal{F}_\theta(N')_+\} \tag{8}$$

We recall the following integral representation of positive distributions:

Theorem 2.1.[9] *Let $\Phi \in \mathcal{F}'_\theta(N')_+$, then there exists a unique Radon measure μ_Φ on X', such that*

$$\Phi(f) = \int_{X'} f(y + i0) d\mu_\Phi(y) \; ; \; f \in \mathcal{F}_\theta(N').$$

Conversely, let μ be a finite, positive Borel measure on X'. Then μ represents a positive distribution in $\mathcal{F}'_\theta(N')_+$ if and only if μ is supported by some X_{-p}, $p \in \mathbb{N}^*$, and there exists some $m > 0$ such that

$$\int_{X_{-p}} e^{\theta(m|y|_{-p})} d\mu(y) < \infty. \tag{9}$$

Remark 2.1. Note that the test function space $\mathcal{F}_\theta(N')$ can be generalized in two infinite dimensional variables test functions space denoted by $\mathcal{F}_{(\theta_1,\theta_2)}(N'_1 \oplus N'_2)$ (see Ref. 6), where θ_1, θ_2 are two Young functions and N_1, N_2 are two nuclear Fréchet spaces.

2.2. *Basic results about Large deviation theory and the Gibbs conditioning principle*

In this section, we give some useful results that will be needed.

We denote by $\mathcal{P}_\theta(X')$ the space of probability measures on the set $\mathcal{F}'_\theta(N')_{1,+}$ defined by

$$\mathcal{F}'_\theta(N')_{1,+} = \{\Psi \in \mathcal{F}'_\theta(N')_+ : \langle\langle \Psi, \mathbf{1} \rangle\rangle = 1\}.$$

Let $\Phi \in \mathcal{F}'_\theta(N')_{1,+}$ be a fixed distribution. Denote by $\mu = \mu_\Phi$ the associated measure (see Theorem 2.1). For every integer n, we define the empirical distribution functional

$$\begin{aligned} L_n : \quad & (X')^n \to \mathcal{F}'_\theta(N')_{1,+} \\ (\sigma_1, \sigma_2, \ldots, \sigma_n) & \mapsto \frac{1}{n} \sum_{i=1}^n \delta_{\sigma_i}, \end{aligned} \tag{10}$$

where δ_{σ_i} denotes the Dirac measure at the point σ_i. We define the measure $\tilde{\mu}_n$ to be the distribution of L_n under the n times product of the measure μ denoted by μ^n. Denote by $\tilde{\mu}_1$ the image measure of μ under the map

$$\begin{aligned} L_1 : X' & \to \mathcal{F}'_\theta(N')_{1,+} \\ \sigma & \mapsto \delta_\sigma. \end{aligned} \tag{11}$$

Then, the Logarithmic generating function (see Refs. 2, 3 and 4) of $\tilde{\mu}_1$ is given by

$$\begin{aligned}\Lambda_{\tilde{\mu}_1}(f) &:= \log\left(\int_{\mathcal{F}'_\theta(N')_{1,+}} e^{\langle\langle f,\xi\rangle\rangle} d\tilde{\mu}_1(\xi)\right) \\ &= \log\left(\int_{X'} e^{f(x)} d\mu(x)\right), \quad f \in \mathcal{F}_\theta(N'),\end{aligned} \tag{12}$$

and we define the Legendre transform of $\Lambda_{\tilde{\mu}_1}$ by

$$\Lambda^*_{\tilde{\mu}_1}(\Psi) := \sup_{f\in\mathcal{F}_\theta(N')} \{\langle\langle f,\Psi\rangle\rangle - \Lambda_{\tilde{\mu}_1}(f)\}, \quad \Psi \in \mathcal{F}'_\theta(N').$$

Note that, by Theorem 2.1, if $\Psi \in \mathcal{F}'_\theta(N')_{1,+}$, we have

$$\Lambda^*_{\tilde{\mu}_1}(\Psi) = \sup_{f\in\mathcal{F}_\theta(N')} \left\{\int_{X'} f d\mu_\Psi - \Lambda_{\tilde{\mu}_1}(f)\right\}. \tag{13}$$

Now, we recall (see Ref. 4) the relative entropy of a probability measure ν with respect to μ by

$$H(\nu|\mu) = \begin{cases} \int_{X'} f \log f d\mu & \text{if} \quad \nu \ll \mu \quad \text{and} \quad f = \frac{d\nu}{d\mu} \\ \infty & \text{otherwise.}\end{cases} \tag{14}$$

where the symbol $\nu \ll \mu$ denotes that ν is absolutely continuous with respect to μ. So, in our framework, a positive distribution Ψ is absolutely continuous with respect to the fixed distribution Φ if and only if there exists $f \in \mathcal{F}_\theta(N')_+$ such that $\Psi = f\Phi$, and therefore the entropy for distributions is given by

$$\begin{aligned}H(\Psi|\Phi) &= H(\mu_\Psi|\mu_\Phi) \\ &:= \langle\langle \Phi, f\log f\rangle\rangle \\ &= \int_{X'} f(x)\log f(x) d\mu_{\Phi(x)}.\end{aligned}$$

For any distribution $\Psi \in \mathcal{F}'_\theta(N')_{1,+}$ such that the associated measure μ_Ψ is absolutely continuous with respect to the fixed measure $\mu = \mu_\Phi$, we prove in Ref. 1 that the rate function $\Lambda^*_{\tilde{\mu}_1}$ coincides with the entropy

$$\Lambda^*_{\tilde{\mu}_1}(\Psi) = H(\Psi|\Phi). \tag{15}$$

Therefore, equation (15) allows us to extend Sanov's theorem to positive distributions (see Ref. 1 and references therein) as follows:

Theorem 2.2.[1] *Let $\Phi \in \mathcal{F}'_\theta(N')_{1,+}$ be a positive distribution and $\mu_\Phi = \mu$ the associated measure. Let $\tilde{\mu}_n \in \mathcal{P}_\theta(X')$ be the distribution of the function*

L_n under μ^n. Then, $H(.|\mu)$ is a good, convex rate function on $\mathcal{F}'_\theta(N')_{1,+}$ and $\{\tilde{\mu}_n : n \geq 1\}$ satisfies the full large deviation principle with rate function $H(.|\mu)$.

As an application of our extended Sanov's theorem, we study also in Ref. 1 the Gibbs conditioning principle. In fact, let $Y = (Y_i)_{i\in\mathbb{N}}$ be a sequence of random variables defined on the probability space $(\Omega, \mathcal{A}, \mathrm{P})$ with values in X', independent and identically distributed with law μ. Then, the empirical measure is defined by

$$\begin{aligned} L_n^Y : (\Omega, \mathcal{A}, \mathrm{P}) &\to \mathcal{F}'_\theta(N')_{1,+} \\ \omega &\mapsto \frac{1}{n}\sum_{i=1}^{n} \delta_{Y_i(\omega)}. \end{aligned} \tag{16}$$

For all measurable subset A of $\mathcal{F}'_\theta(N')_{1,+}$, we define the set function H given by

$$H(A) = \inf\{H(\Psi|\mu); \Psi \in A\}.$$

Let $(A_\delta)_{\delta>0}$ be a family of increasing borelian subsets of $\mathcal{F}'_\theta(N')_{1,+}$. Let $(F_\delta)_{\delta>0}$ be a family of increasing closed subsets such that $A_\delta \subseteq F_\delta$. Define $F_0 = \bigcap_{\delta>0} F_\delta$ and $A_0 = \bigcap_{\delta>0} A_\delta$ and suppose that the family $(A_\delta)_{\delta>0}$ satisfies the following assumption:

(A) $H(A_0) < \infty$ and for any $\delta > 0$

$$\mathrm{P}(\{\mathrm{L_n^Y} \in \mathrm{A}_\delta\}) > 0, \quad \forall \mathrm{n} \geq 1.$$

Consider the set of the minimizers of the entropy

$$\mathcal{M} = \{\Psi \in F_0 : H(\Psi|\mu) = H(A_0)\}.$$

If we suppose that A_0 is a convex set, then $\mathcal{M}$ is reduced to a unique element Ψ_* of the space $\mathcal{F}'_\theta(N')_{1,+}$, i. e.

$$\mathcal{M} = \{\Psi_*\}. \tag{17}$$

Denote by

$$\mu^n_{Y^k|A_\delta} = \mathrm{P}((\mathrm{Y}_1, ..., \mathrm{Y_k}) \in .|\mathrm{L_n^Y} \in \mathrm{A}_\delta),$$

the law of $Y^k := (Y_1, Y_2, ..., Y_k)$ conditional upon the event $\{L_n^Y \in A_\delta\}$, then we have the following result.

Theorem 2.3.[1] *Assume that (**A**) is fulfilled and that the constraint set A_0 is convex. Then $\mu^n_{Y^k|A_\delta}$ converge weakly in $\mathcal{F}'_\theta(N')_{1,+}$ to ν_*^k, when $n \to \infty$ and $\delta \to 0$, where ν_* denotes the measure associated to the distribution Ψ_* defined in (17).*

3. The Non-Interacting case

Let $U : X' \to [0, \infty)$ be a positive test function in $\mathcal{F}_\theta(N')_+$. Define the energy functional

$$E_U : \quad \mathcal{F}'_\theta(N')_{1,+} \to [-1, \infty]$$
$$\Psi \mapsto E_U(\Psi) = \langle\langle \Psi, U \rangle\rangle - 1. \tag{18}$$

In this section, we are interested in the particular case where the measurable sets A_δ have the special following forms:

$$A_\delta = \{\Psi : |E_U(\Psi)| \leq \delta\}. \tag{19}$$

Sine U is a test function A_δ are closed sets, and we consider $F_\delta = A_\delta$. Then, consider the constraint

$$\{L_n^Y \in A_\delta\} = \{|E_U(L_n^Y)| \leq \delta\} = \{|\frac{1}{n}\sum_{i=1}^{n} U(Y_i) - 1| \leq \delta\}.$$

Our goal is to show that, in this case, the positive distribution Ψ_* defined in (17) is a Gibbs measure γ_β. Namely, for any reference measure μ and any $\beta \geq 0$, the Gibbs measure γ_β is defined as

$$\frac{d\gamma_\beta}{d\mu} = \frac{e^{-\beta U(x)}}{Z_\beta}, \tag{20}$$

where $Z_\beta = \int_{X'} e^{-\beta U(x)} d\mu(x)$ is the normalizing constant.

Remark 3.1. Let μ be the fixed reference measure. Then, the Gibbs measure γ_β defined in (20) is an element of $\mathcal{F}'_\theta(N')_{1,+}$. In fact for any test function $\varphi \in \mathcal{F}_\theta(N')$

$$\langle\langle \gamma_\beta, \varphi \rangle\rangle = \frac{1}{Z_\beta} \langle\langle \mu, e^{-\beta U(x)} \varphi \rangle\rangle. \tag{21}$$

Since $\beta \geq 0$, U is a positive function and $\mu \in \mathcal{F}'_\theta(N')_{1,+}$, the equality (21) shows that the Gibbs measures γ_β belong to $\mathcal{F}'_\theta(N')_{1,+}$.

The following lemma is the key to show that the unique solution of the optimization problem

$$\inf_{F_0 = \{\Psi : \langle\langle \Psi, U \rangle\rangle = 1\}} H(\Psi|\mu) \tag{22}$$

is a Gibbs measure.

Lemma 3.1. *Assume that $\mu(\{x : U(x) < 1\}) > 0$ and $E_U(\mu) > 0$. Then, there exists a unique $\beta^* \in (0, \infty)$ such that the Gibbs measure γ_{β^*} satisfies $E_U(\gamma_{\beta^*}) = 0$.*

Proof. It is obvious to see that the function

$$\beta \in (0,\infty) \mapsto \log Z_\beta, \tag{23}$$

is a C^∞ function. Then, by the dominated convergence theorem, we have

$$\langle\!\langle \gamma_\beta, U \rangle\!\rangle = -\frac{d}{d\beta} \log Z_\beta,$$

and

$$\frac{d}{d\beta} \langle\!\langle \gamma_\beta, U \rangle\!\rangle = -\int_{X'} (U - \langle\!\langle \gamma_\beta, U \rangle\!\rangle)^2 d\gamma_\beta < 0.$$

Hence, the function $\beta \in (0,\infty) \mapsto \langle\!\langle \gamma_\beta, U \rangle\!\rangle$ is strictly decreasing and continuous. To prove the uniqueness and existence of γ_{β^*} it is enough to show that

$$\lim_{\beta\to+\infty} E_U(\gamma_\beta) < 0. \tag{24}$$

By assumption, there exists $u_0 \in]0,1[$ such that $\mu(\{x : U(x) < u_0\}) > 0$. Then, for $\beta > 0$, we have

$$\int_{X'} e^{-\beta U(x)} d\mu(x) \geq e^{-\beta u_0} \mu(\{x : U(x) \in [0, u_0)\}),$$

and

$$\begin{aligned} &\textstyle\int_{X'} (U(x) - u_0) e^{-\beta U(x)} d\mu(x) \\ &\quad\leq e^{-\beta u_0} \int_{X'} (U(x) - u_0) e^{-\beta(U(x)-u_0)} \mathbf{1}_{\{x:U(x)>u_0\}} d\mu(x) \\ &\quad\leq \frac{e^{-\beta u_0}}{\beta} \sup_{y\geq 0} \{y e^{-y}\}. \end{aligned}$$

Therefore, there exists a constant $C < \infty$ such that

$$\langle\!\langle \gamma_\beta, U \rangle\!\rangle = u_0 + \frac{\int_{X'} (U(x) - u_0) e^{-\beta U(x)} d\mu(x)}{Z_\beta} \leq u_0 + \frac{C}{\beta}.$$

Now, we consider the limit when $\beta \to \infty$ and we obtain the limit (24). □

The main result of this section is the following.

Theorem 3.1. *Suppose that $\mu(\{x : U(x) < 1\}) > 0$ and $E_U(\mu) > 0$. Then, the set $\mathcal{M}$ of the minimizers of the entropy is reduced to the unique Gibbs measure γ_{β^*}. Moreover, for all measurable set $\Gamma \subset \mathcal{F}'_\theta(N')_{1,+}$, we have*

$$\lim_{\delta\to 0} \lim_{n\to\infty} \mathrm{P}((Y_1, ..., Y_k) \in \Gamma | L_n^Y \in A_\delta) = \gamma_{\beta^*}^k(\Gamma) \tag{25}$$

Proof. By the preceding lemma, the obtained Gibbs measure γ_{β^*} belongs to the set F_0. We obtain

$$\begin{aligned}
H(\gamma_{\beta^*}|\mu) &= \int_{X'} \frac{d\gamma_{\beta^*}}{d\mu} \log \frac{d\gamma_{\beta^*}}{d\mu} d\mu \\
&= \int_{X'} \log \frac{e^{-\beta^* U(x)}}{Z_{\beta^*}} d\gamma_{\beta^*} \\
&= \int_{X'} (-\beta^* U(x) - \log Z_{\beta^*}) d\gamma_{\beta^*} \\
&= -\beta^* \langle\langle \gamma_{\beta^*}, U \rangle\rangle - \log Z_{\beta^*} < \infty. \qquad (26)
\end{aligned}$$

Therefore, $\inf_{\Psi \in F_0} H(\Psi|\mu) < \infty$. Since $H(.|\mu)$ is strictly convex, it follows that there exists a unique $\nu_0 \in F_0$ such that $\mathcal{M} = \{\nu_0\}$, therefore we have

$$H(\nu_0|\mu) \leq H(\gamma_{\beta^*}|\mu).$$

Then,

$$-H(\nu_0|\gamma_{\beta^*}) \geq -H(\nu_0|\gamma_{\beta^*}) + H(\nu_0|\mu) - H(\gamma_{\beta^*}|\mu).$$

By the equivalence of the measures μ and γ_{β^*}, we have

$$\begin{aligned}
H(\nu_0|\mu) - H(\nu_0|\gamma_{\beta^*}) &= - \int_{X'} \log \frac{d\mu}{d\gamma_{\beta^*}} d\nu_0 \\
&= - \int_{X'} \log\left(Z_{\beta^*} e^{\beta^* U(x)}\right) d\nu_0 \\
&= - \log Z_{\beta^*} - \beta^* \langle\langle \nu_0, U \rangle\rangle. \qquad (27)
\end{aligned}$$

Therefore, it follows that

$$\begin{aligned}
-H(\nu_0|\gamma_{\beta^*}) &\geq \beta^*(\langle\langle \gamma_{\beta^*}, U \rangle\rangle - \langle\langle \nu_0, U \rangle\rangle) \\
&= \beta^*(1 - \langle\langle \nu_0, U \rangle\rangle).
\end{aligned}$$

Then, $H(\nu_0|\gamma_{\beta^*}) \leq 0$ since $E_U(\nu_0) \leq 0$ (because $\nu_0 \in F_0$) and $\beta^* \geq 0$. Therefore, $\nu_0 = \gamma_{\beta^*}$ and $\mathcal{M} = \{\gamma_{\beta^*}\}$.

$\square$

4. The interacting case

In this section, we investigate the case where interaction is present. This case is interesting from a physical point of view. Let us consider a real valued two-variables test function $U \in \mathcal{F}_{(\theta,\theta)}(N' \times N')$.[6] Now, define in this case the energy functional

$$\begin{aligned} E_U : \mathcal{F}'_\theta(N')_{1,+} &\to [-1, \infty] \\ \Psi &\mapsto E_U(\Psi) = \langle\langle \Psi \otimes \Psi, U \rangle\rangle - 1, \end{aligned} \tag{28}$$

where by definition of the tensor product of measures

$$\langle\langle \Psi \otimes \Psi, U \rangle\rangle = \int_{X' \times X'} U(x,y) d\Psi(x) d\Psi(y).$$

In the sequel, we denote by A_δ the set

$$A_\delta = \{\Psi : |E_U(\Psi)| \leq \delta\}, \quad \delta > 0.$$

Since $\langle\langle \Psi \otimes \Psi, U \rangle\rangle$ represents the duality between $\mathcal{F}'_\theta(N' \times N')$ and $\mathcal{F}_\theta(N' \times N')$ (see Ref. 6 for more details), the continuity of the map $\Psi \mapsto \langle\langle \Psi \otimes \Psi, U \rangle\rangle$ follows immediately. Therefore the sets A_δ are closed.

In the next, we would like to prove that, as in the non interacting case, the set $\mathcal{M}$ of the minimizers of the Hamiltonian defined in (31) is consisting of a unique Gibbs measure γ_β satisfying

$$\frac{d\gamma_\beta}{d\mu} = \frac{e^{-\beta \langle\langle \gamma_\beta, U_x \rangle\rangle}}{Z_\beta}, \tag{29}$$

where U_x is the one variable test function given by $U_x(y) = U(x,y)$ and the normalizing constant Z_β denotes the partition function

$$Z_\beta = \int_{X'} e^{-\beta \langle\langle \gamma_\beta, U_x \rangle\rangle} d\mu(x). \tag{30}$$

For this purpose, we introduce the following assumptions:
Assumption (1) For any distributions Ψ_1, $\Psi_2 \in \mathcal{F}'_\theta(N')_{1,+}$ such that $H(\Psi_i|\mu) < \infty$, $i = 1, 2$, we have

$$\langle\langle \Psi_1 \otimes \Psi_2, U \rangle\rangle \leq \frac{1}{2}(\langle\langle \Psi_1 \otimes \Psi_1, U \rangle\rangle + \langle\langle \Psi_2 \otimes \Psi_2, U \rangle\rangle).$$

Assumption (2) For the fixed probability measure μ, we have

$$\int_{X' \times X'} U(x,y) d\mu(x) d\mu(y) \geq 1.$$

Assumption (3) There exists a positive distribution $\Psi \in \mathcal{F}'_\theta(N')_{1,+}$ such that $H(\Psi|\mu) < \infty$ and $\langle\langle \Psi \otimes \Psi, U \rangle\rangle < 1$.

To prove the existence of Gibbs measures, it is natural to define the following interaction Hamiltonian

$$H_\beta(\Psi) = H(\Psi|\mu) + \frac{\beta}{2} \langle\langle \Psi \otimes \Psi, U \rangle\rangle, \quad \beta \in [0, \infty). \tag{31}$$

Now, we give the following series of results which are the key properties of the Gibbs measures that are related to the Hamiltonian $H_\beta(.)$:

Lemma 4.1. *Assume that Assumption (1) is fulfilled. Then, for each $\beta \geq 0$, there exists a unique minimizer of $H_\beta(.)$ denoted $\gamma_\beta \in \mathcal{F}'_\theta(N')_{1,+}$ such that*

$$\frac{d\gamma_\beta}{d\mu} = \frac{e^{-\beta\langle\langle \gamma_\beta, U_x\rangle\rangle}}{Z_\beta}. \tag{32}$$

Proof. Let $\beta \geq 0$ be fixed. Then, by definition, we have

$$H(\Psi|\mu) \leq H_\beta(\Psi), \quad \Psi \in \mathcal{F}'_\theta(N')_{1,+}.$$

Further, $H(.|\mu)$ is a good rate function, and by the continuity of the map $\Psi \mapsto \langle\langle \Psi \otimes \Psi, U\rangle\rangle$, $H_\beta(.)$ is also a good rate function. Moreover, $H_\beta(\mu) < \infty$. Hence, there exists $\Psi_0 \in \mathcal{F}'_\theta(N')_{1,+}$ such that

$$H_\beta(\Psi_0) = \inf\{H_\beta(\Psi) : \Psi \in \mathcal{F}'_\theta(N')_{1,+}\} < \infty.$$

Now, using Assumption (1) together with the convexity of the function $H(.|\mu)$, we obtain

$$H_\beta\left(\frac{\Psi_1 + \Psi_2}{2}\right) \leq \frac{1}{2}H_\beta(\Psi_1) + \frac{1}{2}H_\beta(\Psi_2),$$

if $H_\beta(\Psi_i) < \infty$ for $i = 1, 2$. Then, for all integers k, n with $1 \leq k \leq 2^n$, we extend the preceding inequality by iterations to cover $\frac{k}{2^n}\Psi_1 + (1 - \frac{k}{2^n})\Psi_2$ and we have the convexity of the function $H_\beta(.)$ as a consequence of its lower semi-continuity. Moreover, $H_\beta(.)$ is strictly convex by the strict convexity of $H(.|\mu)$. Therefore, let γ_β denotes the associated measure to the unique minimizer γ_β of $H_\beta(.)$.

Let $f = \frac{d\gamma_\beta}{d\mu}$ and consider the set $A = \{x : f(x) = 0\}$. We would like to prove that $\mu(A) = 0$. Assume that $a = \mu(A) > 0$ and define the following probability measure $\Psi \in \mathcal{F}'_\theta(N')_{1,+}$ by

$$d\Psi(x) = \frac{1}{a}\mathbf{1}_A d\mu(x).$$

Note that for all $t \in [0, 1]$, the distribution $\Psi_t = t\Psi + (1-t)\gamma_\beta \in \mathcal{F}'_\theta(N')_{1,+}$ defines a probability measure. The supports of Ψ and γ_β are disjoint, then by a direct computation, we have

$$\begin{aligned} 0 &\leq \tfrac{1}{t}\left(H_\beta(\Psi_t) - H_\beta(\gamma_\beta)\right) \\ &= H_\beta(\Psi) - H_\beta(\gamma_\beta) + \log t \\ &+ \tfrac{1-t}{t}\log(1-t) - \tfrac{\beta}{2}(1-t)\langle\langle(\gamma_\beta - \Psi)^{\otimes 2}, U\rangle\rangle. \end{aligned}$$

On the other hand, since $H(\Psi|\mu) = -\log a < \infty$ then $H_\beta(\Psi) < \infty$.

Now, when considering the limit $t \downarrow 0$, the preceding inequality results with a contradiction. Therefore, it remains to prove that (32) holds. So, let us fix a test function $\phi \in \mathcal{F}_\theta(N')_+$, $\phi \neq 0$ and $\delta = 2|||\phi\|_{\theta,m,p} > 0$. For all $t \in]-\delta, \delta[$, let $\Psi_t \in \mathcal{F}'_\theta(N')_{1,+}$ be the probability measure given by

$$\frac{d\Psi_t}{d\gamma_\beta} = 1 + t(\phi - \langle\langle \gamma_\beta, \phi \rangle\rangle).$$

Note that the function $t \mapsto H_\beta(\Psi_t)$ is differentiable and possesses a minimum at $t = 0$. Then, it follows that $dH_\beta(\Psi_t)/dt = 0$ at $t = 0$ which leads to

$$\begin{aligned} 0 &= \textstyle\int_{X'} (\phi - \langle\langle \gamma_\beta, \phi \rangle\rangle) (\log f + \beta \langle\langle \gamma_\beta, U_x \rangle\rangle) \, d\gamma_\beta \\ &= \textstyle\int_{X'} \phi f \, (\log f + \beta \langle\langle \gamma_\beta, U_x \rangle\rangle - H(\gamma_\beta|\mu) - \beta \langle\langle \gamma_\beta \otimes \gamma_\beta, U \rangle\rangle) \, d\mu. \end{aligned}$$

Therefore, (32) holds since ϕ is arbitrary, $f > 0, \mu$-a.e. and $H(\gamma_\beta|\mu)$ and $\beta\langle\langle \gamma_\beta \otimes \gamma_\beta, U \rangle\rangle$ are finite constants. □

Lemma 4.2. *Assume that assumption (1) is fulfilled. Then, the function*

$$g : \beta \in [0, \infty) \mapsto \langle\langle \gamma_\beta \otimes \gamma_\beta, U \rangle\rangle$$

is continuous.

Proof. Define a sequence $\{\beta_n\}_{n \geq 1}$ such that $\beta_n \to \beta \in [0, \infty)$. Then, the sequence $\{\beta_n\}$ is bounded and therefore there exists a compact level set C of $H(.|\mu)$ such that $\{\gamma_{\beta_n}\} \subset C$. Hence, the sequence $\{\gamma_{\beta_n}\}$ has one limit point Ψ. By the characterization of the measure γ_{β_n}, we have

$$H_\beta(\Psi) \leq \liminf_{n \to \infty} H_{\beta_n}(\gamma_{\beta_n}) \leq \liminf_{n \to \infty} H_{\beta_n}(\gamma_\beta) = H_\beta(\gamma_\beta).$$

Finally, using Lemma 4.1, the sequence $\{\gamma_{\beta_n}\}$ converges to γ_β and the continuity of $\beta \mapsto g(\beta)$ follows. □

Lemma 4.3. *Assume that the assumptions (1), (2) and (3) are fulfilled and define*

$$\beta^* := \inf\{\beta \geq 0 : g(\beta) \leq 1\}. \tag{33}$$

Then, $\beta^* < \infty$ and $g(\beta^*) = 1$.

Proof. Let Ψ fulfilling Assumption (3). Then, by Lemma 4.1, we have

$$g(\beta) \leq \frac{2}{\beta} H_\beta(\gamma_\beta) \leq H(\Psi|\mu) + \langle\langle \Psi \otimes \Psi, U \rangle\rangle, \quad \forall \beta > 0,$$

since $H(.|\mu) \geq 0$. Hence,

$$\limsup_{\beta \to \infty} g(\beta) \leq \langle\!\langle \Psi \otimes \Psi, U \rangle\!\rangle < 1.$$

Now, by Assumption (2), $\gamma_0 = \mu$ and $g(0) \geq 1$. Finally, Lemma 4.2 leads to $\beta^* < \infty$ and $g(\beta^*) = 1$. □

So we obtain the next result.

Theorem 4.1. *Assume that the assumptions (1), (2) and (3) are fulfilled. Then,* $\mathcal{M} = \{\gamma_{\beta^*}\}$ *and for every measurable subset* $\Gamma \subset \mathcal{F}'_\theta(N')_{1,+}$, *we have*

$$\lim_{\delta \to 0} \lim_{n \to \infty} \mathrm{P}((Y_1, ..., Y_k) \in \Gamma | L_n^Y \in A_\delta) = \gamma_{\beta^*}^k(\Gamma) \tag{34}$$

where β^* *is as defined in (33) and* γ_{β_*} *denotes the Gibbs measure which corresponds to the unique minimizer of* $H_\beta(.)$.

Proof. Since the functional E_U is continuous. Then, we can take $F_\delta = A_\delta$ and therefore $F_0 = A_0 = \{\Psi : \langle\!\langle \Psi \otimes \Psi, U \rangle\!\rangle = 1\}$. Note that $\gamma_{\beta*} \in F_0$ yields to $\inf_{\Psi \in F_0} H(\Psi|\mu) < \infty$ and $\mathcal{M} \neq \emptyset$. Moreover, by a direct computation, we observe that for $\Psi \in F_0$ and $H(\Psi|\mu) < \infty$,

$$\begin{aligned}
&H(\Psi|\mu) - H(\Psi|\gamma_{\beta^*}) - H(\gamma_{\beta^*}|\mu) \\
&= \beta^* \left(\langle\!\langle \gamma_{\beta^*} \otimes \gamma_{\beta^*}, U \rangle\!\rangle - \langle\!\langle \gamma_{\beta^*} \otimes \Psi, U \rangle\!\rangle\right) \\
&\geq \tfrac{\beta^*}{2} \left(\langle\!\langle \gamma_{\beta^*} \otimes \gamma_{\beta*}, U \rangle\!\rangle - \langle\!\langle \Psi \otimes \Psi, U \rangle\!\rangle\right),
\end{aligned}$$

where the inequality holds by Assumption (1). We note that

$$\langle\!\langle \gamma_{\beta^*} \otimes \gamma_{\beta*}, U \rangle\!\rangle = \langle\!\langle \Psi \otimes \Psi, U \rangle\!\rangle = 1,$$

since the distributions γ_{β^*} and Ψ are elements of F_0 and therefore

$$H(\Psi|\mu) - H(\Psi|\gamma_{\beta^*}) - H(\gamma_{\beta^*}|\mu) \geq 0.$$

Thus, for $\Psi \in \mathcal{M}$, we have

$$-H(\Psi|\gamma_{\beta^*}) \geq -H(\Psi|\gamma_{\beta^*}) + H(\Psi|\mu) - H(\gamma_{\beta^*}|\mu) \geq 0.$$

which leads to $\Psi = \gamma_{\beta^*}$. Hence, $\mathcal{M} = \{\gamma_{\beta^*}\}$ and the conditional law in this interacting case, given by the equality (34), follows from Theorem 2.3. □

References

1. S. Chaari, F. Cipriano, S. Gheryani and H. Ouerdiane, *Sanov's theorem for white noise distributions and application to the Gibbs conditioning principle*; Acta Appl. Math., Vol **104**, (2008) 313-324.
2. S. Chaari, F. Cipriano and H. Ouerdiane, *Large deviations for infinite dimensional analytical distributions*; Adv. Theo. Appl. Math., Vol **1**, No. 3, (2006) 173-187.
3. A. Dembo and O. Zeitouni, *Large deviations techniques and applications*; Second edition. Applications of Mathematics (New York), 38. Springer-Verlag, New York, (1998).
4. J. D. Deuschel and D. W. Stroock, *Large deviations; AMS Chelsea Publishing. Providence, Rhode Island, (2001).*
5. *R. Gannoun, R. Hachaichi, H. Ouerdiane, and A. Rezgui, Un théorème de dualité entre espace de fonctions Holomorphes à croissance exponentielle*; J. Funct. Analysis **171**, (2000) 1-14.
6. U. C. Ji, N. Obata and H. Ouerdiane, *Analytic Characterization of Generalized Fock Space Operators as Two-variable Entire Functions with Growth Condition*, World Scientific, vol. 5, No. 3, (2002), 395-407.
7. H. Ouerdiane, *Fonctionnelles analytiques avec conditions de croissance et application à l'analyse gaussienne*; Japanese Journal of Math. **20**, (1994) 187–198.
8. H. Ouerdiane and N. Privault, *Asymptotic estimates for white noise distributions*; C. R. Acad. Sci. Paris, Ser. I **338**, (2004) 799-804.
9. H. Ouerdiane and A. Rezgui, *Répresentation intégrale de fonctionnelles analytiques*; Stochastic Processes, Physics and Geometry: New Interplays. A volume in honor of S. Albeverio, Canadian Math. Society. Conference Proceedings Series, (1999).

MARKOV TRIPLETS ON CAR ALGEBRAS

JÓZSEF PITRIK

Department for Mathematical Analysis, Budapest University of Technology and Ecomomics,

Budapest, Egry J. utca 1, H-1521, Hungary

E-mail: pitrik@math.bme.hu

We investigate the equivalence of Markov triplets and the states which saturates the strong subadditivity of von Neumann entropy with equality for CAR algebras.

Keywords: von Neumann entropy, CAR algebra, Markov triplet, strong subadditivity

1. Introduction

The notion of Markov property was invented by L. Accardi in the noncommutative (or quantum probabilistic) setting.[1,5] This Markov property is based on a completely positive, identity preserving map, so-called quasi-conditional expectation and it was formulated in the tensor product of matrix algebras. A state of a tensor product system is Markovian if and only if the von Neumann entropy increase is constant. This property and a possible definition of the Markov condition was suggested in Ref. 17. A remarkable property of the von Neumann entropy is the strong subadditivity[10,12,16,19] which plays an important role in the investigations of quantum system's correlations. The above mentioned constant increase of the von Neumann entropy is the same as the equality for the strong subadditivity of von Neumann entropy. Moreover the exact structure of such states was also established.[10,11]

However a pivotal example of quantum composite systems is tensor product of Hilbert spaces, the definition of Markov property has been given under a very general setting that is not limited to the most familiar case of tensor-product systems.[4] That is, it does not require in priciple any specific algebraic location among systems imbedded in the total system. A very important example from this point of view the algebra of the Canonical Anticommutation Relation or briefly CAR algebra, that serve as the description of fermion lattice systems. Fermions are particles with half-integer spin, satisfy Fermi-Dirac statistics, which is the algebraic statement of the Pauli exclusion principle. The requirement that the wave-function of n indistinguishable particles should be antisymmetric of the position variables ensures that the wave-function is zero if any two particles sit at the same point. This anti-symmetry can be ensured if we use creation and annihilation operators obeying the following anti-commutation relations. For $I \subset \mathbb{Z}$, let $\mathcal{A}(I)$ be the CAR algebra, i.e. a unital C^*-algebra generated by the elements $\{a_i : i \in I\}$ satisfying the anticommutation relations

$$\begin{aligned} a_i a_j + a_j a_i &= 0, \\ a_i a_j^* + a_j^* a_i &= \delta_{i,j} \mathbf{1} \end{aligned}$$

for $i, j \in I$. From the first relation we have $(a_i^*)^2 = 0$ for all i, which expresses that we cannot create two particles in the same state. When the set I is countable, the CAR algebra is isomorphic to the C^*-infinite tensor product $\overline{\otimes_I M_2(\mathbb{C})}^{C^*}$, but the isomorphism does not preserve the natural localization. The elements of the disjoint subsystems do not commute in contrast to the tensor product case. In spite of these difficulties the strong subadditivity of von Neumann entropy also holds for CAR algebras.[7,14,22] Moreover a similar equivalence relation of the Markov property and the strong additivity of von Neumann entropy was showed for a relevant subset of states, for so called even states.[13] In this paper we give a slight extension of this result for any states, with some constraints for fixed point algebras.

2. The CAR algebra

In this section we summarize known properties of the algebra of the canonical anticommutation relation. The works[6,7,9] contain all what we need with other details.

Definition 2.1. For $I \subset \mathbb{Z}$, the unital C*-algebra $\mathcal{A}(I)$ generated by the elements satisfying the **canonical anticommutation relations**, i.e.

$$a_i a_j + a_j a_i = 0, \tag{1}$$

$$a_i a_j^* + a_j^* a_i = \delta_{i,j}\mathbf{1} \tag{2}$$

for $i, j \in I$ is called a **CAR algebra**.

The operators a^* and a are often called **creator** and **annihilator**, respectively. It is easy to see that $\mathcal{A}(I)$ is the linear span of the identity and monomials of the form

$$A_{i(1)}A_{i(2)}\ldots A_{i(k)}, \tag{3}$$

where $i(1) < i(2) < \cdots < i(k)$ and each factor $A_{i(j)}$ is one of the four operators $a_{i(j)}, a^*_{i(j)}, a_{i(j)}a^*_{i(j)}$ and $a^*_{i(j)}a_{i(j)}$. The CAR algebra $\mathcal{A}$ is defined by

$$\mathcal{A} \equiv \overline{\bigvee_{l\in L} \mathcal{A}(\{l\})}^{C^*}.$$

It is known that for $I = \{1, 2, \ldots, n\}$, $\mathcal{A}(I)$ is isomorphic to a matrix algebra $M_{2^n}(\mathbb{C}) \simeq \overset{1}{\overbrace{M_2(\mathbb{C})}} \otimes \cdots \otimes \overset{n}{\overbrace{M_2(\mathbb{C})}}$. An explicit isomorphism is given by the so-called **Jordan-Wigner isomorphism**. Namely, the relations

$$\begin{aligned} e_{11}^{(i)} &:= a_i a_i^* & \qquad e_{12}^{(i)} &:= V_{i-1} a_i \\ e_{21}^{(i)} &:= V_{i-1} a_i^* & \qquad e_{22}^{(i)} &:= a_i^* a_i \end{aligned}$$

$$V_i := \prod_{j=1}^{i} (I - 2a_j^* a_j)$$

determine a family of mutually commuting 2×2 matrix units for $i \in I$. Since

$$a_i = \prod_{j=1}^{i-1} \left(e_{11}^{(j)} - e_{22}^{(j)}\right) e_{12}^{(i)},$$

the above matrix units generate $\mathcal{A}(I)$ and give an isomorphism between $\mathcal{A}(I)$ and $M_2(\mathbb{C}) \otimes \cdots \otimes M_2(\mathbb{C})$:

$$e_{i_1 j_2}^{(1)} e_{i_2 j_2}^{(2)} \ldots e_{i_n j_n}^{(n)} \longleftrightarrow e_{i_1 j_1} \otimes e_{i_2 j_2} \otimes \cdots \otimes e_{i_n j_n}. \tag{4}$$

(Here e_{ij} stand for the standard matrix units in $M_2(\mathbb{C})$.) Let $J \subset \mathbb{Z}$. There exists a unique automorphism Θ_J of $\mathcal{A}(\mathbb{Z})$ such that

$$\Theta_J(a_i) = -a_i \text{ and } \Theta_J(a_i^*) = -a_i^* \quad (i \in J)$$
$$\Theta_J(a_i) = a_i \text{ and } \Theta_J(a_i^*) = a_i^* \quad (i \notin J).$$

In particular, we write Θ instead of $\Theta_{\mathbb{Z}}$. Θ_J is inner i.e. there exists a v_J self-adjoint unitary in $\mathcal{A}(J)$ given by

$$v_J \equiv \prod_{i\in J} v_i, \qquad v_i \equiv a_i^* a_i - a_i a_i^* \tag{5}$$

such that $\Theta_J(A) = (\mathrm{Ad}v_J)A \equiv v_J A v_J^*$ for any $a \in \mathcal{A}(J)$. The **odd** and **even parts** of $\mathcal{A}_I$ are defined as

$$\mathcal{A}(I)^+ := \{A \in \mathcal{A}(I) : \Theta_I(A) = A\}, \tag{6}$$
$$\mathcal{A}(I)^- := \{A \in \mathcal{A}(I) : \Theta_I(A) = -A\}. \tag{7}$$

Remark that $\mathcal{A}(I)^+$ is a subalgebra but $\mathcal{A}(I)^-$ is not. The **graded commutation relation** for CAR algebras is known: if $A \in \mathcal{A}(K)$ and $B \in \mathcal{A}(L)$ where $K \cap L = \emptyset$, then

$$AB = \epsilon(A, B)BA \tag{8}$$

where

$$\epsilon(A, B) = \begin{cases} -1 \text{ if A and B are odd} \\ +1 \qquad \text{otherwise.} \end{cases} \tag{9}$$

The parity automorphism is the special case of the action of the **gauge group** $\{\alpha_\theta : 0 \le \theta < 2\pi\}$ with

$$\alpha_\theta(a_i) = e^{-i\theta} a_i.$$

An element $a \in \mathcal{A}$ is **gauge-invariant** if $\alpha_\theta(a) = a$ for all $0 \le \theta < 2\pi$. A state ϕ on the CAR algebra $\mathcal{A}$ is called **even state** if it is Θ-invariant:

$$\phi(\Theta(A)) = \phi(A) \tag{10}$$

for all $a \in \mathcal{A}$. Note that $\phi(A) = 0$ for all $A \in \mathcal{A}^-$ is equivalent to the condition that ϕ is even state of $\mathcal{A}$. Let I and J be two disjoint subsets of $\mathbb{Z}$. We say that ϕ is a **product state** with respect to $\mathcal{A}(I)$ and $\mathcal{A}(J)$, if

$$\phi(AB) = \phi(A)\phi(B) \tag{11}$$

holds for all $A \in \mathcal{A}(I)$ and $B \in \mathcal{A}(J)$. If a state ϕ of the joint system $\mathcal{A}(I \cup J)$ (which is the same as the C*-algebra generated by $\mathcal{A}(I)$ and $\mathcal{A}(J)$) coincides with ϕ_I on $\mathcal{A}(I)$ and ω_J on $\mathcal{A}(J)$, i.e.

$$\phi(A) = \phi_I(A), \qquad A \in \mathcal{A}(I) \tag{12}$$

$$\phi(B) = \omega_J(B), \qquad B \in \mathcal{A}(J), \tag{13}$$

then ϕ is called the **joint extension** of ϕ_I and ω_J. As a special case, if

$$\phi(AB) = \phi_I(A)\omega_J(B) \equiv \phi_I \wedge \omega_J(AB), \tag{14}$$

then $\phi = \phi_I \wedge \omega_J$ is called a **product state extension** of ϕ_I and ω_J. A product state extension does not exist unconditionally. Indeed, suppose that both A and B are odd elements. If the product state extension of ϕ_I and ω_J exist, then

$$\begin{aligned} \phi_I(A)\omega_J(B) &= \phi_I \wedge \omega_J(AB) = \overline{\phi_I \wedge \omega_J((AB)^*)} \\ &= \overline{\phi_I \wedge \omega_J(B^*A^*)} = -\overline{\phi_I \wedge \omega_J(A^*B^*)} \\ &= -\overline{\phi_I(A^*)\omega_J(B^*)} = -\phi_I(A)\omega_J(B), \end{aligned}$$

where we have used (8). This shows that at least one of two states must be even, i.e. must vanish on odd elements. This result was generalized in[8] in the following form:

Theorem 2.1. *Let $I_1, I_2, \ldots$ be an arbitrary (finite or infinite) number of mutually disjoint subsets of $\mathbb{Z}$ and ϕ_i be a given state of $\mathcal{A}(I_i)$ for each i.*

(1) A product state extension of ϕ_i, $i = 1, 2, \ldots$ exists if and only if all states ϕ_i except at most one are even. It is unique if it exists. It is even if and only if all ϕ_i are even.

(2) Suppose that all ϕ_i are pure. If there exists a joint extension of ϕ_i, $i = 1, 2, \ldots$, then all states except at most one have to be even. If this is the case, the joint extension is uniquely given by the product state extension and is a pure state.

A state τ is called **tracial state** if $\tau(AB) = \tau(BA)$ for all $A, B \in \mathcal{A}$. We remark that the existence of a tracial state follows from the isomorphism

(4) immediately. A tracial state τ is an even product state.[7] The **right shift automorphism** γ on $\mathcal{A}$ is defined by $\gamma(a_i) = a_{i+1}$ and $\gamma(a_i^*) = a_{i+1}^*$ for all $a_i, a_i^* \in \mathcal{A}$, $i \in \mathbb{Z}$. A state ϕ on $\mathcal{A}$ is **translation invariant** if $\phi \circ \gamma = \phi$ holds. It is important to know that any translation invariant state is automatically even.[9] Let $\mathcal{B} \subset \mathcal{A}$ be a subset of a C*-algebra $\mathcal{A}$. The **commutant** of $\mathcal{B}$ is defined by

$$\mathcal{B}' = \{A \in \mathcal{A} : AB = BA, \quad \forall B \in \mathcal{B}\}. \tag{15}$$

It is a unital subalgebra of $\mathcal{A}$ and plays an important role in our investigations. The commutants in the CAR algebra are given by the following theorem.[7]

Theorem 2.2. *For a finite $I \subset \mathbb{Z}$,*

(1) $\mathcal{A}(I)' \cap \mathcal{A} = \mathcal{A}(I^c)^+ + v_I \mathcal{A}(I^c)^-$

(2) $(\mathcal{A}(I)^+)' \cap \mathcal{A} = \mathcal{A}(I^c) + v_I \mathcal{A}(I^c)$,

where I^c denotes the complementer set of I.

The fundamental object of the Markov property is the conditional expectation. Now we investigate the existence of the conditional expectation in the CAR algebra case. Recall the following definition.

Definition 2.2. By a (Umegaki) **conditional expectation** $E : \mathcal{A} \to \mathcal{B} \subset \mathcal{A}$ we mean a norm-one projection of the C^*-algebra $\mathcal{A}$ onto the C^*-subalgebra $\mathcal{B}$.

One can check that the map E is automatically a completely positive identity-preserving $\mathcal{B}$-bimodule map. For CAR algebras the existence of conditional expectation which preserves the tracial state τ follows from generalities about conditional expectations or the isomorphism (4). Inspite of these, it is useful to have a construction for E_J^I following the original proof.[7]

Lemma 2.1. *Let $J \subset I$. Then $\mathcal{A}(J) \subset \mathcal{A}(I)$ and there exists a unique conditional expectation $E_J^I : \mathcal{A}(I) \to \mathcal{A}(J)$ which preserves the trace, i.e.*

$\tau \circ E^I_J = \tau.$

Proof. The C*-algebra generated by the commuting subalgebras $\mathcal{A}(I)$ and $\mathcal{A}(I \setminus J)^+$ is isomorphic to their tensor product. We have a conditional expectation

$$F_1 : \mathcal{A}(I) \to \mathcal{A}(J) \otimes \mathcal{A}(I \setminus J)^+, \quad F_1(A) = \frac{1}{2}(A + \Theta_{I\setminus J}(A)) \tag{16}$$

and another

$$F_2 : \mathcal{A}(J) \otimes \mathcal{A}(I \setminus J)^+ \to \mathcal{A}(J), \quad F_2(A \otimes B) = \tau(B)\, A. \tag{17}$$

The composition $F_2 \circ F_1$ is E^I_J. □

To have an example, assume that $I = [1,4]$, $J = [1,2]$ and consider the action of the above conditional expectations on terms like (3). F_1 keeps $a_1 a_2^* a_2 a_3^* a_4$ fixed and F_2 sends it to $a_1 a_2^* a_2 \tau(a_3^*)\tau(a_4) = 0$. Moreover, E^I_J sends $a_1 a_2^* a_2 a_3 a_3^* a_4^* a_4$ to $a_1 a_2^* a_2 \tau(a_3 a_3^*)\tau(a_4^* a_4)$. It is important to make here a remark. For arbitrary subsets $J_1, J_2 \subset I$

$$E^I_{J_1} | \mathcal{A}(J_2) = E^{J_2}_{J_1 \cap J_2} \tag{18}$$

holds. This means that we have a commuting square:

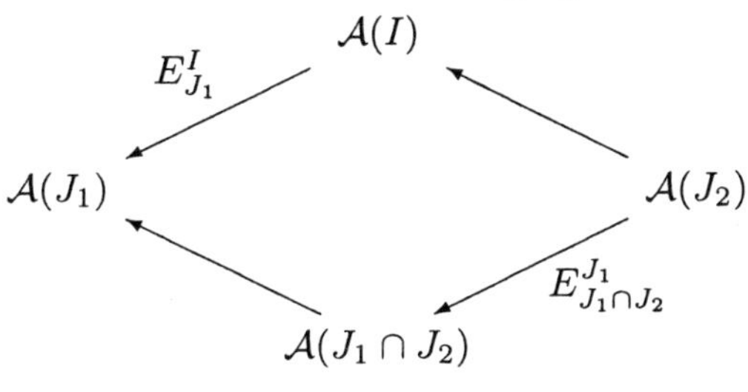

3. The strong subadditivity of the von Neumann entropy and the Markov property

Consider a noncommutative probability space $(\mathcal{A}, \phi)$, where $\mathcal{A}$ is a finite C*-algebra, and ϕ is a normal, faihful state on $\mathcal{A}$ given with a density matrix ρ. (Often we use the same notation for the state and for its density matrix.) The **von Neumann entropy** is defined by the formula

$$S(\phi) \equiv S(\rho) := -\mathrm{Tr}\, \rho \log \rho. \tag{19}$$

Usually, the logarithms are taken to base two. From the definition it is clear that $S(\rho) \geq 0$ and the equality holds if and only if ρ is a one-rank projection, i.e. the state is pure. As the von Neumann entropy is the trace of a continous function of the density matrix, hence it is a continous function on the states. To deduce some properties of the von Neumann entropy, it is useful to introduce the concept of relative entropy. Assume that ρ and σ are density matrices of the states ϕ and ω, respectively on a Hilbert space $\mathcal{H}$, then their **relative entropy** is defined by

$$S(\phi||\omega) \equiv S(\rho||\sigma) = \begin{cases} \operatorname{Tr} \rho(\log \rho - \log \sigma) & \text{if } \operatorname{supp} \rho \subset \operatorname{supp} \sigma, \\ +\infty & \text{otherwise.} \end{cases} \tag{20}$$

As $S(\rho||\sigma) \geq 0$ with equality if and only if $\rho = \sigma$, it is natural to consider the relative entropy as some kind of a distance measure on the set of states, even though it is not symmetric in its arguments and does not satisfy the triangle inequality. The von Neumann entropy and the relative entopy play an important role in the investigations of quantum systems's correlations. The von Neumann entropy is subadditive, i.e.

$$S(\phi_{12}) \leq S(\phi_1) + S(\phi_2),$$

where ϕ_{12} is a normal state of the composite system of $\mathcal{A}_{12} = \mathcal{A}_1 \otimes \mathcal{A}_2$, and the equality holds if and only if ϕ_{12} is product of its marginals, i.e. $\phi_{12} = \phi_1 \otimes \phi_2$, that is the noncommutative analogue of the independent random variables. We also have the remarkable strong subadditivity property.[12] Let $\mathcal{A}_1, \mathcal{A}_2, \mathcal{A}_3$ be subalgebras of $\mathcal{B}(\mathcal{H})$, representing three quantum systems and set $\mathcal{A}_{123} = \mathcal{A}_1 \otimes \mathcal{A}_2 \otimes \mathcal{A}_3$, $\mathcal{A}_{12} = \mathcal{A}_1 \otimes \mathcal{A}_2$, $\mathcal{A}_{23} = \mathcal{A}_2 \otimes \mathcal{A}_3$ as their several compositions. For a state ϕ_{123} of $\mathcal{A}_{123}$ we denote by ϕ_{12}, ϕ_{23} and ϕ_2 its restrictions to $\mathcal{A}_{12}$, $\mathcal{A}_{23}$ and $\mathcal{A}_2$, respectively. The strong subadditivity asserts, that

$$S(\phi_{123}) + S(\phi_2) \leq S(\phi_{12}) + S(\phi_{23}).$$

It is a natural to ask, whether the strong subadditivity of von Neumann entropy also holds for CAR algebras? The root of the problem is the difference between the three fold tensor product system and the CAR algebra from the point of view of the commutation of the subsystems. Indeed, however when the set I is countable, the CAR algebra is isomorphic to the C^*-infinite tensor product $\overline{\otimes_I M_2(\mathbb{C})}^{C^*}$ as we saw, but the isomorphism does not preserve the natural localization. The elements of the disjoint subsystems do not commute in contrast to the tensor product case. In spite of these difficulties the strong subadditivity of von Neumann entropy also holds for CAR algebras:[7,14,22] let I and J be two arbitrary subsets of $\mathbb{Z}$ and denote $\mathcal{A}(I \cup J)$, $\mathcal{A}(I)$, $\mathcal{A}(J)$ and $\mathcal{A}(I \cap J)$ the CAR algebras corresponding to the sets $I \cup J$, I, J and $I \cap J$, respectively as usual with the states $\rho_{I\cup J}$, ρ_I, ρ_J and $\rho_{I\cap J}$. Then

$$S(\rho_I) + S(\rho_J) \geq S(\rho_{I\cap J}) + S(\rho_{I\cup J}) \tag{21}$$

holds. It is interesting to find the states which saturates the strong subadditivity of von Neumann entropy with equality. The following theorem gives an equivalent condition for the equality.[16,18–20,22]

Theorem 3.1. *Assume that $\rho_{I\cup J}$ is invertible. The equality holds in the strong subadditivity of entropy if and only if the following equivalent conditions hold*

(1) $\log \rho_{I\cup J} - \log \rho_J = \log \rho_I - \log \rho_{I\cap J}$

(2) ${\rho_{I\cup J}}^{it} {\rho_J}^{-it} = {\rho_I}^{it} {\rho_{I\cap J}}^{-it}, \qquad t \in \mathbb{R}.$

We can reformulate the strong subadditivity of von Neumann entropy (21) by the relative entropy as

$$S(\rho_{I\cup J}||\rho_J) \geq S(\rho_I||\rho_{I\cap J}) \tag{22}$$

or equivalently with the help of the conditional expectation $E_I^{I\cup J}$

$$S(\rho_{I\cup J}||\rho_J) \geq S(E_I^{I\cup J}(\rho_{I\cup J})||E_I^{I\cup J}(\rho_J)). \tag{23}$$

Uhlmann's monotonicity theorem tell us, that if $\alpha : \mathcal{A}_1 \to \mathcal{A}_2$ is a unital Schwarz mapping between C*-algebras, i.e. α is linear and fulfill the $\alpha(A^*A) \geq \alpha(A)^*\alpha(A)$ so-called Schwarz inequality for all $A \in \mathcal{A}_1$, then for states ω and ϕ of $\mathcal{A}_2$ the inequality:

$$S(\omega \circ \alpha||\phi \circ \alpha) \leq S(\omega||\phi)$$

or equivalently with the dual map T

$$S(\rho||\sigma) \geq S(T(\rho)||T(\sigma))$$

holds, where ρ and σ are the density matrices of ω and ϕ, respectively. As $E_I^{I\cup J}$ is a completely positive unital trace preserving map, it satisfies the Schwarz inequality, hence the strong subadditivity of von Neumann entropy follows from the Uhlmann's theorem immediately. For this reason we characterize the equality case in the Uhlmann's theorem:[15,18,20]

Theorem 3.2. *Let T be a coarse graining, i.e. a trace preserving, 2-positive linear map. For states ρ and σ,*

$$S(\rho||\sigma) = S(T(\rho)||T(\sigma))$$

if and only if there exists a coarse graining $\hat{T}$ such that

$$\hat{T}T(\rho) = \rho, \qquad \hat{T}T(\sigma) = \sigma.$$

To give the definition of Markov triplets, we need recall the definition of the quasi-conditional expectation in the context of the CAR algebras.[4]

Definition 3.1. Let I and J be two arbitrary subsets of $\mathbb{Z}$. Consider a triplet $\mathcal{A}(I \setminus J) \subset \mathcal{A}(I) \subset \mathcal{A}(I \cup J)$ of CAR subalgebras. A **quasi-conditional expectation** w.r.t the given triplet, is a completely positive, identity-preserving linear map $\gamma : \mathcal{A}(I \cup J) \to \mathcal{A}(I)$ such that

$$\gamma(xy) = x\gamma(y), \qquad x \in \mathcal{A}(I \setminus J), y \in \mathcal{A}(I \cup J). \tag{24}$$

We also have $\gamma(yx) = \gamma(y)x$ as γ is a real map. The definition of Markov triplet is the following.

Definition 3.2. Let I and J be two arbitrary subsets of $\mathbb{Z}$. A state $\phi_{I \cup J}$ on the CAR algebra $\mathcal{A}(I \cup J)$ is called a **Markov triplet** corresponding to the localization $\{\mathcal{A}(I \setminus J), \mathcal{A}(I), \mathcal{A}(I \cup J)\}$ if there exists a quasi-conditional expectation γ w.r.t the triplet $\mathcal{A}(I \setminus J) \subset \mathcal{A}(I) \subset \mathcal{A}(I \cup J)$ satisfying

$$\phi_I \circ \gamma = \phi_{I \cup J}, \tag{25}$$

$$E(\mathcal{A}(J)) \subset \mathcal{A}(I \cap J). \tag{26}$$

Here the subalgebras $\mathcal{A}(I \setminus J)$, $\mathcal{A}(I)$ and $\mathcal{A}(I \cup J)$ symbolize the past, the present and the future, respectively. For (25) we also say that the state $\phi_{I \cup J}$ is compatible with the map γ. Condition (26) is the Markov property which possesses the usual interpretation: the future does not depend on the past but only the present.

It has been shown that the Markovianity is tightly related to the strong subadditivity of von Neumann entropy. Namely, a state of a three-composed tensor-product system is a Markov state if and only if it takes the equality for the strong subadditivity inequality of entropy.[10,18,20] Moreover a translation invariant quantum Markov state of the quantum spin algebra has a constant entropy increment at each step by the strong additivity, see *Proposition 11.5* in Ref. 16. Our goal to investigate this situation in the CAR case: we prove the equivalence between the Markov states and the states which fulfill the strong additivity of von Neumann entropy in the CAR algebras whenever the states are even. This result was also showed in Ref. 13 in other context.

Theorem 3.3. *Let I and J be two arbitrary subsets of $\mathbb{Z}$. Let $\phi_{I\cup J}$ be an even state on the CAR algebra $\mathcal{A}(I \cup J)$ with the density matrix $\rho_{I\cup J}$. Then $\phi_{I\cup J}$ is a Markov state corresponding to the localization $\{\mathcal{A}(I\setminus J), \mathcal{A}(I), \mathcal{A}(I\cup J)\}$ if and only if it saturates the strong subadditivity inequality of entropy , ie.*

$$S(\rho_I) + S(\rho_J) = S(\rho_{I\cap J}) + S(\rho_{I\cup J}), \tag{27}$$

where ρ_J, ρ_I and $\rho_{I\cap J}$ are the density matrices of the appropriate restrictions of $\phi_{I\cup J}$.

Proof. At first let suppose that we (27) holds or by expressing with the relative entropy we have

$$S(\rho_{I\cup J}||\rho_J) = S(\rho_I||\rho_{I\cap J}). \tag{28}$$

Let's define the map

$$\gamma(X) = {\rho_{I\cap J}}^{-1/2} E_I^{I\cup J}({\rho_J}^{1/2} X {\rho_J}^{1/2}) {\rho_{I\cap J}}^{-1/2}, \qquad X \in \mathcal{A}(I\cup J). \tag{29}$$

It is clear, that $\gamma : \mathcal{A}(I\cup J) \to \mathcal{A}(I)$. We show that γ is a quasi-conditional expectation with respect to the desired triplet, which preserves the even state $\phi_{I\cup J}$, that is $\phi_{I\cup J}$ is a Markov state. It is obvious that γ is completely positive and preserves the identity. We show that $\mathcal{A}(I \setminus J) \subset \mathrm{Fix}(\gamma)$, the fixpoint algebra of γ. For any $X \in \mathcal{A}(I \setminus J) \subset \mathcal{A}(I)$ we have

$$\begin{aligned}\gamma(X) &= {\rho_{I\cap J}}^{-1/2} E_I^{I\cup J}({\rho_J}^{1/2} X {\rho_J}^{1/2}) {\rho_{I\cap J}}^{-1/2} \\ &= X {\rho_{I\cap J}}^{-1/2} E_I^{I\cup J}(\rho_J) {\rho_{I\cap J}}^{-1/2} = X,\end{aligned}$$

where we have used that $\rho_{I\cap J}$ commutes with any element of $\mathcal{A}(I \setminus J)$ since $\phi_{I\cup J}$ and its all restrictions are even, and $E_I^{I\cup J}(\rho_J) = \rho_{I\cap J}$ by the commuting square property. So, get that γ leaves the elements of the algebra $\mathcal{A}(I \setminus J)$ fixed. We remark that

$$\mathcal{A}(I \setminus J)^+ \subset \mathrm{Fix}(\gamma), \tag{30}$$

always, even if $\phi_{I\cup J}$ is not an even state. If $X \in \mathcal{A}(I \setminus J)$ and $Y \in \mathcal{A}(I \cup J)$ then

$$\gamma(XY) = {\rho_{I\cap J}}^{-1/2} E_I^{I\cup J}({\rho_J}^{1/2} XY {\rho_J}^{1/2}) {\rho_{I\cap J}}^{-1/2} \tag{31}$$

$$= X {\rho_{I\cap J}}^{-1/2} E_I^{I\cup J}({\rho_J}^{1/2} Y {\rho_J}^{1/2}) {\rho_{I\cap J}}^{-1/2} = X\gamma(Y) \tag{32}$$

which shows the modular property. We remark again, that $\gamma(XY) = X\gamma(Y)$ holds for all $X \in \mathcal{A}(I \setminus J)^+$, $Y \in \mathcal{A}(I \cup J)$, even if $\phi_{I\cup J}$ is not even. For any $X \in \mathcal{A}(J)$

$$\begin{aligned}\gamma(X) &= E_I^{I\cup J}({\rho_{I\cap J}}^{-1/2} {\rho_J}^{1/2} X {\rho_J}^{1/2} {\rho_{I\cap J}}^{-1/2}) \\ &= E_I^{I\cup J}\left(E_J^{I\cup J}({\rho_{I\cap J}}^{-1/2} {\rho_J}^{1/2} X {\rho_J}^{1/2} {\rho_{I\cap J}}^{-1/2})\right) \\ &= E_{I\cap J}^{I\cup J}({\rho_{I\cap J}}^{-1/2} {\rho_J}^{1/2} X {\rho_J}^{1/2} {\rho_{I\cap J}}^{-1/2}),\end{aligned}$$

by the commuting square property, that is $\gamma(\mathcal{A}(J)) \subset \mathcal{A}(I \cap J)$ holds. These properties show that γ is a quasi-conditional expectation with respect to the triple

$$\mathcal{A}(I \setminus J) \subset \mathcal{A}(I) \subset \mathcal{A}(I \cup J).$$

Our assumption (28), according to the Theorem 3.1, is equivalent with

$${\rho_{I\cup J}}^{it} {\rho_J}^{-it} = {\rho_I}^{it} {\rho_{I\cap J}}^{-it}, \qquad t \in \mathbb{R},$$

or by the analytic continuation for $t = -i$ we have

$$\rho_{I\cup J} {\rho_J}^{-1} = \rho_I {\rho_{I\cap J}}^{-1}. \tag{33}$$

With the help of this relation we get for any $X \in \mathcal{A}(I \cup J)$

$$\begin{aligned}\phi_I(\gamma(X)) &= \tau\left(\rho_I E_I^{I\cup J}({\rho_{I\cap J}}^{-1/2}{\rho_J}^{1/2}X{\rho_J}^{1/2}{\rho_{I\cap J}}^{-1/2})\right)\\ &= \tau\left(E_I^{I\cup J}({\rho_I}^{1/2}{\rho_{I\cap J}}^{-1/2}{\rho_J}^{1/2}X{\rho_J}^{1/2}{\rho_{I\cap J}}^{-1/2}{\rho_I}^{1/2})\right)\\ &= \tau\left({\rho_I}^{1/2}{\rho_{I\cap J}}^{-1/2}{\rho_J}^{1/2}X{\rho_J}^{1/2}{\rho_{I\cap J}}^{-1/2}{\rho_I}^{1/2}\right)\\ &= \tau\left({\rho_{I\cup J}}^{1/2}X{\rho_{I\cup J}}^{1/2}\right) = \phi_{I\cup J}(X),\end{aligned}$$

which means that $\phi_I \circ \gamma = \phi_{I\cup J}$, so $\phi_{I\cup J}$ is a Markov state.

For the converse statement let consider a Markov state $\phi_{I\cup J}$ ie.

$$\phi_I \circ F = \phi_{I\cup J} \tag{34}$$

and

$$F(\mathcal{A}(J)) \subset \mathcal{A}(I \cap J) \tag{35}$$

where F is a quasi-conditional expectation with respect to the triple $\mathcal{A}(I \setminus J) \subset \mathcal{A}(I) \subset \mathcal{A}(I \cup J)$. Let denote F^* the dual map of F with respect to the Hilbert-Schmidt scalar product $\langle X, Y\rangle = \tau(X^*Y)$. In this case for any $X \in \mathcal{A}(I \cup J)$ we have

$$\begin{aligned}\phi_{I\cup J}(X) &= \tau(\rho_{I\cup J}X) = \langle \rho_{I\cup J}, X\rangle\\ &= \phi_I(F(X)) = \tau(\rho_I F(X))\\ &= \langle \rho_I, F(X)\rangle = \langle F^*(\rho_I), X\rangle\end{aligned}$$

which shows that

$$F^*(\rho_I) = \rho_{I\cup J}. \tag{36}$$

Now suppose that $X \in \mathcal{A}(J)$. We get

$$\begin{aligned}\phi_{I\cup J}(X) &= \phi_J(X) = \tau(\rho_J X)\\ &= \langle \rho_J, X\rangle = \phi_I(F(X))\\ &= \phi_{I\cap J}(F(X)) = \tau(\rho_{I\cap J}F(X))\\ &= \langle \rho_{I\cap J}, F(X)\rangle = \langle F^*(\rho_{I\cap J}), X\rangle\end{aligned}$$

where we used that (34) and (35). The computation above shows that

$$F^*(\rho_{I\cap J}) = \rho_J \tag{37}$$

also holds. As F^* is a dual of a quasi-conditional expectation, it is completely positive and trace preserving, i.e. it is a coarse graining, so with the equations (36) and (37) F^* fulfill the the necessary and sufficient conditions of the Theorem 3.2. We proved that we have equality in the strong subadditivity of von Neumann entropy for all Markovian state without any restriction for its evenness. □

From the proof it turns out attending to (30) and the remark after (31), that we can leave the condition of the evenness of our state, if we require a stronger condition to our localization:

Corollary 3.1. *Let I and J be two arbitrary subsets of $\mathbb{Z}$. Let $\phi_{I\cup J}$ be any state on the CAR algebra $\mathcal{A}(I\cup J)$ with the density matrix $\rho_{I\cup J}$. Then $\phi_{I\cup J}$ is a Markov state corresponding to the localization $\{\mathcal{A}(I\setminus J)^+, \mathcal{A}(I), \mathcal{A}(I\cup J)\}$ if and only if it saturates the strong subadditivity inequality of entropy, ie.*

$$S(\rho_I) + S(\rho_J) = S(\rho_{I\cap J}) + S(\rho_{I\cup J}),$$

where ρ_J, ρ_I and $\rho_{I\cap J}$ are the density matrices of the appropriate restrictions of $\phi_{I\cup J}$.

Corollary 3.2. *An even state ϕ on $\mathcal{A}$ is quantum Markov state, ie. for each $n\in\mathbb{N}$, there exists a quasi-conditional expectation E_n w.r.t the triplet $\mathcal{A}(n-1]) \subset \mathcal{A}(n]) \subset \mathcal{A}(n+1])$ satisfying*

$$\phi_{n]} \circ E_n = \phi_{n+1]},$$

$$E_n\left(\mathcal{A}\left([n,n+1]\right)\right) \subset \mathcal{A}\left(\{n\}\right),$$

if and only if

$$S\left(\phi_{n+1]}\right) + S\left(\phi_{\{n\}}\right) = S\left(\phi_{[n,n+1]}\right) + S\left(\phi_{n]}\right)$$

for all n.

Proof. For a fixed n, with the choice $I = n]$, $J = [n, n+1]$ the quantum Markov state become a Markov triplet. By using the theorem above for all n we get the statement. □

Acknowledgments

The author is grateful to L. Accardi, V.P. Belavkin, B. Dóra and D. Petz for the useful discussions. The work was partially supported by the Hungarian Research Grant OTKA TS049835.

References

1. L. Accardi, *Funkcional. Anal. i Prilozen.* **9**, 1–8 (1975).
2. L. Accardi, *Funct. Anal. Appl.* **8**, 1–8. (1975).
3. L. Accardi, in *Quantum probability and applications to the quantum theory of irreversible processes* (Lecture Notes in Math., Springer, 1984).
4. L. Accardi, F. Fidaleo and F. Mukhamedov, *Infin. Dim. Anal. Quantum Probab. Relat. Top.* **10**, 165–183 (2007).
5. L. Accardi and A. Frigerio, *Proc. R. Ir. Acad.* **83**, 251–263 (1983).
6. R. Alicki and M. Fannes, *Quantum dynamical systems*, (Oxford University Press, Oxford, 2001).
7. H. Araki and H. Moriya, *Rev. Math. Phys* **15**, 93–198 (2003).
8. H. Araki and H. Moriya, *Comm. Math. Phys.* **237**, 105–122 (2003).
9. O. Bratteli and D.W. Robinson, *Operator algebras and quantum statistical mechanics I, II* (Springer, 1979, 1981).
10. P. Hayden, R. Jozsa, D. Petz and A. Winter, *Comm. Math. Phys.* **246**, 359–374 (2004).
11. A. Jenčová and D. Petz, *Comm. Math. Phys.* **263**, 259–276 (2006).
12. E. H. Lieb and M. B. Ruskai, *J. Math. Phys.* **14**, 1938–1941 (1973).
13. H. Moriya, *J. Math. Phys.* **47**, 033510 (2006).
14. H. Moriya, in *Recent trends in infinite-dimensional noncommutative analysis*

(Japanese) (Kyoto, 1997). Surikaisekikenkyusho Kokyuroku **1035**, 128–132 (1998).

15. M. Mosonyi and D. Petz, *Lett. Math. Phys.* **68**, 19–30 (2004).
16. M. Ohya and D. Petz, *Quantum entropy and its use*, (Springer-Verlag, Heidelberg, 1993).
17. D. Petz, *Riv. di Math. Pura ed Appl.* **14**, 33–42 (1994).
18. D. Petz, *Rev. Math. Phys.* **15**, 79–91 (2003).
19. D. Petz, *Quantum Information Theory and Quantum Statistics*, (Springer-Verlag, Heidelberg, 2008).
20. D. Petz, *Quart. J. Math. Oxford* **39**, 475–483 (1984).
21. J. Pitrik, *J. Math. Phys.* **48**, 112110 (2007).
22. J. Pitrik and V.P. Belavkin, arXiv:math-ph/0602035 (2006)

Quantum Fokker-Planck Models: Limiting Case in the Lindblad Condition

F. FAGNOLA

Dipartimento di Matematica, Politecnico di Milano,

Piazza Leonardo da Vinci 32, I-20133 Milano, Italy

E-mail: franco.fagnola@polimi.it

L. NEUMANN

Institute for Analysis and Scientific Computing, TU Wien,

Wiedner Hauptstr. 8, A-1040 Wien, Austria

E-mail: Lukas.Neumann@tuwien.ac.at

In this article we study stationary states and the long time asymptotics for the quantum Fokker–Planck equation. We continue the investigation of an earlier work in which we derived convergence to a steady state if the Lindblad condition $D_{pp}D_{qq} - D_{pq}^2 \geq \gamma^2/4$ is satisfied with strict inequality. Here we extend our results to the limiting case that turns out to be more difficult because irreducibility of the quantum Markov semigroup does not follow from triviality of the generalized commutator with position and momentum operators.

Keywords: Quantum Markov Semigroups, Quantum Fokker Planck, steady state, large-time convergence.

1. Quantum Fokker–Planck model

This paper is concerned with the long-time asymptotics of quantum Fokker–Planck (QFP) models, a special type of open quantum systems that models the quantum mechanical charge-transport including diffusive effects, as needed, *e.g.*, in the description of quantum Brownian motion, quantum optics, and semiconductor device simulations. We shall consider two equivalent descriptions, the Wigner function formalism and the density matrix formalism. We continue our analysis that we commenced in [2].

In the quantum kinetic Wigner picture a quantum state is described by the real valued Wigner function $w(x,v,t)$, where $(x,v) \in \mathbb{R}^2$ denotes the position–velocity phase space. Its time evolution in a harmonic confinement potential $V_0(x) = \omega^2 \frac{x^2}{2}$ with $\omega > 0$ is given by the Wigner Fokker–Planck equation

$$\begin{aligned} \partial_t w &= \omega^2 x \partial_v w - v \partial_x w + Qw \,, \\ Qw &= 2\gamma \partial_v (vw) + D_{pp} \Delta_v w + D_{qq} \Delta_x w + 2D_{pq} \partial_v \partial_x w \,. \end{aligned} \tag{1}$$

The (real valued) diffusion constants D_{pp}, D_{pq}, D_{qq} and the friction $\gamma > 0$ satisfy the Lindblad condition

$$\Delta := D_{pp} D_{qq} - D_{pq}^2 - \gamma^2/4 \geq 0 \,, \tag{2}$$

and $D_{pp}, D_{qq} \geq 0$. In fact (2) together with $\gamma > 0$ implies $D_{pp}, D_{qq} > 0$. We assume that the particle mass and $\hbar$ are scaled to 1. This equation has been partly derived in [7]. Well–posedness [3,4,6], the classical limit [5] and long time asymptotics for purely harmonic oscillator potential [17] have been studied. For some applications we refer the reader to [9,10]. More references can be found in [1] or [16].

This equation can be equivalently studied in the Heisenberg-picture. The corresponding evolution equation on the space of bounded operators is given by[2]

$$\frac{\mathrm{d}A_t}{\mathrm{d}t} = \mathcal{L}(A_t) \,,$$

subject to initial conditions $A_{t=0} = A_0$. The generator $\mathcal{L}$ of the evolution semigroup $\mathcal{T}$ is given by

$$\begin{aligned} \mathcal{L}(A) &= \frac{i}{2} \left[p^2 + \omega^2 q^2 + 2V(q), A \right] + i\gamma \left\{ p, [q, A] \right\} \\ &\quad - D_{qq}[p,[p,A]] - D_{pp}[q,[q,A]] + 2D_{pq}[q,[p,A]] \,, \quad A \in \mathcal{B}(\mathsf{h}) \,. \end{aligned}$$

It can be written in (generalised) GKSL form like

$$\mathcal{L}(A) = i[H, A] - \frac{1}{2}\sum_{\ell=1}^{2} (L_\ell^* L_\ell A - 2L_\ell^* A L_\ell + A L_\ell^* L_\ell) \tag{3}$$

with the "adjusted" Hamiltonian

$$H = \frac{1}{2}\left(p^2 + \omega^2 q^2 + \gamma(pq + qp)\right) + V(q)\,,$$

and the Lindblad operators L_1 and L_2 given by

$$L_1 = \frac{-2D_{pq} + i\gamma}{\sqrt{2D_{pp}}} p + \sqrt{2D_{pp}}\, q\,, \qquad L_2 = \frac{2\sqrt{\Delta}}{\sqrt{2D_{pp}}} p\,. \tag{4}$$

Note that here we use the external potential $U(q) = \omega^2 q^2/2 + V(q)$. The harmonic oscillator potential is the simplest way of ensuring confinement to guarantee the existence of a non trivial steady state. $V(q)$ is a perturbation potential, assumed to be twice continuously differentiable and satisfy

$$|V'(x)| \le g_V \left(1 + |x|^2\right)^{\alpha/2}\,, \tag{5}$$

with $g_V > 0$ and $0 \le \alpha < 1$.

2. Previous results

In Ref. 2 we proved the existence of the minimal Quantum Markov semigroup (QMS) for the Lindbladian (3). We will only sketch the result here. First note that all operators can be defined on the domain of the Number operator $N := (p^2 + q^2 - 1)/2$,

$$\mathrm{Dom}(N) = \left\{ u \in \mathsf{h} \,\middle|\, Nu \in \mathsf{h} \right\} = \left\{ u \in \mathsf{h} \,\middle|\, p^2 u,\, q^2 u \in \mathsf{h} \right\}.$$

For details on domain problems we refer to [2].
We consider the operator G, defined on $\mathrm{Dom}(N)$, by

$$G = -\frac{1}{2}(L_1^* L_1 + L_2^* L_2) - iH = -\left(D_{qq} + \frac{i}{2}\right)p^2 - \left(D_{pp} + \frac{i\omega^2}{2}\right)q^2 \\ + \left(D_{pq} - \frac{i\gamma}{2}\right)(pq + qp) + \frac{\gamma}{2} - iV(q)\,. \tag{6}$$

It can be checked that the domain of the adjoint operator G^* is again $\mathrm{Dom}(N)$. The operators G and G^* are dissipative and thus G generates a strongly continuous contraction semigroup $(P_t)_{t\ge 0}$ on h.

Since the *formal* mass preservation holds we can apply results from Ref. 12 to construct $\mathcal{T}$, the minimal QMS associated with G and the L_ℓ's. Moreover applying results form Refs. 8 and 12 we proved the following theorem.

Theorem 2.1.[2] *Suppose that the potential V is twice differentiable and satisfies the growth condition (5). Then the minimal semigroup associated with the closed extensions of the operators G, L_1, L_2 is Markov and admits a normal invariant state.*

Note that this also implies the existence of the predual semigroup $\mathcal{T}_*$ on $\mathcal{J}_1$, the set of positive trace–class operators (*i.e.* density metrices).

The next step in our analysis is the proof of irreducibility. This implies that any initial density matrix, in the evolution, gives a positive mass on any subspace of h and allows us to apply powerful convergence results.

A QMS $\mathcal{T}$ on $\mathcal{B}(\mathsf{h})$ is called *irreducible* if the only subharmonic projections[13] Π in h (*i.e.* projections satisfying $\mathcal{T}_t(\Pi) \geq \Pi$ for all $t \geq 0$) are the trivial ones 0 or $\mathbf{1}$. If a projection Π is subharmonic, the total mass of any normal state σ with support in Π (*i.e.* such that $\Pi\sigma\Pi = \Pi\sigma = \sigma\Pi = \sigma$), remains concentrated in Π during the evolution. As an example, the support projection of a normal stationary state for a QMS is subharmonic.[13] Thus if a QMS is irreducible and has a normal invariant state, then its support projection must be $\mathbf{1}$, *i.e.* it must be *faithful*. Subharmonic projections are characterised by the following theorem.

Theorem 2.2.[13] *A projection Π is subharmonic for the QMS associated with the operators G, L_ℓ if and only if its range $\mathcal{X}$ is an invariant subspace for all the operators P_t of the contraction semigroup generated by G (i.e. $\forall t \geq 0 : P_t\mathcal{X} \subseteq \mathcal{X}$) and $L_\ell\,(\mathcal{X} \cap Dom(G)) \subseteq \mathcal{X}$ for all ℓ's.*

The application to our model yields the following Theorem. A sketch of the proof will be given in the beginning of the next section.

Theorem 2.3.[2] *Suppose that $\Delta > 0$. Then the QMS $\mathcal{T}$ associated with (the closed extensions of) the operators G, L_ℓ given by (6) and (4) is irreducible and thus all normal invariant states are faithful.*

We denote by $\{ H, L_1, L_1^*, L_2, L_2^* \}'$ the *generalized commutant*, *i.e.* the set of all operators that commute with H as well as with L_1,L_1^*,L_2 and L_2^*. Now since the semigroup has a faithful invariant state a combination of result by Frigerio,[15] Fagnola and Rebolledo[11,14] gives (under some technical conditions that can be checked for our model[2]) a criterium for convergence towards the steady state. If $\{ H, L_1, L_1^*, L_2, L_2^* \}' = \{ L_1, L_1^*, L_2, L_2^* \}' = \mathbb{C}\mathbf{1}$ then $\mathcal{T}_{*t}(\sigma)$ converges as $t \to \infty$ towards a unique invariant state in the trace norm. From $\gamma > 0$ we conclude that L_1 and L_1^* are linearly independent and thus $\{ L_1, L_1^*, L_2, L_2^* \}'$ contains operators commuting with both q and p. This yields $\mathbb{C}\mathbf{1} = \{ L_1, L_1^*, L_2, L_2^* \}' \supseteq \{ H, L_1, L_1^*, L_2, L_2^* \}'$ and leads to

Corollary 2.1. *Let $\gamma > 0$ and $V \in C^2(\mathbb{R})$ satisfy (5). If the QMS associated with G and L_ℓ is irreducible (by Thm. 2.3 this holds true if $\Delta > 0$) then it has a unique faithful normal invariant state ρ. Moreover, for all normal initial states σ, we have*

$$\lim_{t\to\infty} \mathcal{T}_{*t}(\sigma) = \rho$$

in the trace norm.

Note that in the limiting case $\Delta = 0$ the irreducibility can indeed fail:

Proposition 2.1.[2] *Let $V = 0$, $\Delta = 0$, and $0 < \gamma < \omega$. Under the conditions*

$$D_{pq} = -\gamma D_{qq} \quad \text{and} \quad D_{pp} = \omega^2 D_{qq} \,. \tag{7}$$

the semigroup is not irreducible. It admits a steady state that is not faithful.

3. Irreducibility for $\Delta = 0$

In this section we will show that the semigroup is irreducible if the conditions (7) are violated. In doing so we also extend our convergence result.

The interesting case when conditions (7) hold but perturbation potential is different from zero is postponed to a later work. We conjecture that the semigroup becomes irreducible as soon as $V \neq 0$.

First we sketch the idea of the proof of irreducibility in the case $\Delta > 0$. By Theorem 2.2 a projection is subharmonic if its range $\mathcal{X}$ is invariant for G as well as for L_1 and L_2. Since L_1 and L_2 are linearly independent if $\Delta > 0$ we know that $\mathcal{X}$ has to be invariant for p and q. Thus it is also invariant for the creation and annihilation operators a and $a^\dagger$. Now if the closed subspace $\mathcal{X}$ is nonzero it includes an eigenvector of the Number operator. Since it is invariant under both, the creation and the annihilation operator, it has to be the whole space. Now the only subharmonic projections are the trivial ones and the semigroup is irreducible. A precise proof becomes more involved due to domain problems and can be found in [2]. This proof breaks down if $\Delta = 0$ since in this case $L_2 = 0$. Thus we look for an operator that leaves $\mathcal{X}$ invariant and can replace L_2 in the above strategy.

Since $\mathcal{X}$ is G and L_1 invariant, the most natural choice for such an operator should be a polynomial in (the non-commuting) G and L_1. We do all calculations on $\mathcal{C}_c^\infty$ disregarding commutator domains, *i.e.* understanding $[\cdot\,,\cdot]$ as $\overline{[\cdot\,,\cdot]}$. All operators can be extended to $\mathrm{Dom}(N)$ as in Ref. [2].

Lemma 3.1. *Let $\Delta = 0$. The following identities hold on $\mathcal{C}_c^\infty$:*

$$[G, L_1] = B + \frac{-2D_{pq} + i\gamma}{\sqrt{2D_{pp}}} V'(q) \tag{8}$$

where

$$B = \frac{2\gamma(-2D_{pq} + i\gamma) - 2D_{pp}}{\sqrt{2D_{pp}}} p + \omega^2 \frac{-2D_{pq} + i\gamma}{\sqrt{2D_{pp}}} q.$$

The operator B is linearly dependent of L_1 if and only if the identities (7) *hold. In this case*

$$B = \omega^2 \frac{-2D_{pq} + i\gamma}{2D_{pp}} L_1. \tag{9}$$

Proof. Since $L_2 = 0$ we have $G = -\frac{1}{2}L_1^* L_1 - iH$. A straightforward but rather lengthy calculation using the CCR $[q, p] = i$ leads to formula (8). Two operators $xp + yq$, $zp + wq$ (with $x, y, z, w \in \mathbb{C} - \{0\}$) are linearly *dependent* if and only if $x/z = y/w$. Therefore B and L_1 are linearly dependent if and only if

$$\frac{2\gamma(-2D_{pq}+i\gamma)-2D_{pp}}{-2D_{pq}+i\gamma} = \omega^2\frac{(-2D_{pq}+i\gamma)}{2D_{pp}} \tag{10}$$

Clearly $\Delta = 0$ is equivalent to $4D_{pp}D_{qq} = (-2D_{pq}+i\gamma)(-2D_{pq}-i\gamma)$, *i.e.*

$$\frac{2D_{pp}}{-2D_{pq}+i\gamma} = \frac{-2D_{pq}-i\gamma}{2D_{qq}}.$$

Therefore (10) can be written in the form

$$2\gamma + \frac{2D_{pq}+i\gamma}{2D_{qq}} = \omega^2\frac{(-2D_{pq}+i\gamma)}{2D_{pp}}.$$

The imaginary part of the left and right-hand side coincide if and only if $D_{pp} = \omega^2 D_{qq}$. Then the real parts coincide if and only if $D_{pq} = -\gamma D_{qq}$.

Now, if the identities $D_{pp} = \omega^2 D_{qq}$ and $D_{pq} = -\gamma D_{qq}$ hold, then we can write B as

$$B = \frac{-2D_{pq}+i\gamma}{\sqrt{2D_{pp}}}\left(\left(2\gamma - \frac{2D_{pp}}{-2D_{pq}+i\gamma}\right)p + \omega^2 q\right).$$

Writing

$$2\gamma - \frac{2D_{pp}}{-2D_{pq}+i\gamma} = 2\gamma + \frac{2D_{pq}+i\gamma}{2D_{qq}} = \omega^2\frac{-2D_{pq}+i\gamma}{2D_{pp}}$$

we find the identity (9). □

Theorem 3.1. *Let $\Delta = 0$ and $D_{pq} \neq -\gamma D_{qq}$. Moreover assume that V is twice continuously differentiable with V'' bounded.*

The QMS $\mathcal{T}$ associated with (the closed extensions of) the operators G, L_1 given by (6) and (4) is irreducible.

Proof. We only point out the difference with respect to the proof of in the case $\Delta > 0$ in [2]. The proof will proceed in tree steps. First we show that the range $\mathcal{X}$ of a subharmonic protection has to be invariant under the multiplication operator $V''(q)$. In step two we use this to show that $\mathcal{X}$ has to be invariant under an operator of the form $q(1+zV''(q))$ for some $z \in \mathbb{C}$ with $\Im(z) \neq 0$. In step tree we conclude by a technical argument that this ensures invariance of $\mathcal{X}$ under multiplication by q and complete the proof.

Step 1: The subspace $\mathcal{X}$ has to be invariant under the double commutator $[\,[G, L_1], L_1\,]$ *i.e.*, more precisely

$$[\,[G, L_1], L_1\,]\,(\mathcal{X} \cap \mathrm{Dom}(N^n)) \subseteq \mathcal{X} \cap \mathrm{Dom}(N^{n+2})$$

for all $n \geq 0$. A straightforward computation shows that

$$[\,[G, L_1], L_1\,] = z\mathbb{1} + i\frac{(-2D_{pq} + i\gamma)^2}{2D_{pp}} V''(q)$$

for some $z \in \mathbb{C}$. Therefore, by the density of $\mathcal{X} \cap \mathrm{Dom}(N^2)$ in $\mathcal{X}$, and boundedness of the self-adjoint multiplication operator $V''(q)$, we have

$$V''(q)(\mathcal{X}) \subseteq \mathcal{X}.$$

Step 2: We first calculate the commutator $[G, [G, L_1]]$.
To shorten the notation we set $\alpha := \frac{-2D_{pq}+i\gamma}{\sqrt{2D_{pp}}}$. With this abbreviation $\Delta = 0$ becomes $|\alpha|^2 = \alpha\overline{\alpha} = 2D_{qq}$, and we have

$$[G, L_1] = (2\gamma\alpha - \sqrt{2D_{pp}})p + \omega^2\alpha q + \alpha V'(q)\ .$$

Straightforward calculations yield

$$\begin{aligned}[G, [G, L_1]] = &\left[(\alpha\omega^2(i\alpha\overline{\alpha} - 1) - i\alpha\sqrt{2D_{pp}}(2\gamma\alpha - \sqrt{2D_{pp}})\right] p+ \\ &\left[i\sqrt{2D_{pp}}\alpha^2\omega^2 - 2i(D_{pp} + i\omega^2/2)(2\gamma\alpha - \sqrt{2D_{pp}})\right] q+ \\ i(\alpha\overline{\alpha}/2 + i/2)\alpha\{p, V''(q)\} &+ (2\gamma\alpha - \sqrt{2D_{pp}})V'(q) + i\sqrt{2D_{pp}}\alpha^2 V''(q)q\ ,\end{aligned}$$

where $\{p, V''(q)\}$ denotes the anticommutator.
Note that $\mathcal{X}$ is invariant under L_1 and by Step 1 also under the multiplication operator $V''(q)$. Thus it has to be invariant under the anticommutator

$$\{L_1, V''(q)\} = \alpha\{p, V''(q)\} + 2\sqrt{2D_{pp}}qV''(q)\ .$$

We can remove the term proportional to $\{p, V''\}$ from the double commutator by adding a suitable multiple of $\{L_1, V''\}$. The term proportional to V' can be eliminated by a multiple of $[G, L_1]$ and finally we use L_1 to cancel the term with the momentum operator. Doing the tedious algebra leads to

$$[G,[G,L_1]] + c_1\{L_1,V''\} + c_2[G,L_1] + c_3L_1 =$$
$$(-2\gamma\alpha + \sqrt{2D_{pp}})\left[\left(\omega^2 + (-2\gamma\alpha + \sqrt{2D_{pp}})\sqrt{2D_{pp}}/\alpha^2\right) q + qV''(q)\right] ,$$

for explicit constants c_1, c_2, $c_3 \in \mathbb{C}$.
Since $\mathcal{X}$ is invariant for all operators on the left hand side of the above equation (and the coefficient has absolute value different from zero) it is also invariant for $q(y + V'')$ with $y = \omega^2 + (-2\gamma\alpha + \sqrt{2D_{pp}})\sqrt{2D_{pp}}/\alpha^2$. The real and imaginary parts of y are given by

$$\Re(y) = \frac{2D_{pp}}{(4D_{pq}^2 + \gamma^2)^2}\left[4\gamma D_{pq}(4D_{pq}^2 + \gamma^2) + 2D_{pp}(4D_{pq}^2 - \gamma^2)\right] + \omega^2$$
$$\Im(y) = \frac{2D_{pp}}{(4D_{pq}^2 + \gamma^2)^2}\left[2\gamma^2(4D_{pq}^2 + \gamma^2) + 8\gamma D_{pq}D_{pp}\right] .$$

Note that $\Im(y) = 0$ if and only if $D_{pq} = -\gamma D_{qq}$, as can be seen by using $\Delta = 0$ in the equation above. The condition for the real part to be zero, $D_{pq}/D_{qq} = -\gamma \pm \sqrt{\gamma^2 - \omega^2 + \gamma^2/(4D_{qq}^2)}$, is more difficult to see but direct calculations yield that when $\Im(y) = 0$, then $\Re(y) = 0$ if and only if $D_{pp} = \omega^2 D_{qq}$. Thus $|y|$ is zero exactly if B and L_1 are linearly dependent. Since from our assumptions $D_{pq} \neq -\gamma D_{qq}$ we can invert y and see that $\mathcal{X}$ is invariant for an operator $q(1 + zV'')$ with $\Im(z) = -\Im(y)/|y|^2 \neq 0$.

Step 3: Note that

$$|1 + zV''(x)|^2 = (1 + \Re(z)V''(x))^2 + (\Im(z))^2(V''(x))^2$$

and $1 + zV''(x)$ is non-zero for all $x \in \mathbb{R}$ because there is no x such that $1+\Re(z)V''(x) = 0 = V''(x)$ (recall $\Im(z) \neq 0$). Moreover, for the same reason there is no sequence $(x_n)_{n\geq 1}$ of real numbers such that $1+\Re(z)V''(x_n)$ and $V''(x_n)$ both vanish as n goes to infinity. It follows that

$$\inf_{x\in\mathbb{R}} |1 + zV''(x)|^2 > 0$$

and $1 + zV''$ has a bounded inverse. This is given by spectral calculus of normal operators by

$$(1+zV'')^{-1} = \int_0^\infty \mathrm{e}^{-t(1+zV'')}\mathrm{d}t$$

and, since $\mathcal{X}$ is invariant under all powers $(1+zV'')^n$, it is invariant under $\mathrm{e}^{-t(1+zV'')}$ and also under the resolvent operator $(1+zV'')^{-1}$.
Now, for all $u \in \mathcal{X} \cap \mathrm{Dom}(G)$ we have $(1+zV''(q))^{-1}u = v \in \mathcal{X} \cap \mathrm{Dom}(q^2)$ and thus

$$qu = (q(1+zV''(q)))\,(1+zV''(q))^{-1}u = (q(1+zV''(q)))\,v \in \mathcal{X}.$$

It follows that $\mathcal{X}$ is q–invariant. Since $\mathcal{X}$ is also L_1 invariant it has to be p invariant. Thus it is invariant under the creation operator $a = (q+ip)/\sqrt{2}$ and the annihilation operator $a^\dagger = (q-ip)/\sqrt{2}$ and $\mathcal{X}$ has to be either zero or coincide with the whole space (see [2]). □

Acknowledgement:
The authors would like to thank A. Arnold for fruitful discussions.

References

1. A. Arnold: *Mathematical Properties of Quantum Evolution Equations*, Quantum Transport - Modelling, Analysis and Asymptotics, (LNM 1946), G. Allaire, A. Arnold, P. Degond, Th.Y. Hou, Springer, Berlin (2008)
2. A. Arnold, F. Fagnola and L. Neumann: *Quantum Fokker–Planck models: the Lindblad and Wigner approaches*, Proceedings of the 28'th Conf. on Quantum Prob. and Relat. Topics, J. C. García, R. Quezada and S. B. Sontz (Editors), Quantum Probability and White Noise Analysis **23**, (2008), 23–48.
3. A. Arnold, J.L. López, P.A. Markowich, and J. Soler: *An Analysis of Quantum Fokker-Planck Models: A Wigner Function Approach*, Rev. Mat. Iberoam. **20**(3) (2004), 771–814.
   Revised: `http://www.math.tuwien.ac.at/~arnold/papers/wpfp.pdf` .
4. A. Arnold, and C. Sparber: *Quantum dynamical semigroups for diffusion*

models with Hartree interaction, Comm. Math. Phys. **251**(1) (2004), 179–207.

5. R. Bosi: *Classical limit for linear and nonlinear quantum Fokker-Planck systems*, preprint 2007.
6. J.A. Cañizo, J.L. López, and J. Nieto: *Global L^1 theory and regularity for the 3D nonlinear Wigner-Poisson-Fokker-Planck system*, J. Diff. Equ. **198** (2004), 356–373.
7. F. Castella, L. Erdös, F. Frommlet, and P. Markowich: *Fokker-Planck equations as Scaling Limit of Reversible Quantum Systems*, J. Stat. Physics **100**(3/4) (2000), 543–601.
8. A.M. Chebotarev and F. Fagnola: *Sufficient conditions for conservativity of quantum dynamical semigroups*, J. Funct. Anal. **153** (1998), 382–404.
9. A. Donarini, T. Novotný, and A.P. Jauho: *Simple models suffice for the single-dot quantum shuttle*, New J. of Physics **7** (2005), 237–262.
10. M. Elk and P. Lambropoulos: *Connection between approximate Fokker-Planck equations and the Wigner function applied to the micromaser*, Quantum Semiclass. Opt. **8** (1996), 23–37.
11. F. Fagnola and R. Rebolledo: *The approach to equilibrium of a class of quantum dynamical semigroups*, Infin. Dimens. Anal. Quantum Probab. Relat. Top. **1**(4) (1998), 561–572.
12. F. Fagnola and R. Rebolledo: *On the existence of stationary states for quantum dynamical semigroups*, J. Math. Phys. **42**(3) (2001), 1296–1308.
13. F. Fagnola and R. Rebolledo: *Subharmonic projections for a quantum Markov semigroup*, J. Math. Phys. **43**(2) (2002), 1074–1082.
14. F. Fagnola and R. Rebolledo: *Algebraic Conditions for Convergence to a Steady State*, Infin. Dimens. Anal. Quantum Probab. Relat. Top. **11** (2008), 467–474.

15. A. Frigerio: *Quantum dynamical semigroups and approach to equilibrium*, Lett. Math. Phys. **2**(2) (1977/78), 79–87.

16. A. Isar, A. Sandulescu, H. Scutaru, E. Stefanescu, and W. Scheid: *Open quantum Systems*, Int. J. Mod. Phys. E, **3**(2) (1994), 635–719.

17. C. Sparber, J.A. Carrillo, J. Dolbeault, and P.A. Markowich: *On the long time behavior of the quantum Fokker-Planck equation*, Monatsh. Math. **141**(3) (2004), 237–257.

GENERALIZED EULER HEAT EQUATION

Abdessatar BARHOUMI

Department of Mathematics

Higher School of Sci. and Tech. of Hammam-Sousse

University of Sousse, Sousse, Tunisia

E-mail: abdessatar.barhoumi@ipein.rnu

Habib OUERDIANE

Department of Mathematics

Faculty of Sciences of Tunis

University of Tunis El-Manar, Tunis, Tunisia

E-mail: habib.ouerdiane@fst.rnu.tn

Hafedh RGUIGUI

Department of Mathematics

Faculty of Sciences of Tunis

University of Tunis El-Manar, Tunis, Tunisia

E-mail: hafedh.rguigui@yahoo.fr

We extend some results about second quantization and K-Gross Laplacian on nuclear algebra of entire functions. Then, we investigate the solution of a initial-value problem associated to a suitable generalized Euler Operator.

Keywords: Differential second quantization, Euler heat equation, Fourier-Gauss transform, K-Gross Laplacian.

1. Introduction

Gross[12] and Piech[21] initiated the study of the infinite dimensional Laplacians (the Gross Laplacian Δ_G and the number operator N, resp.) on infinite dimensional abstract Wiener space, as the infinite dimensional analogue of the finite dimensional Laplacian. In Ref. 19, Kuo has studied the heat equation associated with the Gross Laplacian in white noise analysis setting. In Ref. 17, Kang has studied the heat equation associated with the number operator N in white noise analysis setting. In this paper we will investigate the existence of a solution of the Cauchy problem associated with the generalized Euler operator Δ_E in the basis of nuclear algebras of entire functions. The operator $\Delta_E := \Delta_G + N$ studied in Refs. 5, 6 and 22 is the infinite dimensional analogue of the so-called *Euler operator* which is defined as the first order differential operator $\sum_{i=1}^{d} x_i \frac{\partial}{\partial x_i}$ on $\mathbb{R}^d$ (see Ref. 10).

We define

$$\Delta_E(K, B) = \Delta_G(\frac{1}{2}K) + N(B)$$

for $K \in \mathcal{L}(N, N')$ and $B \in \mathcal{L}(N, N)$ and we call it also the (infinite dimensional) *generalized Euler operator*. Then we show that under some appropriate conditions, $\Delta_E(K, B)$ is the generator of a one-parameter group transformation. By using the $\mathcal{G}_{K,B}$-transform studied in Refs. 2, 5 and 14 we investigate the existence of a solution of the following *generalized Euler heat equation*:

$$\frac{\partial u(t)}{\partial t} = \Delta_E(K, B)u(t), \quad u(0) = \varphi \in \mathcal{F}_\theta(N').$$

We expect some other interesting features concerning regularity of the solutions as well as their quantum extensions. These topics are, however, somehow beyond the scop of this paper and we hope to discuss them elsewhere.

The paper is organized as follows. In Section 2 we briefly recall well-known results on nuclear algebras of entire holomorphic functions. In Section 3, we recall some basic useful results for the transformations $\mathcal{G}_{A,B}$ in white noise theory within the framework of nuclear algebras of entire functions. In Section 4, we study the K-Gross Laplacian $\Delta_G(K)$, the second quantization $\Gamma(A)$ and the differential second quantization $N(A)$ within the framework of nuclear algebras of entire functions. In Section 5, we investigate the solution of a initial-value problem associated to the sum of the differential second quantization and the K-Gross Laplacian as a generalization of the so-called Euler equation.

2. Preliminaries

In this Section we shall briefly recall some of the concepts, notations and known results on nuclear algebras of entire functions.[7,9,11,19,20] Let H be a real separable (infinite dimensional) Hilbert space with inner product $\langle \cdot, \cdot \rangle$ and norm $|\cdot|_0$. Let $A \geq 1$ be a positive self-adjoint operator in H with Hilbert-Schmidt inverse. Then there exist a sequence of positive numbers $1 < \lambda_1 \leq \lambda_2 \leq \cdots$ and a complete orthonormal basis of H, $\{e_n\}_{n=1}^{\infty} \subseteq \mathrm{Dom}(A)$ such that

$$Ae_n = \lambda_n e_n \, , \qquad \sum_{n=1}^{\infty} \lambda_n^{-2} = \left\|A^{-1}\right\|_{HS}^2 < \infty.$$

For every $p \in \mathbb{R}$ we define:

$$|\xi|_p^2 := \sum_{n=1}^{\infty} \langle \xi, e_n \rangle^2 \lambda_n^{2p} = |A^p \xi|_0^2 \, , \qquad \xi \in H.$$

The fact that, for $\lambda > 1$, the map $p \mapsto \lambda^p$ is increasing implies that:

(i) for $p \geq 0$, the space X_p, of all $\xi \in H$ with $|\xi|_p < \infty$, is a Hilbert space with norm $|\cdot|_p$ and, if $p \leq q$, then $X_q \subseteq X_p$;

(ii) denoting X_{-p} the $|\cdot|_{-p}$-completion of H ($p \geq 0$), if $0 \leq p \leq q$, then $X_{-p} \subseteq X_{-q}$.

This construction gives a decreasing chain of Hilbert spaces $\{X_p\}_{p \in \mathbb{R}}$ with natural continuous inclusions $i_{q,p} : X_q \hookrightarrow X_p$ ($p \leq q$). Defining the countably Hilbert nuclear space (see e.g. Ref. 11):

$$X := \operatorname*{projlim}_{p \to \infty} X_p \cong \bigcap_{p \geq 0} X_p$$

the strong dual space X' of X is:

$$X' := \operatorname*{indlim}_{p \to \infty} X_{-p} \cong \bigcup_{p \geq 0} X_{-p}$$

and the triple

$$X \subset H \equiv H' \subset X' \tag{1}$$

is called a real standard triple.[20] The complexifications of X_p, X and H respectively will be denoted

$$N_p := X_p + iX_p \, ; \quad N := X + iX \, ; \quad \mathcal{H} := H + iH. \tag{2}$$

Notice that $\{e_n\}_{n=1}^{\infty}$ is also a complete orthonormal basis of $\mathcal{H}$. Thus the complexification of the standard triple (1) is:

$$N \subset \mathcal{H} \subset N'.$$

When dealing with complex Hilbert spaces, we will always assume that the scalar product is linear in the second factor and the duality $\langle N', N\rangle$, also denoted $\langle\cdot,\cdot\rangle$, is defined so to be compatible with the inner product of $\mathcal{H}$.

For $n \in \mathbb{N}$ we denote by $N^{\widehat{\otimes} n}$ the n-fold symmetric tensor product of N equipped with the π-topology and by $N_p^{\widehat{\otimes} n}$ the n-fold symmetric Hilbertian tensor product of N_p. We will preserve the notation $|\cdot|_p$ and $|\cdot|_{-p}$ for the norms on $N_p^{\widehat{\otimes} n}$ and $N_{-p}^{\widehat{\otimes} n}$, respectively.

From Ref. 9 we recall the following background. Let θ be a Young function, i.e., it is a continuous, convex, and increasing function defined on $\mathbb{R}_+$ and satisfies the condition $\lim_{x\to\infty}\theta(x)/x = \infty$. We define the conjugate function θ^* of θ by

$$\theta^*(x) = \sup_{t\geq 0}\big(tx - \theta(t)\big), \quad x \geq 0.$$

For a Young function θ, we denote by $\mathcal{F}_\theta(N')$ the space of holomorphic functions on N' with exponential growth of order θ and of minimal type. Moreover, for each $p \in \mathbb{Z}$ and $m > 0$, define $\mathrm{Exp}(N_p, \theta, m)$ to be the space of entire functions f on N_p satisfying the condition:

$$\|f\|_{\theta,p,m} = \sup_{x\in N_p} |f(x)|e^{-\theta(m|x|_p)} < \infty.$$

Then the space $\mathcal{F}_\theta(N')$ can be represented as

$$\mathcal{F}_\theta(N') = \bigcap_{p\in\mathbb{N},\, m>0} \mathrm{Exp}(N_{-p}, \theta, m),$$

and is equipped with the projective limit topology. The space $\mathcal{F}_\theta(N')$ is called the space of *test functions* on N'.

For $p \in \mathbb{N}$ and $m > 0$, we define the Hilbert spaces

$$F_{\theta,m}(N_p) = \Big\{\vec{\varphi} = (\varphi_n)_{n=0}^\infty\,;\, \varphi_n \in N_p^{\widehat{\otimes} n},\, \sum_{n=0}^\infty \theta_n^{-2} m^{-n} |\varphi_n|_p^2 < \infty\Big\},$$

where

$$\theta_n = \inf_{r>0} e^{\theta(r)}/r^n, \rangle n \in \mathbb{N}. \tag{3}$$

Put

$$F_\theta(N) = \bigcap_{p\in\mathbb{N}, m>0} F_{\theta,m}(N_p).$$

The space $F_\theta(N)$ equipped with the projective limit topology is a nuclear Frechét space.[9]

It was proved in Ref. 9 that the *Taylor map* defined by

$$T:\ \varphi \longmapsto \Big(\frac{1}{n!}\varphi^{(n)}(0)\Big)_{n=0}^{\infty}$$

is a topological isomorphism from $\mathcal{F}_\theta(N')$ onto $F_\theta(N)$. For $\vec{\varphi} = (\varphi_n)_{n\geq 0} \in F_\theta(N)$ we write $\varphi \sim (\varphi_n)_{n\geq 0}$ for short. The following estimate is useful.

Lemma 2.1. *(See Ref. 9) Let* $\varphi \sim (\varphi_n)_{n\geq 0}$ *in* $\mathcal{F}_\theta(N')$. *Then, for any* $n \geq 0$, $p \geq 0$ *and* $m > 0$, *there exist* $q > p$ *such that*

$$|\varphi_n|_p \leq e^n \theta_n m^n \|i_{q,p}\|_{HS}^n \|\varphi\|_{\theta,q,m}. \tag{4}$$

The Borel σ-algebra on X' will be denoted by $\mathcal{B}(X')$. It is well-known[11] that $\mathcal{B}(X')$ coincides with the σ-algebra generated by the cylinder subsets of X'. Let μ be the standard Gaussian measure on $(X', \mathcal{B}(X'))$, i.e., its characteristic function is given by

$$\int_{X'} e^{i\langle y,\xi\rangle}\, d\mu(y) = e^{-|\xi|_0^2/2}, \quad \xi \in X.$$

3. The $\mathcal{G}_{A,B}$−transform

In this section we shall briefly recall some of results studied in Ref. 2. For locally convex spaces $\mathfrak{X}$ and $\mathfrak{Y}$ we denote by $\mathcal{L}(\mathfrak{X}, \mathfrak{Y})$ the set of all continuous linear operators from $\mathfrak{X}$ into $\mathfrak{Y}$. Let B, $C \in \mathcal{L}(N, N)$ and put $A = C^*C$. The $\mathcal{G}_{A,B}$−transform is defined by

$$\mathcal{G}_{A,B}\varphi(y) = \int_{X'} \varphi(C^*x + B^*y) d\mu(x)\,, \quad y \in N',\ \varphi \in \mathcal{F}_\theta(N'). \tag{5}$$

Theorem 3.1.[2] *Let* B, $C \in \mathcal{L}(N, N)$ *and put* $A = C^*C$, *then* $\mathcal{G}_{A,B}$ *is a continuous linear operator from* $\mathcal{F}_\theta(N')$ *into itself.*

We denote by $\tau(K)$ the corresponding distribution to $K \in \mathcal{L}(N, N')$ under the canonical isomorphism $\mathcal{L}(N, N') \cong (N \otimes N)'$, i.e.

$$\langle \tau(K), \xi \otimes \eta\rangle = \langle K\xi, \eta\rangle, \quad \xi,\ \eta \in N.$$

In particular, $\tau(I)$ is the usual trace τ. It can be easily shown that

$$\tau(K) = \sum_{j=0}^{\infty} (K^* e_j) \otimes e_j,$$

where K^* is the adjoint of K with respect to the dual pairing $\langle N', N\rangle$ and the infinite sum is in the sense of the strong topology on $(N \otimes N)'$.

Proposition 3.1.[2] *Let $B, C \in \mathcal{L}(N, N)$ and put $A = C^*C$, then for any $\varphi(y) = \sum_{n=0}^{\infty}\langle y^{\otimes n}, \varphi_n\rangle \in \mathcal{F}_\theta(N')$, we have*

$$\mathcal{G}_{A,B}\varphi(y) = \sum_{n=0}^{\infty}\langle y^{\otimes n}, g_n\rangle,$$

where g_n is given by

$$g_n = (B)^{\otimes n}\left(\sum_{l=0}^{\infty}\frac{(n+2l)!}{n!\,2^l\,l!}(\tau(A))^{\otimes l}\widehat{\otimes}_{2l}\varphi_{n+2l}\right).$$

Proposition 3.2.[2] *Let B, C_1, C_2, $D \in \mathcal{L}(N, N)$ and denote $A_1 = C_1^*C_1$, $A_2 = C_2^*C_2$, then*

$$\mathcal{G}_{A_2,D}\mathcal{G}_{A_1,B} = \mathcal{G}_{A_1+B^*A_2B,DB}.$$

In particular, if B is invertible, then the operator $\mathcal{G}_{A,B}$ is invertible and

$$\mathcal{G}_{A,B}^{-1} = \mathcal{G}_{-(B^*)^{-1}AB^{-1},B^{-1}}.$$

Proposition 3.3.[2] *Let $B, C \in \mathcal{L}(N, N)$ such that B is invertible and put $C^*C = A$. Then $\mathcal{G}_{A,B}$ realize a topological isomorphisms from $\mathcal{F}_\theta(N')$ into itself.*

4. Generalized Euler operator

Let be given $A \in \mathcal{L}(N, N)$ and $K \in \mathcal{L}(N, N')$. For $\varphi(x) = \sum_{n=0}^{\infty}\langle x^{\otimes n}, \varphi_n\rangle \in \mathcal{F}_\theta(N')$, we define three operators $\Delta_G(K)$, $\Gamma(A)$ and $N(A)$ on $\mathcal{F}_\theta(N')$ as follows.

$$\Delta_G(K)\varphi(x) = \sum_{n=0}^{\infty}(n+2)(n+1)\langle x^{\otimes n}, \tau(K)\widehat{\otimes}_2\varphi_{n+2}\rangle, \tag{6}$$

$$\Gamma(A)\varphi(x) = \sum_{n=0}^{\infty} \langle x^{\otimes n}, A^{\otimes n}\varphi_n\rangle, \tag{7}$$

$$N(A)\varphi(x) = \sum_{n=0}^{\infty} \langle x^{\otimes n}, \gamma_n(A)\varphi_n\rangle,$$

where $\gamma_n(A)$ is given by $\gamma_0(A) = 0$ and

$$\gamma_n(A) = \sum_{k=0}^{n-1} I^{\otimes k} \otimes A \otimes I^{\otimes(n-1-k)}, \quad n \geq 1.$$

$\Delta_G(K)$, $\Gamma(A)$ and $N(A)$ are called the *K-Gross Laplacian*, the *second quantization* and the *differential second quantization* of A, respectively. Notice that, $\Delta_G(I) \equiv \Delta_G$ is the usual Gross Laplacian, $N(I) \equiv N$ is the standard number operator and $\Gamma(B) \equiv \mathcal{G}_{0,B}$. For various related studies on these operators we refer Refs. 3, 5, 6, 14.

The so-called *Euler operator* is defined as the first order differential operator $\sum_{i=1}^{d} x_i \frac{\partial}{\partial x_i}$ on $\mathbb{R}^d$ (see Ref.[10]). Its infinite dimensional analogue is well-known[5,22] to be $\Delta_E := \Delta_G + N$.

For $K \in \mathcal{L}(N, N')$ and $B \in \mathcal{L}(N, N)$, we define

$$\Delta_E(K, B) = \Delta_G(\frac{1}{2}K) + N(B)$$

and we call it also the (infinite dimensional) *generalized Euler operator.*

For completeness the following result is given with proof ultimately connected to our setting.

Proposition 4.1. *For any $A \in \mathcal{L}(N, N)$ and $K \in \mathcal{L}(N, N')$, $\Delta_G(K)$, $\Gamma(A)$ and $N(A)$ are three continuous linear operators from $\mathcal{F}_\theta(N')$ into itself.*

Proof. Consider $\varphi(x) = \sum_{n=0}^{\infty} \langle x^{\otimes n}, \varphi_n\rangle \in \mathcal{F}_\theta(N')$. We have

$$\Delta_G(K)\varphi(x) = \sum_{n=0}^{\infty} (n+2)(n+1) \left\langle x^{\otimes n}, \tau(K)\widehat{\otimes}_2\varphi_{n+2}\right\rangle.$$

Then, for $p \geq 0$,

$$|\Delta_G(K)\varphi(x)| \leq \sum_{n=0}^{\infty} (n+2)(n+1)|\tau(K)|_{-p}\, |x|_{-p}^n\, |\varphi_{n+2}|_p.$$

From (5), for any $n \geq 1$, we have the following inequality

$$|\varphi_n|_p \leq e^n\, \theta_n m^n \|\varphi\|_{\theta,-q,m} \|i_{q,p}\|_{HS}^n, \quad \forall m \geq 0,\, q > p.$$

It follows that

$$|\Delta_G(K)\varphi(x)| \leq |\tau(K)|_{-p}\|\varphi\|_{\theta,-q,m} \times \sum_{n=0}^{\infty}(n+2)(n+1)\,|x|_{-p}^{n}\,e^{n+2}\,\theta_{n+2}m^{n+2}\|i_{q,p}\|_{HS}^{n+2}.$$

Thus, using the inequalities

$$\theta_{p+q} \leq 2^{p+q}\theta_p\,\theta_q, \quad n^2 \leq 2^{2n}$$

we get, for any $m' > 0$,

$$|\Delta_G(K)\varphi(x)|\,e^{-\theta(m'|x|_{-p})} \leq C\,|\tau(K)|_{-p}\|\varphi\|_{\theta,-q,m},$$

where

$$C = \sum_{n=0}^{\infty}\theta_2\,\{8\frac{em}{m'}\|i_{q,p}\|_{HS}\}^{n+2}.$$

Therefore for $q > p$ and $m > 0$ such that $8\frac{em}{m'}\|i_{q,p}\|_{HS} < 1$, we can conclude that $C < \infty$. This completes the proof of the statement for $\Delta_G(K)$.

Let be given $p \geq 0$. We have

$$|\Gamma(A)\varphi(x)| \leq \sum_{n=0}^{\infty}|x|_{-p}^{n}\|A\|^n|\varphi_n|_p.$$

From (5) there exist $q > p$ and $m' > 0$ such that

$$|\varphi_n|_p \leq e^n\|i_{q,p}\|_{HS}^{n}(m')^n\theta_n\|\varphi\|_{\theta,-q,m'}. \tag{8}$$

Then, for $m > 0$,

$$\begin{aligned}|\Gamma(A)\varphi(x)| &\leq \sum_{n=0}^{\infty}|x|_{-p}^{n}\|A\|^n|\varphi_n|_p \\ &\leq \|\varphi\|_{\theta,-q,m'}\sum_{n=0}^{\infty}\left(\frac{m'}{m}e\|A\|\|i_{q,p}\|_{HS}\right)^n(m|x|_{-p})^n\theta_n \\ &\leq \|\varphi\|_{\theta,-q,m'}\sum_{n=0}^{\infty}\left(\frac{m'}{m}e\|A\|\|i_{q,p}\|_{HS}\right)^n e^{\theta(m|x|_{-p})}.\end{aligned}$$

Therefore, for $m > 0$ such that $\frac{m'}{m}e\|A\|\|i_{q,p}\|_{HS} < 1$ we deduce

$$\|\Gamma(A)\varphi\|_{\theta,-p,m} \leq \|\varphi\|_{\theta,-q,m'}\sum_{n=0}^{\infty}\left(\frac{m'}{m}e\|A\|\|i_{q,p}\|_{HS}\right)^n$$

which follows the proof of the statement for $\Gamma(A)$.

For $p \geq 0$ and $m > 0$ we have

$$|N(A)\varphi(x)|e^{-\theta(m|x|_{-p})} \leq \sum_{n=0}^{\infty} |\gamma_n(A)\varphi_n|_p |x|_{-p}^n e^{-\theta(m|x|_{-p})}.$$

Then, by using the obvious inequalities

$$|\gamma_n(A)\varphi_n|_p \leq n\|A\||\varphi_n|_p \leq 2^n\|A\||\varphi_n|_p\,, \quad \theta_n(m|x|_{-p})^n e^{-\theta(m|x|_{-p})} < 1$$

and (5), there exist $m' > 0$ such that

$$\|N(A)\varphi\|_{\theta,-p,m} \leq \|\varphi\|_{\theta,-q,m'}\|A\| \sum_{n=0}^{\infty} \left(2e\|i_{q,p}\|_{HS}\frac{m'}{m}\right)^n.$$

Thus, with the assumption $\frac{m'}{m}2e\|i_{q,p}\|_{HS} < 1$, we complete the proof of the statement for $N(A)$. □

From the above Proposition, we notice that the generalized Euler operator $\Delta_E(K, B) = \Delta_G(\frac{1}{2}K) + N(B)$ is the sum of two continuous linear operators, so it is also a continuous linear operator from $\mathcal{F}_\theta(N')$ into itself.

We denote by $GL(\mathcal{F}_\theta(N'))$ the group of all linear homeomorphisms from $\mathcal{F}_\theta(N')$ onto itself.

Lemma 4.1. *Let $A \in \mathcal{L}(N, N)$, then $\{\Gamma(e^{tA})\}_{t\in\mathbb{R}}$ is a regular one-parameter subgroup of $GL(\mathcal{F}_\theta(N'))$ with infinitesimal generator $N(A)$.*

Proof. We have

$$\begin{aligned}(e^{tA})^{\otimes n} &= \left[(I + tA) + \sum_{l=2}^{\infty}\frac{(tA)^l}{l!}\right]^{\otimes n} \\ &= I^{\otimes n} + tnI^{\otimes(n-1)}\widehat{\otimes}A + \sum_{j=0}^{n-2} C_n^j I^{\otimes j}\widehat{\otimes}(t^{n-j}A^{\otimes(n-j)}) \\ &\quad + \sum_{k=0}^{n-1} C_n^k (I + tA)^{\otimes k}\widehat{\otimes}\left(t^2A^2\sum_{l=0}^{\infty}\frac{(tA)^l}{(l+2)!}\right)^{\otimes(n-k)}.\end{aligned}$$

Then, for $\varphi(x) = \sum_{n=0}^{\infty}\langle x^{\otimes n}, \varphi_n\rangle \in \mathcal{F}_\theta(N')$, one can write

$$\Gamma(e^{tA})\varphi(x) = \varphi(x) + tN(A)\varphi(x) + t^2\varepsilon(t, A)\varphi(x)$$

where $\varepsilon(t,A)$ is given by

$$\varepsilon(t,A)\varphi(x) = \sum_{n=2}^{\infty}\left\langle x^{\otimes n}, \left(\sum_{j=0}^{n-2} t^{n-j-2} C_n^j A^{\otimes(n-j)} \widehat{\otimes} I^{\otimes j}\right)\varphi_n\right\rangle$$

$$+\sum_{n=1}^{\infty}\left\langle x^{\otimes n}, \left(\sum_{k=0}^{n-1} C_n^k t^{2n-2k-2}(I+tA)^{\otimes k}\widehat{\otimes}\left(A^2\sum_{l=0}^{\infty}\frac{(tA)^l}{(l+2)!}\right)^{\otimes(n-k)}\right)\varphi_n\right\rangle.$$

Hence

$$\left(\frac{\Gamma(e^{tA})\varphi-\varphi}{t} - N(A)\varphi\right) \sim (t\varepsilon_n^1)_{n\geq 1} + (t\varepsilon_n^2)_{n\geq 2}$$

where

$$\varepsilon_n^1 = \left(\sum_{k=0}^{n-1} C_n^k t^{2n-2k-2}(I+tA)^{\otimes k}\widehat{\otimes}\left(A^2\sum_{l=0}^{\infty}\frac{(tA)^l}{(l+2)!}\right)^{\otimes(n-k)}\right)\varphi_n$$

and

$$\varepsilon_n^2 = \left(\sum_{j=0}^{n-2} t^{n-j-2} C_n^j A^{\otimes(n-j)} \widehat{\otimes} I^{\otimes j}\right)\varphi_n.$$

Now, for $|t|\leq 1$, we have,

$$\begin{aligned}
|\varepsilon_n^1|_p &\leq |\varphi_n|_p \sum_{k=0}^{n-1} C_n^k (1+\parallel A \parallel)^k \left(\parallel A \parallel^2 \sum_{l=0}^{\infty}\frac{(\parallel A \parallel)^l}{(l+2)!}\right)^{n-k} \\
&\leq |\varphi_n|_p \sum_{k=0}^{n-1} C_n^k (1+\parallel A \parallel)^k \left(\parallel A \parallel^2 e^{\parallel A \parallel}\right)^{n-k} \\
&\leq |\varphi_n|_p \left(1+\parallel A \parallel + \parallel A \parallel^2 e^{\parallel A \parallel}\right)^n.
\end{aligned}$$

Similarly, for $|t|<1$, we have,

$$|\varepsilon_n^2|_p \leq |\varphi_n|_p \sum_{j=0}^{n-2} C_n^j \parallel A \parallel^{n-j} \leq |\varphi_n|_p (1+\parallel A \parallel)^n.$$

Then, from (5), for $p\geq 0$, $m>0$, there exist $q>p$, $m'>0$ such that

$$\left|\frac{\Gamma(e^{tA})\varphi(x)-\varphi(x)}{t} - N(A)\varphi(x)\right|$$

$$\leq |t|\|\varphi\|_{\theta,-q,m'}\sum_{n=0}^{\infty}\left\{\left(2+2\|A\|+\|A\|^2e^{\|A\|}\right)^2\frac{m'}{m}e\|i_{q,p}\|_{HS}\right\}^n \theta_n(m|x|_{-p})^n$$

$$\leq |t|\|\varphi\|_{\theta,-q,m'}\sum_{n=0}^{\infty}\left\{\left(2+2\|A\|+\|A\|^2e^{\|A\|}\right)^2\frac{m'}{m}e\|i_{q,p}\|_{HS}\right\}^n e^{\theta(m|x|_{-p})}.$$

Under the assumption $(2+2\|A\|+\|A\|^2e^{\|A\|})^2\frac{m'}{m}e\|i_{q,p}\|_{HS} < 1$, we obtain

$$\left\|\frac{\Gamma(e^{tA})\varphi-\varphi}{t} - N(A)\varphi\right\|_{\theta,-p,m}$$

$$\leq |t|\|\varphi\|_{\theta,-q,m'}\sum_{n=0}^{\infty}\left\{\left(2+2\|A\|+\|A\|^2e^{\|A\|}\right)^2\frac{m'}{m}e\|i_{q,p}\|_{HS}\right\}^n \tag{9}$$

which complete the proof. □

In the following Section we investigate the generalized Euler operator. We show that, under some conditions, $\Delta_E(K,B)$ is the generator of a one-parameter group transformation and we use the $\mathcal{G}_{K,B}$-transform to investigate the existence of a solution of the following *generalized Euler heat equation*:

$$\frac{\partial u(t)}{\partial t} = \Delta_E(K,B)u(t), \quad u(0)=\varphi\in\mathcal{F}_\theta(N'). \tag{10}$$

5. Generalized Euler heat equation

Let $K\in\mathcal{L}(N,N')$. For $T\in\mathcal{L}(N\otimes N, N\otimes N)$, we define a distribution $\tau(K)\circ T\in(N\otimes N)'$ by

$$\langle\tau(K)\circ T, f\otimes g\rangle := \langle\tau(K), T(f\otimes g)\rangle\,, \quad f,\, g\in N.$$

By the kernel theorem, there exist a unique operator $\mathfrak{K}_{K,T}\in\mathcal{L}(N,N')$ with kernel $\tau(K)\circ T$, i.e.,

$$\left\langle\mathfrak{K}_{K,T}f, g\right\rangle = \langle\tau(K)\circ T, f\otimes g\rangle\,, \quad f,\, g\in N.$$

Example 5.1. Let $B\in\mathcal{L}(N,N)$ and put $T=B\otimes I$. Then, $\tau(K)\circ(B\otimes I)$

is given by

$$\begin{aligned}\langle \tau(K) \circ (B \otimes I), f \otimes g \rangle &= \langle \tau(K), (Bf) \otimes g \rangle \\ &= \langle (KB)f, g \rangle \\ &= \langle \tau(KB), f \otimes g \rangle\end{aligned}$$

i.e., $\tau(K) \circ (B \otimes I) = \tau(KB)$, or equivalently $\mathfrak{K}_{K,B\otimes I} = KB$.

Example 5.2. Let $B \in \mathcal{L}(N,N)$ and put

$$T_B := (e^B)^{\otimes 2} - I^{\otimes 2}.$$

Then, for any $f, g \in N$, we have

$$\begin{aligned}\langle \tau(K) \circ T_B, f \otimes g \rangle &= \left\langle \tau(K), \left((e^B)^{\otimes 2} - I^{\otimes 2}\right)(f \otimes g)\right\rangle \\ &= \left\langle \tau(K), (e^B f) \otimes (e^B g)\right\rangle - \langle \tau(K), f \otimes g \rangle \\ &= \left\langle (Ke^B)f, e^B g\right\rangle - \langle Kf, g \rangle \\ &= \left\langle (e^{B^*} K e^B - K)f, g \right\rangle.\end{aligned}$$

Then, $\tau(K) \circ T_B$ is the kernel of the operator $\mathfrak{K}_{K,T_B} \in \mathcal{L}(N, N')$ given by

$$\mathfrak{K}_{K,T_B} = e^{B^*} K e^B - K.$$

Now, for $K \in \mathcal{L}(N, N')$, $B \in \mathcal{L}(N,N)$ and $\alpha \in \mathbb{C}\backslash\{0\}$, we define the transformation

$$\Upsilon_K^B(t) = \mathcal{G}_{\frac{1}{2\alpha}\mathfrak{K}_{K,T_{tB}}, e^{tB}}, \quad t \in \mathbb{R}.$$

Lemma 5.1. *Let be given $K \in \mathcal{L}(N, N')$ and $B \in \mathcal{L}(N,N)$. Then, the family $\left\{\Upsilon_K^B(t)\right\}_{t\in\mathbb{R}}$ is a one-parameter transformation subgroup in $GL(\mathcal{F}_\theta(N'))$.*

Proof. It is obvious that $\Upsilon_K^B(0) = \mathcal{G}_{0,I} = I$. Now we shall prove that

$$\Upsilon_K^B(t)\,\Upsilon_K^B(s) = \Upsilon_K^B(t+s)\,, \quad t,\, s \in \mathbb{R}.$$

By using Proposition 3.4 we can write

$$\begin{aligned}\Upsilon_K^B(t)\,\Upsilon_K^B(s) &= \mathcal{G}_{\frac{1}{2\alpha}\mathfrak{K}_{K,T_{tB}},e^{tB}}\mathcal{G}_{\frac{1}{2\alpha}\mathfrak{K}_{K,T_{sB}},e^{sB}} \\ &= \mathcal{G}_{C_{s,t},D_{s,t}},\end{aligned}$$

where $D_{s,t} = e^{(s+t)B}$ and $C_{s,t}$ is given by

$$\begin{aligned}C_{s,t} &= \frac{1}{2\alpha}\left(\mathfrak{K}_{K,T_{sB}} + e^{sB^*}\mathfrak{K}_{K,T_{tB}}e^{sB}\right) \\ &= \frac{1}{2\alpha}\left\{(e^{sB^*}Ke^{sB} - K) + e^{sB^*}(e^{tB^*}Ke^{tB} - K)e^{sB}\right\} \\ &= \frac{1}{2\alpha}\left(e^{(s+t)B^*}Ke^{(s+t)B} - K\right) \\ &= \frac{1}{2\alpha}\mathfrak{K}_{K,T_{(t+s)B}}.\end{aligned}$$

This completes the proof. □

Now, having the transformation group $\left\{\Upsilon_K^B(t)\right\}_{t\in\mathbb{R}}$, we are ready to give explicitly the solution of the Euler heat equation (10).

Theorem 5.1. *Let $K \in \mathcal{L}(N, N')$ and $B \in \mathcal{L}(N, N)$ satisfying $\tau(K)\circ(B\otimes I) = \alpha\tau(K)$ for some $\alpha \in \mathbb{C}\backslash\{0\}$. Then*

$$u(t) = \Upsilon_K^B(t)\varphi \in \mathcal{F}_\theta(N')$$

is the unique solution of the generalized Euler heat equation (10).

Proof. To prove the statement, we shall prove that $\{\Upsilon_K^B(t)\}_{t\in\mathbb{R}}$ is a differentiable one-parameter transformation group with infinitesimal generator $\Delta_E(K,B)$, i.e., for any $p \geq 0$, $m > 0$ and $\varphi \in \mathcal{F}_\theta(N')$, we have

$$\lim_{t\to 0}\left\|\frac{\Upsilon_K^B(t)\varphi - \varphi}{t} - \Delta_E(K,B)\varphi\right\|_{\theta,-p,m} = 0.$$

Let $p \in \mathbb{R}$, $m > 0$ and $\varphi \sim (\varphi_n)_{n\geq 0} \in F_\theta(N)$ be given. Then we have

$$\frac{\Upsilon_K^B(t)\varphi-\varphi}{t} - \Delta_E(K,B)\varphi$$
$$\sim \left\{\left[\frac{(e^{tB})^{\otimes n} - I^{\otimes n}}{t} - n(B\widehat{\otimes}I^{\otimes(n-1)})\right]\varphi_n\right\}$$
$$+ \left\{(n+2)(n+1)\left[\frac{(e^{tB})^{\otimes n}\tau_t(K)}{2t}\widehat{\otimes}_2\varphi_{n+2} - \frac{\tau(K)}{2}\widehat{\otimes}_2\varphi_{n+2}\right]\right\}$$
$$+ \left\{\sum_{l=2}^{\infty}\frac{(n+2l)!}{n!\,l!\,2^l}(e^{tB})^{\otimes n}(\frac{(\tau_t(K))^{\otimes l}}{t}\widehat{\otimes}_{2l}\varphi_{n+2l})\right\}, \tag{11}$$

where

$$\tau_t(K) = \frac{1}{2\alpha}\tau(K)\circ T_{tB}.$$

From (11) we have

$$\left\|\frac{\Upsilon_K^B(t)\varphi-\varphi}{t} - \Delta_E(K,B)\varphi\right\|_{\theta,-p,m}$$
$$\leq \left\|\frac{\Gamma(e^{tA})\varphi-\varphi}{t} - N(A)\varphi\right\|_{\theta,-p,m} + \left\|F_1^{(t)}\right\|_{\theta,-p,m} + \left\|F_2^{(t)}\right\|_{\theta,-p,m}$$

where

$$F_1^{(t)}(x) = \sum_{n=0}^{\infty}\left\langle x^{\otimes n}, (n+2)(n+1)\left[\frac{(e^{tB})^{\otimes n}\tau_t(K)}{2t}\widehat{\otimes}_2\varphi_{n+2} - \frac{\tau(K)}{2}\widehat{\otimes}_2\varphi_{n+2}\right]\right\rangle$$

and

$$F_2^{(t)}(x) = \sum_{n=0}^{\infty}\left\langle x^{\otimes n}, \sum_{l=2}^{\infty}\frac{(n+2l)!}{n!\,l!\,2^l}(e^{tB})^{\otimes n}(\frac{(\tau_t(K))^{\otimes l}}{t}\widehat{\otimes}_{2l}\varphi_{n+2l})\right\rangle.$$

From (9) clearly we have

$$\lim_{t\to 0}\left\|\frac{\Gamma(e^{tA})\varphi-\varphi}{t} - N(A)\varphi\right\|_{\theta,-p,m} = 0.$$

It remains to prove

$$\lim_{t\to 0}\left(\|F_1^{(t)}\|_{\theta,-p,m} + \|F_2^{(t)}\|_{\theta,-p,m}\right) = 0.$$

Step 1. We shall prove $\lim_{t\to 0}\|F_1^{(t)}\|_{\theta,-p,m} = 0$.
Observe that

$$\left|\left[\frac{(e^{tB})^{\otimes n}\tau_t(K)}{2t}\widehat{\otimes}_2\varphi_{n+2} - \frac{\tau(K)}{2}\widehat{\otimes}_2\varphi_{n+2}\right]\right|_p$$

$$\leq \left|(e^{tB})^{\otimes n}\left[\frac{\tau_t(K)}{t}\widehat{\otimes}_2\varphi_{n+2} - \tau(K)\widehat{\otimes}_2\varphi_{n+2}\right]\right|_p$$

$$+ \left|\left[(e^{tB})^{\otimes n} - I^{\otimes n}\right]\tau(K)\widehat{\otimes}_2\varphi_{n+2}\right|_p.$$

Then, by using the condition $\tau(KB) = \tau(\alpha K)$ which is equivalent to the condition $\tau(K) \circ (B \otimes I) = \alpha\tau(K)$, for any $t \neq 0$, we have

$$\left|(e^{tB})^{\otimes n}\left[\frac{\tau_t(K)}{t}\widehat{\otimes}_2\varphi_{n+2} - \tau(K)\widehat{\otimes}_2\varphi_{n+2}\right]\right|_p$$

$$\leq 2|t||\tau(K)|_{-p}\|B\|e^{(n+2)|t|\|B\|}|\varphi_{n+2}|_p.$$

On the other hand, we have

$$\begin{aligned}
|((e^{tB})^{\otimes n} - I^{\otimes n})\varphi_n|_p &\leq \sum_{k=0}^{n-1}\left|((e^{tB})^{\otimes(n-1-k)} \otimes (e^{tB} - I) \otimes I^{\otimes k})\varphi_n\right|_p \\
&\leq |t|\|B\|e^{|t|\|B\|}\sum_{k=0}^{n-1} e^{|t|\|B\|(n-1-k)}|\varphi_n|_p \\
&\leq |t|\|B\|e^{|t|\|B\|}n\left(e^{|t|\|B\|}\right)^n|\varphi_n|_p \\
&\leq |t|\|B\|e^{|t|\|B\|}\left(2e^{|t|\|B\|}\right)^n|\varphi_n|_p.
\end{aligned}$$

Then we deduce

$$\left|\left[(e^{tB})^{\otimes n} - I^{\otimes n}\right]\tau(K)\widehat{\otimes}_2\varphi_{n+2}\right|_p$$

$$\leq |t|\|B\||\tau(K)|_{-p}e^{|t|\|B\|}\left(2e^{|t|\|B\|}\right)^n|\varphi_{n+2}|_p.$$

Hence, by using (8) and the obvious inequalities

$$e^{|t|\|B\|} \leq e^{2|t|\|B\|}, \quad e^{n|t|\|B\|} \leq (2e^{|t|\|B\|})^n,$$

for $q > p$ and $m' > 0$, we get

$$\left\|F_1^{(t)}\right\|_{\theta,-p,m} \leq |t|\|\varphi\|_{\theta,-q,m'}\left[3|\tau(K)|_{-p}\|B\|e^{2|t|\|B\|}\left(e\frac{m'}{m}\|i_{q,p}\|_{HS}\right)^2\right]$$

$$\times\sum_{n=0}^{\infty}\left(2e\frac{m'}{m}e^{|t|\|B\|}\|i_{q,p}\|_{HS}\right)^n(m|x|_{-p})^{n+2}\theta_{n+2}e^{-\theta(m|x|_{-p})}.$$

Therefore, we obtain

$$\left\|F_1^{(t)}\right\|_{\theta,-p,m} \leq |t|\|\varphi\|_{\theta,-q,m'} \left[3|\tau(K)|_{-p}\|B\|e^{2|t|\|B\|}\left(e\frac{m'}{m}\|i_{q,p}\|_{HS}\right)^2\right] \times \sum_{n=0}^{\infty}\left(2e\frac{m'}{m}e^{|t|\|B\|}\|i_{q,p}\|_{HS}\right)^n.$$

Under the condition $2e\frac{m'}{m}e^{|t|\|B\|}\|i_{q,p}\|_{HS} < 1$, the last series converges and therefore we deduce

$$\lim_{t\to 0}\left\|F_1^{(t)}\right\|_{\theta,-p,m} = 0$$

as desired.

Step 2. We shall prove $\lim_{t\to 0}\left\|F_2^{(t)}\right\|_{\theta,-p,m} = 0.$

By using the inequality $(n+2k)! \leq 2^{n+4l}n!(l!)^2$, we calculate

$$\left|\sum_{l=2}^{\infty}\frac{(n+2l)!}{n!\,l!\,2^l}(e^{tB})^{\otimes n}\Big(\frac{(\tau_t(K))^{\otimes l}}{t}\widehat{\otimes}_{2l}\varphi_{n+2l}\Big)\right|_p \leq \sum_{l=2}^{\infty} l!\,2^{n+3l}\frac{e^{n|t|\|B\|}}{|t|}|\tau_t(K)|_{-p}^l|\varphi_{n+2l}|_p.$$

Then, using the inequality (8) and $|\tau_t(K)|_{-p} \leq |t|e^{2|t|\|B\|}|\tau(K)|_{-p}$, for any $q \geq p$, $m' > 0$, $|t| < 1$, we have

$$\begin{aligned}\left|F_2^{(t)}(x)\right| &\leq \|\varphi\|_{\theta,-q,m'}\sum_{l=2}^{\infty}\left[64\|K\|(em'\|i_{q,p}\|_{HS}e^{|t|\|B\|})^2\right]^l|t|^{l-1}l!\theta_{2l} \\ &\quad\times\sum_{n=0}^{\infty}(2e\frac{m'}{m}e^{|t|\|B\|}\|i_{q,p}\|_{HS})^n(m|x|_{-p})^n\theta_n \\ &\leq |t|\left\{\sum_{l=2}^{\infty}\left[64\|K\|\left(em'\|i_{q,p}\|_{HS}e^{|t|\|B\|}\right)^2\right]^l l!\theta_{2l}\right\} \\ &\quad\times\left\{\sum_{n=0}^{\infty}(2e\frac{m'}{m}e^{|t|\|B\|}\|i_{q,p}\|_{HS})^n\right\}e^{\theta(m|x|_{-p})}.\end{aligned}$$

Under the assumption

$$\max\left(64\|K\|(em'\|i_{q,p}\|_{HS}e^{|t|\|B\|})^2, 2e\frac{m'}{m}e^{|t|\|B\|}\|i_{q,p}\|_{HS}\right) < 1$$

the last two series converge and therefore we obtain

$$\lim_{t\to 0}\left\|F_2^{(t)}\right\|_{\theta,-p,m} = 0.$$

This completes the proof. □

References

1. A. Barhoumi, H.-H. Kuo and H. Ouerdiane, *Generalized heat equation with noises*, Soochow J. Math. 32 (2006), 113-125 .
2. A. Barhoumi, H. Ouerdiane, H. Rguigui and A. Riahi, *On operator-parameter transforms based on nuclear algebra of entire functions and applications*, preprint.
3. M. Ben Chrouda, M. El Oued and H. Ouerdiane, *Convolution calculus and application to stockastic differential equation*, Soochow J. Math. 28 (2002), 375-388.
4. M. Ben Chrouda and H. Ouerdiane *Algebras of Operators on Holomorphic Functions and Application*, Mathematical Physics, Analysis Geomety 5: 65-76, 2002.
5. D. M. Chung and U.C. Ji, *Transform on white noise functionals with their application to Cauchy problems*, Nagoya Math. J. Vol 147 (1997), 1-23.
6. D. M. Chung and U. C. Ji, *Transformation groups on white noise functionals and their application*, Appl. Math. Optim. 37 No. 2 (1998), 205-223.
7. D. Dineen, *Complex analysis in locally convex space*, Mathematical Studies, North Holland, Amsterdam 75, 1981.
8. R. Gannoun, R. Hachaichi and H. Ouerdiane, *Division de fonctions holomorphes à croissance $\theta-$exponentielle*, Preprint, BiBos No E 00-01-04, (2000).
9. R. Gannoun, R. Hachachi, H. Ouerdiane and A. Rezgui, *Un théorème de dualité entre espaces de fonctions holomorphes á croissance exponentielle*, J. Funct. Anal., Vol. 171 No. 1 (2000), 1-14.
10. I.M. Gel'fand and G.E. Shilov, *Generalized Functions*, Vol.I. Academic Press, Inc., New York, 1968.

11. I.M. Gel'fand and N.Ya. Vilenkin, *Generalized unctions*, Vol. 4, Academic press New York and London 1964.

12. L. Gross, *Abstract Wiener spaces*, Proc. 5-th Berkeley Symp. Math. Stat. Probab. 2 (1967), 31-42.

13. T. Hida, H.-H. Kuo and N. Obata, *Transformations for white noise functionals*, J. Funct. Anal., 111 (1993), 259-277.

14. U. C. Ji, *Integral kernel operators on regular generalized white noise functions.* Bull. Korean Math. Soc. 37 No. 3 (2000), 601-618.

15. U. C. Ji, *Quantum extensions of Fourier-Gauss and Fourier-Mehler transforms*, J. Korean Math. Soc. 45 No.6 (2008), 1785-1801.

16. U. C. Ji, N. Obata and H. Ouerdiane, *Analytic characterization of generalized Fock space operators as two-variable entire functions with growth condition*, World Scientific, Vol. 5 No. 3 (2002), 395-407.

17. S. J. Kang, *Heat and Poisson equations associated with number operator in white noise analysis*, Soochow J. Math. 20 (1994), 45–55.

18. H.-H. Kuo, *The Fourier transform in white noise calculus*, J. Multivariate Analysis 31 (1989), 311-327.

19. H.-H. Kuo, *White noise distrubution theory*, CRC Press, Boca Raton 1996.

20. N. Obata, *White Noise calculs and Fock Space.* Lecture Notes in Math. 1577, Spring, New York 1994.

21. M. A. Piech, *Parabolic equations associated with the number operator*, Trans. Amer. Math. Soc. 194 (1974), 213-222.

22. J. Potthoff and J. A. Yan, *Some results about test and generalized functionals of white noise*, Proc. Singapore Prob. Conf. L.Y. Chen et al.(eds.) (1989), 121-145.

ON QUANTUM DE FINETTI'S THEOREMS

VITONOFRIO CRISMALE and YUN GANG LU

Dipartimento di Matematica, Università degli Studi di Bari

Via E. Orabona, 4 I-70125 Bari, ITALY

E-mail: crismalev@dm.uniba.it

E-mail: lu@dm.uniba.it

We review some new results and developments about quantum De Finetti's type theorems.

1. Preliminaries

In these notes we review the formulation of the De Finetti's type theorems in the sense of Ref. 3, emphasizing the analogies or differences with respect to other results on the same matter. To avoid redundancies we omit most of the proofs, addressing the interested reader to the existing literature.

De Finetti's type theorems deal with the relations between exchangeable and independent families of random variables. The pioneering work on the subject is Ref. 8. There it was proved that any finite joint distribution of an exchangeable (symmetric) sequence of two points valued random variables is a mixture of i.i.d. random variables. This fine result was the starting point for many generalizations whose interest is yet nowadays prominent in Classical (see Ref. 14 for a detailed account), Quantum and Free Probability as well as in Quantum Computation.

A deep generalization was given by Hewitt and Savage:[11] they extended the first result to each exchangeable distribution on the infinite product $X = E \times E \times \ldots$, where E is a compact Hausdorff space. Consequently they characterized the extremal points of the convex of exchangeable stochastic processes in X as the i.i.d. stochastic processes.

If one takes $\mathcal{A} := \mathcal{C}(X)$ and $\mathcal{B} := \mathcal{C}(E)$, i.e. $\mathcal{A} \simeq \mathcal{B} \otimes \mathcal{B} \otimes \ldots$, the Hewitt-Savage's result has a functional analytic formulation: each exchangeable positive functional on $\mathcal{A}$ is uniquely decomposed into a convex combination of product ones. The extension of this statement to arbitrary (*noncommutatives*) C^*- algebras $\mathcal{B}$ and $\mathcal{A} = \mathcal{B} \otimes \mathcal{B} \otimes \ldots$ was firstly obtained by Størmer.[22] His paper gave rise to the emergence of new de Finetti's type results in the setting of Operator Algebras (see, e.g., Ref. 12 for locally normal states, Ref 19 for m-dependent states, Refs. 18, 16, 17 in Free Probability), Quantum Statistical Mechanics[9,10] and Quantum Computation (see Ref. 7 and references therein).

Accardi and Lu[4] obtained also a continuous version of the theorem, thus characterizing the extremal points of the exchangeable increment processes as the (stochastic) independent increment stationary processes.

Another (classical) generalized version of the theorem, which is called extended de Finetti's Theorem and based on Ryll-Nadzewski's work,[20] states the equivalence among exchangeable, spreadable and conditionally i.i.d. sequences of random variables. Very recently Köstler[15] obtained a non-commutative counterpart of such a result. Namely he showed how, dropping the commutativity, spreadability does not imply exchangeability and the conditional independence with respect to the tail (i.e. asymptotic) algebra is not enough to guarantee spreadability.

In the spirit of Ref. 4, the authors in Ref. 3 offered a characterization for the extreme points of the convex set of exchangeable algebraic stochastic processes and, as fashion in the setting of De Finetti's theorems, proved that for any exchangeable state there exists a unique decomposing Radon measure which is concentrated in the interior of the convex of the exchangeable states. The characterization for such extreme points involves various forms of the so-called *singleton condition*, a notion which is common to all the main stochastic independencies (tensor, Fermi, free, monotone, boolean) and thus is almost sufficient for the validity of many central limit theorems; in Ref. 3 it is shown as the singleton condition becomes a substitute of classical independence in the quantum De Finetti theorem.

2. Singleton and exchangeability conditions

An algebraic probability space is a pair $\{\mathcal{A}, \varphi\}$ where $\mathcal{A}$ is a unital $*$–algebra and $\varphi : \mathcal{A} \to \mathbb{C}$ a state. An algebraic stochastic process is a quadruple $\{\{\mathcal{A}, \varphi\}, \mathcal{B}, (j_i)_{i \in \mathcal{I}}\}$ where $\{\mathcal{A}, \varphi\}$ is an algebraic probability space, $\mathcal{I}$ a set, $\mathcal{B}$ a unital $*$–algebra and for any $i \in \mathcal{I}$, $j_i : \mathcal{B} \to \mathcal{A}$ a $*$– homomorphism, which will be called algebraic random variable. When for any $i \in \mathcal{I}$, $j_i(1_{\mathcal{B}}) = 1_{\mathcal{A}}$ (or, equivalently, all the $*$–subalgebras $j_i(\mathcal{B})$ are endowed with the same unit $1_{\mathcal{A}}$), we speak of unital algebraic stochastic process. We shall take $\mathcal{A}$ as minimal, i.e. $\mathcal{A}$ is the $*$–algebra generated by $\{j_i(\mathcal{B})\}_{i\in\mathcal{I}} \cup 1_{\mathcal{A}}$. Since the j_k's are homomorphisms, then in any expectation value of the form

$$\varphi\left(j_{k_q}(b_q)\cdots j_{k_1}(b_1)\right) \tag{1}$$

we can always suppose that

$$k_j \neq k_{j+1}, \quad \forall j = 1, \ldots, q-1 \tag{2}$$

This convention will be ever assumed and any expectation value as (1) will be said to be in standard form if (2) is satisfied.

Definition 2.1. Let $\{\{\mathcal{A}, \varphi\}, \mathcal{B}, (j_i)_{i\in\mathcal{I}}\}$ be an algebraic stochastic process. It satisfies the **singleton** condition (with respect to φ) if for any $n \geqslant 1$, for any choice of $i_1, \ldots, i_n \in I$ and $b_n, \cdots, b_1 \in \mathcal{B}$

$$\varphi\left(j_{i_n}(b_n)\cdots j_{i_1}(b_1)\right) = 0 \tag{3}$$

whenever $\{i_1, \ldots, i_n\}$ has a singleton i_s and $\varphi(j_s(b_s)) = 0$. The same process satisfies the **strict singleton** condition (with respect to φ) if

$$\varphi\left(j_{i_n}(b_n)\cdots j_{i_1}(b_1)\right) = \varphi\left(j_{i_s}(b_s)\right)\varphi\left(j_{i_n}(b_n)\cdots \widehat{j_{i_s}(b_s)}\cdots j_{i_1}(b_1)\right) \tag{4}$$

whenever $\{i_1, \ldots, i_n\}$ has a singleton i_s.

Definition 2.2. The algebraic stochastic process $\{\{\mathcal{A}, \varphi\}, \mathcal{B}, (j_i)_{i\in\mathcal{I}}\}$ satisfies the **block singleton** condition if for any $n \geqslant 1$, $i_1, \ldots, i_n \in I$ and $b_n, \cdots, b_1 \in \mathcal{B}$

$$\begin{aligned}\varphi\left(j_{i_n}(b_n)\cdots j_{i_1}(b_1)\right) &= \varphi\left(j_{i_{s+q}}(b_{s+q})\cdots j_{i_s}(b_s)\right)\\ &\times\varphi\left(j_{i_n}(b_n)\cdots \widehat{j_{i_{s+q}}(b_{s+q})}\cdots \widehat{j_{i_s}(b_s)}\cdots j_{i_1}(b_1)\right)\end{aligned}$$

if, for any $j = 0, \dots, q$, $i_{s+j} \notin \left\{ i_1, \dots \widehat{i_s}, \widehat{i_{s+1}}, \dots \widehat{i_{s+q}}, \dots, i_n \right\}$.

The strict singleton condition implies the singleton condition but the converse is not true in general (see Ref. 3, Example 1). For unital algebraic stochastic processes the strict singleton and singleton conditions are equivalent.

The strict singleton condition comes from the block singleton condition when $q = 0$. As the definitions suggests, a strict singleton state would be not necessarily block singleton, but in literature we have not yet found a counterexample. Hence we will assume the two conditions are different.

Remark 2.1. In the classical case the strict singleton condition is equivalent to stochastic independence.

Algebraic stochastic processes which are tensor, free, boolean, monotone or symmetric projectively independent[5] satisfy the singleton condition. Furthermore in the tensor and free cases, the block singleton condition is verified too (see Ref. 3, Propositions 2.1 and 2.2).

Definition 2.3. Let $\mathcal{S}_0 = \mathcal{S}_0(\mathbb{Z})$ be the group of one–to–one maps $\pi : \mathbb{Z} \to \mathbb{Z}$ such that $\pi(j) = j$ for any $j \in \mathbb{Z}$ but a finite number of points. We say that an algebraic stochastic process $\{\{\mathcal{A}, \varphi\}, \mathcal{B}, (j_k)_{k\in\mathbb{Z}}\}$ is φ–exchangeable if it is $\mathcal{S}_0$–invariant i.e. if for any $q \in \mathbb{N}$, $b_q, \dots, b_1 \in \mathcal{B}$, $k_q, \dots, k_1 \in \mathbb{Z}$ and any $\pi \in \mathcal{S}_0$

$$\varphi\left(j_{k_q}\left(b_q\right) \cdots j_{k_1}\left(b_1\right)\right) = \varphi\left(j_{\pi(k_q)}\left(b_q\right) \cdots j_{\pi(k_1)}\left(b_1\right)\right)$$

In such a case we say also that φ is exchangeable (with respect to the process $\{j_k\}_{k\in\mathbb{Z}}$).

Obviously exchangeability implies all the random variables are identically distributed, i.e. for each $b \in \mathcal{B}$, $k, h \in \mathbb{Z}$

$$\varphi\left(j_k\left(b\right)\right) = \varphi\left(j_h\left(b\right)\right)$$

3. De Finetti type results

Our aim is to characterize the extremal points of the convex of the exchangeable states. In this section we see that block singleton states almost allow to reach our goal. Therefore, in this perspective, the block singleton condition seems to be the counterpart, in the $*$–algebraic setting, of classical independence. Since the block singleton condition implies stochastic independence, the passage from Classical to Quantum Probability consists therefore in strengthening the necessary condition of our desired equivalence. Since we will show the main result is based on the ergodic decomposition theory on C^*–algebras (see Refs. 6 and 21), from now on we assume $\mathcal{A}$ is a C^*–algebra and the cyclic triple associated to $\{\mathcal{A},\varphi\}$ is $\{\mathcal{H}_\varphi,\pi_\varphi,\Phi\}$, where $\mathcal{H}_\varphi$ is a Hilbert space and π_φ is a $*$–representation of $\mathcal{A}$ into $\mathbf{B}(\mathcal{H}_\varphi)$.

Let $\{\{\mathcal{A},\varphi\},\mathcal{B},(j_k)_{k\in\mathbb{Z}}\}$ be an exchangeable algebraic stochastic process and $\{\mathcal{H}_\varphi,\pi_\varphi,\Phi\}$ be the GNS representation of $\{\mathcal{A},\varphi\}$. Arguing as in Proposition 1.4 of Ref. 2, it can be found that there exists a unitary operator $U:\mathcal{H}_\varphi\to\mathcal{H}_\varphi$ and a $*$–automorphism u on $\pi_\varphi(\mathcal{A})$ such that

$$U\Phi=\Phi$$
$$u\left(\pi_\varphi\left(j_k\left(b\right)\right)\right):=U\pi_\varphi\left(j_k\left(b\right)\right)U^*=\pi_\varphi\left(j_{k+1}\left(b\right)\right),\quad\text{for all } b\in\mathcal{B},\ k\in\mathbb{Z} \tag{5}$$

The $*$–automorphism u is called the shift on the $*$-subalgebra $\pi_\varphi(\mathcal{A})$ of $\mathbf{B}(\mathcal{H}_\varphi)$. The following theorem is the main ergodic result and the proof traces out the one given in Ref. 13, Theorem 2.2.1.

Theorem 3.1. *Let $\{\{\mathcal{A},\varphi\},\mathcal{B},(j_k)_{k\in\mathbb{Z}}\}$ be an exchangeable algebraic stochastic process and $\{\mathcal{H}_\varphi,\pi_\varphi,\Phi\}$ be the GNS representation of $\{\mathcal{A},\varphi\}$. The limit*

$$\lim_{N\to\infty}\frac{1}{N}\sum_{l=0}^{N-1}u^l=\mathbb{E}_\infty$$

exists strongly on $\pi_\varphi\left(\mathcal{A}\right)''$ and is equal to the Umegaki conditional expectation E_∞ onto the algebra

$$\mathcal{A}_\infty^\varphi=\mathcal{A}_\infty:=\left\{x\in\pi_\varphi\left(\mathcal{A}\right)'':u(x)=x\right\}$$

For the definition of Umegaki conditional expectation, one can see Ref. 1. As stressed at the beginning of the section, singleton conditions have a prominent position to gain a quantum version of the De Finetti's Theorem. Thus we need now to get a bridge between the notions of singleton states and extremal exchangeable states. The definition below fulfils our wish.

Definition 3.1. An exchangeable state φ on a unital C^*−algebra $\mathcal{A}$ is called **ergodic** if it is extremal.

φ is called **1–ergodic** if the asymptotic algebra $\mathcal{A}_\infty$ is trivial, i.e.

$$\mathcal{A}_\infty = \mathbb{C} \cdot 1$$

The former part of definition above comes from Ref. 21, Definition 3.1.9. The next result clarifies the emergence of the latter.

Theorem 3.2. *For* $\{\{\mathcal{A}, \varphi\}, \mathcal{B}, (j_k)_{k\in\mathbb{Z}}\}$ *exchangeable algebraic stochastic process, the following are equivalent*

(i) φ *is 1–ergodic*

(ii) φ *is a block singleton state.*

Corollary 3.1. *Let* $\{\{\mathcal{A}, \varphi\}, \mathcal{B}, (j_k)_{k\in\mathbb{Z}}\}$ *be an exchangeable algebraic stochastic process. If* φ *is 1–ergodic, then it is a strict singleton state.*

The proof of these result are given in Ref.3, Theorem 3.2 and Corollary 3.3. Looking at the Corollary above, one can ask himself if the converse implication holds. Namely, does a strict singleton state be also 1-ergodic? The answer is in general negative. To get just a flavor of the reason, if one aims to obtain information about the asymptotic algebra, one needs something like the following equality

$$\begin{aligned}&\langle \eta, \mathbb{E}_\infty \left(\pi_\varphi \left(j_{k_n}(b_n) \cdots j_{k_1}(b_1)\right)\right) \xi \rangle \\ &= \langle \eta, \mathbb{E}_\infty \left(\pi_\varphi \left(j_{k_n}(b_n)\right)\right) \cdots \mathbb{E}_\infty \left(\pi_\varphi \left(j_{k_1}(b_1)\right)\right) \xi \rangle \end{aligned} \tag{6}$$

for any $j_{k_n}(b_n) \cdots j_{k_1}(b_1)$, $n \in \mathbb{N}$, $k_1, \ldots, k_n \in \mathbb{N}$, $b_1, \ldots, b_n \in \mathcal{B}$, for any $\xi, \eta \in \pi_\varphi(\mathcal{A})\Phi$.

But, as shown in Ref. 3, the strict singleton condition and exchangeability do not allow to reach it. Hence, giving an affirmative answer to the above question is subordinated to the addiction of some further conditions on the algebra $\mathcal{A}$. In particular, since Lemma 3.1 of Ref. 3 ensures that, under exchangeability and strict singleton condition (6) holds when $\{k_1, \ldots, k_n\}$ is an ordered set, one can suppose the desired condition should be that, for each $n \in \mathbb{N}$, $k_n > \ldots > k_1 \in \mathbb{N}$, $b_1, \ldots, b_n \in \mathcal{B}$, the products $j_{k_n}(b_n) \cdots j_{k_1}(b_1)$ are total in $\mathcal{A}$. In this case one say $\mathcal{A}$ satisfies condition TOD (the totally ordered products are dense). The following result is given in Ref. 3, Proposition 3.3.

Proposition 3.1. *Let $\{\{\mathcal{A}, \varphi\}, \mathcal{B}, (j_k)_{k\in\mathbb{Z}}\}$ be an exchangeable algebraic stochastic process and suppose $\mathcal{A}$ satisfies condition TOD. Then, if φ is a strict singleton state, it is 1–ergodic.*

Recalling we need to obtain some relations between singleton conditions and extremal states and having performed the connections between singleton conditions and the asymptotic algebra, our goal can be reached by finding possible links of this algebra with extremal states. The Lemma of Hewitt and Savage[11] in Probability Theory states that for a family of classical exchangeable random variables, the symmetric σ-algebra coincides with the asymptotic one. A quantum version of such a result, presented in Ref. 3, Proposition 3.4, allows to achieve the above goal.

Namely, take $\{\{\mathcal{A}, \varphi\}, \mathcal{B}, (j_k)_{k\in\mathbb{Z}}\}$ an exchangeable algebraic stochastic process and $\{\mathcal{H}_\varphi, \pi_\varphi, \Phi\}$ the GNS representation of $\{\mathcal{A}, \varphi\}$. If $\left\{\{\mathcal{A}, \varphi\}, \mathcal{B}, (j_k^{(1)})_{k\in\mathbb{Z}}\right\}$, where $j_k^{(1)} := j_{\sigma(k)}$ for all $k \in \mathbb{Z}$, then there exists a unitary isomorphism $U_\sigma : \mathcal{H}_\varphi \to \mathcal{H}_\varphi$ such that $U_\sigma \Phi = \Phi$ and for any $b \in \mathcal{B}$, $k \in \mathbb{Z}$

$$U_\sigma \pi_\varphi(j_k(b)) = \pi_\varphi\left(j_{\sigma(k)}(b)\right) U_\sigma$$

Moreover the map u_σ on $\pi_\varphi(\mathcal{A})$ such that for any $b \in \mathcal{B}$, $k \in \mathbb{Z}$

$$u_\sigma \pi_\varphi(j_k(b)) := U_\sigma \pi_\varphi(j_k(b)) U_\sigma^* = \pi_\varphi\left(j_{\sigma(k)}(b)\right)$$

is a $*$-automorphism. Define the symmetric algebra

$$\mathcal{A}_{\mathcal{S}_0} = \mathcal{A}_{\mathcal{S}_0}^{\varphi} := \left\{x \in \pi_\varphi(\mathcal{A})'' : u_\sigma(x) = x \text{ for all } \sigma \in \mathcal{S}_0\right\}$$

and the closed subspace of $\mathcal{H}_\varphi$

$$E_{\mathcal{S}_0} := \{\xi \in \mathcal{H}_\varphi : U_\sigma \xi = \xi \text{ for all } \sigma \in \mathcal{S}_0\}$$

Proposition 3.2. *(Quantum Hewitt–Savage Lemma) For an exchangeable stochastic process* $\{\{\mathcal{A},\varphi\},\mathcal{B},(j_k)_{k\in\mathbb{Z}}\}$, *in the same notations introduced above, one has*

$$\mathcal{A}_{\mathcal{S}_0}=\mathcal{A}_\infty$$

A known result in Decomposition Theory for C^*–algebras (see Ref. 21, Proposition 3.1.10) states that, if φ is an exchangeable state on $\mathcal{A}$ and $\mathcal{A}_{\mathcal{S}_0}=\mathbb{C}\cdot 1$, then φ is extremal. As a consequence, each 1–ergodic state is ergodic and, in the convex of exchangeable states, all the block singleton ones are extremal. On the contrary ergodicity does not implies 1–ergodicity, as shown in Example 3 of Ref. 3. The equivalence between ergodicity and 1-ergodicity can be obtained by means of a further request on $\mathcal{A}$, namely the $\mathcal{S}_0$–abelianity (see Ref. 21, Definition 3.1.11). This means that for any exchangeable state φ on $\mathcal{A}$, $P_{\mathcal{S}_0}\pi_\varphi(\mathcal{A})P_{\mathcal{S}_0}$ is a family of mutually commutative operators, where $P_{\mathcal{S}_0}$ is the orthogonal projection of $\mathcal{H}_\varphi$ onto $E_{\mathcal{S}_0}$. In the classical case this condition is ever satisfied, since $\mathcal{A}$ is a subalgebra of $\mathbf{L}^\infty(\Omega,\mathcal{F},\mathbf{P})$.

The section ends with the statement of the main theorem, i.e. a quantum version of De Finetti's result. In the following $\mathcal{S}_E$ is the set of exchangeable states.

Theorem 3.3. *Let us suppose* $\mathcal{A}$ *satisfies the* $\mathcal{S}_0$*–abelianity. Then the weak* compact convex of exchangeable states is a Choquet simplex whose extremal points are exactly all the block singleton states. Therefore, for any* $\varphi\in\mathcal{S}_E$, *there exists a unique Radon probability measure* μ *on* $\mathcal{S}_E$ *such that*

$$\varphi(a)=\int_{\mathcal{S}_E}\psi(a)\,d\mu(\psi)\,,\quad a\in\mathcal{A}$$

and for any Baire set Δ *in* $\mathcal{S}_E$ *with* $\mathcal{E}_E\cap\Delta=\emptyset$, $\int_\Delta d\mu(\psi)=0$ *(where* $\mathcal{E}_E$ *is the set of extreme points of* $\mathcal{S}_E$*).*

As a consequence, if $\mathcal{A}$ satisfies the $\mathcal{S}_0$–abelianity and φ is exchangeable and extremal, then φ is a strict singleton state. Moreover, if $\mathcal{A}$ satisfies the condition TOD and φ is an exchangeable and strict singleton state, then φ is extremal.

4. Final remarks

As already observed in Section 1, the literature on non commutative analogues of De Finetti Theorem is very huge. In Refs. 12, 19, 4 one finds results which are similar to Theorem 3.3.

In Ref. 12 the authors proved that, if $\mathcal{B} := \mathbf{B}(\mathcal{H})$, $\mathcal{A} := \mathcal{B}\otimes\mathcal{B}\otimes\mathcal{B}\otimes\cdots$ and φ is an exchangeable, locally normal state on $\mathcal{A}$, then there exists a unique probability Radon measure μ concentrated on the normal states, such that

$$\varphi(a) = \int \psi(a)\, d\mu(\psi), \quad a \in \mathcal{A}$$

Moreover in Ref. 19 it is found that the closed extremal boundary of the compact convex set of m–symmetric states consists of the m–dependent states and for any φ m–symmetric state, there exists a unique probability Radon measure μ on the m–dependent states, such that

$$\varphi(a) = \int \psi(a)\, d\mu(\psi), \quad a \in \mathcal{A}$$

As stressed in Section 1, Köstler[15] gave a noncommutative counterpart of the extended De Finetti Theorem. We again recall that such a result states the equivalence among exchangeable, spreadable and conditionally i.i.d. (with respect to the asymptotic $\sigma-$algebra) sequences of classical random variables. On the contrary, Köstler showed that, taking as an algebraic probability space the pair $(\mathcal{M}, \psi)$, where $\mathcal{M}$ is a von Neumann algebra and ψ a faithful normal state, spreadability does not implies exchangeability and a conditional independence with respect to the asymptotic algebra is weaker than spreadability. As a consequence of such a result, one has that, in the setting of von Neumann algebras, spreadability (and, a fortiori, exchangeability) implies a factorization rule for the elements of the asymptotic algebra. Such a rule, which can be expressed in terms of (6) via the GNS representation, does not hold in the more general case of C^*-algebras. Furthermore, as pointed out in Ref. 3 and recalled in Section 3, even if one takes an exchangeable and strict singleton state φ, (6) is not verified.

References

1. L. Accardi: *An outline of quantum probability*, Uspehi Math. Nauk. (1993).
2. L. Accardi, A. Frigerio, J. Lewis: *Quantum stochastic processes*, Publ. Res. Inst. Math. Sci. **18** no.1 (1982), 97-133.

3. L. Accardi, A. Ben Ghorbal, V. Crismale and Y.G. Lu: *Singleton Conditions and Quantum De Finetti's theorem*, Infin. Dimens. Anal. Quantum Probab. Rel. Top. **11** no.4 (2008), 639-660.
4. L. Accardi, Y.G. Lu: *A continuous version of de Finetti's theorem*, Ann. Probab. **21** no. 3 (1993), 1478-1493.
5. A. Ben Ghorbal, V. Crismale: *Independence arising from interacting Fock spaces and related central limit theorems*, to appear on Probab. Math. Statist.
6. O. Bratteli, D. Robinson: *Operator Algebras and Quantum Statistical Mechanics I*, 2nd edition, Springer (1987).
7. M Christandl, R. König, G. Mitchison, R. Renner: *One-and-a-half de Finetti's Theorems*, Commun. Math. Phys. **273** (2007), 473-498.
8. B. de Finetti: *Funzione caratteristica di un fenomeno aleatorio*, Atti Accad. Naz. Lincei, VI Ser., Mem. Cl. Sci. Fis. Mat. Nat. **4** (1931), 251-299.
9. M. Fannes: *An application of de Finetti's theorem*, Quantum Probability snd Applications III. Proceedings Oberwolfach 1987 (eds. L. Accardi, W. von Waldenfels), Lect. Notes Math., Vol **1303**, Springer-Verlag (1988), 89-102.
10. M. Fannes, J. T. Lewis, A. Verbeure: *Symmetric states of composite systems*, Lett. Math. Phys. **15** (1988), 255-260.
11. E. Hewitt, L. F.Savage: *Symmetric measures on Cartesian products*, Trans. Am. Math. Soc. **80** (1955), 470-501.
12. R. L. Hudson, G. R. Moody: *Locally normal symmetric states and an analog of de Finetti's theorem*, Z. Wahr. Ver. Gebiete **33** (1976), 343-351.
13. R. Jajte: *Strong limit theorems in non-commutative probability*, Springer LNM 1110 (1985).
14. O. Kallenberg: *Probabilistic Symmetries and Invariance Principles*, Probability and Its Applications. Springer-Verlag (2005).
15. C. Köstler: *A noncommutative extended De Finetti's theorem*, arXiv:

0806.3621 v1 [math OA], (2008).

16. C. Köstler: *On Lehner's 'free' noncommutative analogue of de Finetti's theorem*, arXiv: 0806.3632 v1 [math OA], (2008).
17. C. Köstler and R. Speicher: *A noncommutative De Finetti theorem: invariance under quantum permutations is equivalent to freeness with amalgamation*, arXiv: 0807.0677 v1 [math OA], (2008).
18. F. Lehner: *Cumulants in noncommutative probability theory. IV. Noncrossing cumulants: de Finetti's theorem and L^p-inequalities*, J. Funct. Anal. **239** (2006), 214-246.
19. D. Petz: *A de Finetti-type Theorem with m-Dependent States*, Probab. Th. Rel. Fields **85** (1990), 65-72.
20. C. Ryll-Nardzewski: *On stationary sequences of random variables and the de Finetti's equivalence*, Colloq. Math. **4** (1957), 149-156.
21. S. Sakai: *C^*-algebras and W^*-algebras*, Springer (1971).
22. E. Størmer: *Symmetric states on infinite tensor product of C^*-algebras*, J. Funct. Anal. **3** (1969), 48-68.

KOLMOGOROVIAN MODEL FOR EPR-EXPERIMENT

D. AVIS

School of Computer Science and GERAD, McGill University

3480 University Montreal, Quebec, Canada H3A 2A7

P. FISCHER

Informatics and Mathematical Modelling

Technical University of Denmark

DK-2800 Kgs. Lyngby, Denmark

A. HILBERT and A. KHRENNIKOV *

International Center for Mathematical Modeling

in Physics and Cognitive Sciences

University of Växjö, S-35195, Sweden

E-mail: Andrei.Khrennikov@vxu.se

We present Kolmogorovian models for probabilistic data from the EPR-experiment. The crucial point is that these probabilities should be considered as conditional probabilities. This approach provide a possibility to escape contradiction with Bell's inequality.

*This work is supported by the profile of Växjö University – "Mathematical Modelling and System Collaboration".

1. Introduction

The aim of this paper is to provide alternative probabilistic models for the EPR-Bohm-Bell experiment (see books[1,2] for detailed presentation of this experiment from the viewpoint of probability theory) based on conditional expectations. We show that quantum (experimental) statistical data can be consistently combined with a classical probabilistic description.

2. CHSH inequality

We recall the CHSH inequality:

Theorem. *Let* $A^{(i)}(\omega)$ *and* $B^{(i)}(\omega), i = 1, 2,$ *be random variables taking values in* $[-1, 1]$ *and defined on a single probability space* $\mathcal{P}$. *Then the following inequality holds:*

$$| < A^{(1)}, B^{(1)} > + < A^{(1)}, B^{(2)} > + < A^{(2)}, B^{(1)} > - < A^{(2)}, B^{(2)} > | \leq 2. \tag{1}$$

The classical correlation is defined as it is in classical probability theory:

$$< A^{(i)}, B^{(j)} >= \int_\Omega A^{(i)}(\omega) B^{(j)}(\omega) d\mathbf{P}(\omega).$$

Bell proposed the following methodology. To verify an inequality of this type, one should put statistical data collected for four pairs of PBSs settings:

$$\theta_{11} = (\theta_1, \theta_1'), \theta_{12} = (\theta_1, \theta_2'), \theta_{21} = (\theta_2, \theta_1'), \theta_{22} = (\theta_2, \theta_2'),$$

into it. Here $\theta = \theta_1, \theta_2$ and $\theta' = \theta_1', \theta_2'$ are selections of angles for orientations of respective PBSs.

Following Bell, the selection of the angle θ_i determines the random variable

$$A^{(i)}(\omega) \equiv a_{\theta_i}(\omega).$$

There are two detectors coupled to the PBS with the θ-orientation: "up-spin" (or "up-polarization") detector and "down-spin" (or "down-polarization") detector. A click of the up-detector assigns to the random variable $a_\theta(\omega)$ the value +1 and a click of the down-detector assigns to it the value -1. However, since a lot of photons disappear without any click, it is also permitted for random variables to take the value zero in the case of no detection. Therefore in Bell's framework it is sufficient to consider $a_\theta(\omega)$ taking values $-1, 0, +1$.

In the same way selection of the angle θ' determines

$$B^{(i)}(\omega) \equiv b_{\theta'_i}(\omega),$$

where $b_{\theta'_i}(\omega)$ takes values $-1, 0, +1$.

It seems that Bell's random model is not proper for the EPR-Bohm-Bell experiment. Bell's description does not take into account probabilities of choosing pairs of angles (orientations of PBSs) $\theta_{11}, \ldots, \theta_{22}$. Thus his model provides only incomplete probabilistic description. This allows to include probabilities of choosing experimental settings $\mathbf{P}(\theta_{ij})$ into the model; this way completing it.

In the next section we shall provide such a complete probabilistic description of the EPR-Bohm-Bell experiment. We point out that random variables of our model (which will be put into the CHSH inequality) does not coincide with Bellian variables.

3. Proper random experiment

a). There is a source of entangled photons.

b). There are four PBSs and corresponding pairs of detectors. PBSs are labelled as $i = 1, 2$ and $j = 1, 2$ [a].

c). Directly after source there is a distribution device which opens at each instance of time, $t = 0, \tau, 2\tau, \ldots$ ways to only two (of four) optical fibers going to the corresponding two PBSs. For simplicity, we suppose that the detector at each end is chosen independently at random with probability 1/2. Therefore each pair $(i, j) : (1, 1), (1, 2), (2, 1), (2, 2)$ can be opened with equal probability:

$$\mathbf{P}(i, j) = 1/4.$$

We now define some random variables. To simplify considerations, we consider the *ideal experiment with 100% detectors efficiency.* Thus in Bell's framework random variables $a_\theta(\omega)$ and $b_{\theta'}(\omega)$ should take only values ± 1. The zero-value will play a totally different role in our model.

1) $A^{(i)}(\omega) = \pm 1, i = 1, 2$ if the corresponding (up or down) detector is coupled to *ith* PBS fires;

2) $A^{(i)}(\omega) = 0$ if the i-th channel is blocked. In the same way we define random variables $B^{(j)}(\omega)$ corresponding to PBSs $j = 1, 2$.

[a]It is just the form of labelling which is convenient to form pairs (i, j).

Of course, the correlations of these random variables satisfy CHSH inequality. Thus if such an experiment were performed and if CHSH inequality were violated, we should seriously think about e.g. quantum non-locality or the death of realism.

However, to see that CHSH inequality for $< A^{(i)}, B^{(j)} >$- correlations does not contradict to experimental data, we could use statistical data which has been collected for experiments with fixed pairs $\theta_{ij} = (\theta_i, \theta'_j)$ of orientations of PBS. We only need to express correlations of Bell's variables $< a_{\theta_i}, b_{\theta'_j} >$ via correlations $< A^{(i)}, B^{(j)} >$.

4. Frequency analysis

Suppose that our version of EPR-Bohm-Bell experiment was repeated $M = 4N$ times and each pair (i,j) of optical fibers was opened only N times.

The random variables took values

$$A^{(i)} = A_1^{(i)}, \ldots, A_M^{(i)}, i = 1, 2, B^{(j)} = B_1^{(j)}, \ldots, B_M^{(j)}, j = 1, 2.$$

Then by the law of large numbers [b]:

$$< A^{(i)}, B^{(j)} >= \lim_{M \to \infty} \frac{1}{M} \sum_{k=1}^{M} A_k^{(i)} B_k^{(j)}.$$

We remark that, for each pair of gates (i, j), only N pairs $(A_k^{(i)}, B_k^{(j)})$ have both components non zero. Thus

$$< A^{(i)}, B^{(j)} >= \lim_{N \to \infty} \frac{1}{4N} \sum_{l=1}^{N} A_{k_l}^{(i)} B_{k_l}^{(j)},$$

where summation is with respect to only pairs of values with both nonzero components.

Thus the quantities $< A^{(i)}, B^{(j)} >$ are not estimates for the $< a_{\theta_i}, b_{\theta'_j} >$ obtained in physical experiments. The right estimates are given by

$$\frac{1}{N} \sum_{l=1}^{N} A_{k_l}^{(i)} B_{k_l}^{(j)}.$$

---

[b] We assume that different trials are independent. Thus the law of large numbers is applicable

Hence the CHSH inequality for random variables $A^{(i)}, B^{(j)}$ induces the following inequality for "traditional Bellian random variables":

$$| < a_{\theta_1}, b_{\theta'_1} > + < a_{\theta_1}, b_{\theta'_2} > + < a_{\theta_2}, b_{\theta'_1} > - < a_{\theta_2}, b_{\theta'_2} > | \leq 8. \quad (2)$$

It is not violated for known experimental data for entangled photons. Moreover, this inequality provides a trivial constraint on correlations: each correlation of Bellian variables is majorated by 1, hence, their linear combination with $\pm$-signs is always bounded above by 4.

5. Consistent joint probability space

We now construct a joint probability space for the EPR-Bohm-Bell experiment. This is a general construction for combining of probabilities produced by a few incompatible experiments. We have probabilities $p_{ij}(\epsilon, \epsilon'), \epsilon, \epsilon' = \pm 1$, to get $a_{\theta_i} = \epsilon, b_{\theta'_j} = \epsilon'$ in the experiment with the fixed pair of orientations (θ_i, θ'_j). From QM we know that

$$p_{ij}(\epsilon, \epsilon) = \frac{1}{2} \cos^2 \frac{\theta_i - \theta'_j}{2}, p_{ij}(\epsilon, -\epsilon) = \frac{1}{2} \sin^2 \frac{\theta_i - \theta'_j}{2}. \quad (3)$$

However, this special form of probabilities is not important for us. Our construction of unifying Kolmogorov probability space works well for any collection of probabilities $p_{ij} : \sum_{\epsilon,\epsilon'} p_{ij}(\epsilon, \epsilon') = 1$. We remark that $p_{ij}(\epsilon, \epsilon')$ determine automatically marginal probabilities:

$$p_i(\epsilon) = \sum_{\epsilon'} p_{ij}(\epsilon, \epsilon'),$$

$$p_j(\epsilon') = \sum_{\epsilon} p_{ij}(\epsilon, \epsilon').$$

In the EPR-Bohm-Bell experiment they are equal to 1/2. Let us now consider the set $\{-1, 0, +1\}^4$: the set of all vectors

$$\omega = (\omega_1, \omega_2, \omega_3, \omega_4), \omega_l = \pm 1, 0.$$

It contains 3^4 points. Now we consider the following subset Ω of this set:

$$\omega = (\epsilon_1, 0, \epsilon'_1, 0), (\epsilon_1, 0, 0, \epsilon'_2), (0, \epsilon_2, \epsilon'_1, 0), (0, \epsilon_2, 0, \epsilon'_2).$$

It contains 16 points. We define the following probability measure on Ω :

$$\mathbf{P}(\epsilon_1, 0, \epsilon'_1, 0) = \frac{1}{4} p_{11}(\epsilon_1, \epsilon'_1), \mathbf{P}(\epsilon_1, 0, 0, \epsilon'_2) = \frac{1}{4} p_{12}(\epsilon_1, \epsilon'_2)$$

$$\mathbf{P}(0,\epsilon_2,\epsilon_1',0)=\frac{1}{4}p_{21}(\epsilon_2,\epsilon_1'),\mathbf{P}(0,\epsilon_2,0,\epsilon_2')=\frac{1}{4}p_{22}(\epsilon_2,\epsilon_2').$$

We remark that we really have

$$\sum_{\epsilon,\epsilon_1'}\mathbf{P}(\epsilon_1,0,\epsilon_1',0)+\sum_{\epsilon_1,\epsilon_2'}\mathbf{P}(\epsilon_1,0,0,\epsilon_2')+\sum_{\epsilon_2,\epsilon_1'}\mathbf{P}(0,\epsilon_2,\epsilon_1',0)+\sum_{\epsilon_2,\epsilon_2'}\mathbf{P}(0,\epsilon_2,0,\epsilon_2')=$$

$$\frac{1}{4}\left[\sum_{\epsilon,\epsilon_1'}p_{11}(\epsilon_1,\epsilon_1')+\sum_{\epsilon_1,\epsilon_2'}p_{12}(\epsilon_1,\epsilon_2')+\sum_{\epsilon_2,\epsilon_1'}p_{21}(\epsilon_2,\epsilon_2')+\sum_{\epsilon_2,\epsilon_2'}p_{22}(\epsilon_2,\epsilon_2')\right]=1.$$

We now define random variables $A^{(i)}(\omega),B^{(j)}(\omega)$:

$$A^{(1)}(\epsilon_1,0,\epsilon_1',0)=A^{(1)}(\epsilon_1,0,0,\epsilon_2')=\epsilon_1,A^{(2)}(0,\epsilon_2,\epsilon_1',0)=A^{(2)}(0,\epsilon_2,0,\epsilon_2')=\epsilon_2;$$

$$B^{(1)}(\epsilon_1,0,\epsilon_1',0)=B^{(1)}(0,\epsilon_2,\epsilon_1',0)=\epsilon_1',B^{(2)}(\epsilon_1,0,0,\epsilon_2')=B^{(2)}(0,\epsilon_2,0,\epsilon_2')=\epsilon_2'.$$

We find two dimensional probabilities

$$\mathbf{P}(\omega\in\Omega:A^{(1)}(\omega)=\epsilon_1,B^{(1)}(\omega)=\epsilon_1')=\mathbf{P}(\epsilon_1,0,\epsilon_1',0)=\frac{1}{4}p_{11}(\epsilon_1,\epsilon_1'),\ldots,$$

$$\mathbf{P}(\omega\in\Omega:A^{(2)}(\omega)=\epsilon_2,B^{(2)}(\omega)=\epsilon_2')=\frac{1}{4}p_{22}(\epsilon_2,\epsilon_2').$$

We also consider two random variables η_A and η_B which are responsible for the selection of gates by the two parties respectively:

$$\eta_A(\epsilon_1,0,0,\epsilon_2')=\eta_A(\epsilon_1,0,\epsilon_1',0)=1,\eta_A(0,\epsilon_2,0,\epsilon_2')=\eta_A(0,\epsilon_2,\epsilon_1',0)=2,$$

$$\eta_B(\epsilon_1,0,\epsilon_1',0)=\eta_B(0,\epsilon_2,\epsilon_1',0)=1,\eta_B(\epsilon_1,0,0,\epsilon_2')=\eta_B(0,\epsilon_2,0,\epsilon_2')=2,$$

Both η_A and η_B are independent and uniformly distributed by our experimental assumptions.

We create a consistent probabilistic model by means of *conditional expectation.* Let $(\Omega,\mathcal{F},\mathbf{P})$ be an arbitrary probability space and let $\Omega_0\subset\Omega,\Omega_0\in\mathcal{F},\mathbf{P}(\Omega_0)\neq 0$. We also consider an arbitrary random variable $\xi:\Omega\to\mathbf{R}$. Then

$$E(\xi|\Omega_0)=\int_\Omega\xi(\omega)d\mathbf{P}_{\Omega_0}(\omega),$$

where the conditional probability is defined by the Bayes' formula:

$$\mathbf{P}_{\Omega_0}(U)\equiv\mathbf{P}(U|\Omega_0)=\mathbf{P}(U\cap\Omega_0)/\mathbf{P}(\Omega_0).$$

Let us come back to our unifying probability space. Take $\Omega_0 \equiv \Omega_{ij} = \{\omega \in \Omega : \eta_A(\omega) = i, \eta_B(\omega) = j\}$. By uniformity and independence we have $\mathbf{P}(\Omega_{ij}) = \mathbf{P}(\eta_A(\omega) = i)\mathbf{P}(\eta_B(\omega) = j) = 1/4$. Thus

$$E(A^{(i)}B^{(j)}|\eta_A = i, \eta_B = j) = \int_\Omega A^{(i)}(\omega)B^{(j)}(\omega)d\mathbf{P}_{\Omega ij}(\omega)$$

$$= 4\int_{\Omega_{ij}} A^{(i)}(\omega)B^{(j)}(\omega)d\mathbf{P}(\omega)$$

$$= 4\int_\Omega A^{(i)}(\omega)B^{(j)}(\omega)d\mathbf{P}(\omega) = 4 < A^{(i)}, B^{(j)} >=< a_{\theta_i}, b_{\theta'_j} > .$$

Thus QM-correlations can be represented as conditional expectations:

$$< a_{\theta_i}, b_{\theta'_j} >= E(A^{(i)}B^{(j)}|\eta_A = i, \eta_B = j). \tag{4}$$

6. Two-valued random variables

We showed in the last section how to give a complete probabilistic description of an EPR-Bohm-Bell experiment with random variables $A^{(1)}$,$A^{(2)}$, $B^{(1)}$, $B^{(2)}$, η_A and η_B. In that description the $A^{(i)}$, $B^{(j)}$ took three values: ± 1 and 0. In this section we show that it is also possible to do this when the $A^{(i)}$, $B^{(j)}$ take only the values ± 1.

By way of illustration, let us take the standard idealized EPR-Bohm-Bell experiment described in the beginning of the previous section with fixed orientations $\theta_1 = \pi/4$, $\theta_2 = 0$, $\theta'_1 = \pi/8$, $\theta'_2 = 3\pi/8$. The probabilities of the experimental outcome $a_{\theta_i} = \epsilon$, $b_{\theta'_j} = \epsilon'$ are given by (3) and yield the expected values

$$< a_{\theta_1}, b_{\theta'_1} >=< a_{\theta_1}, b_{\theta'_2} >=< a_{\theta_2}, b_{\theta'_1} >= \frac{1}{\sqrt{2}}, < a_{\theta_2}, b_{\theta'_2} >= -\frac{1}{\sqrt{2}} \tag{5}$$

Therefore we have

$$< a_{\theta_1}, b_{\theta'_1} > + < a_{\theta_1}, b_{\theta'_2} > + < a_{\theta_2}, b_{\theta'_1} > - < a_{\theta_2}, b_{\theta'_2} > \quad = \quad 2\sqrt{2}, \tag{6}$$

obtaining the Tsirelson bound[3] on the maximum quantum "violation" of the CHSH inequality.

We construct a Kolmogorov probability space $\mathcal{P} = (\Omega, \mathcal{F}, \mathbf{P})$ with sixteen outcomes and six random variables: $A^{(1)}$, $A^{(2)}$, $B^{(1)}$, $B^{(2)}$, η_A and η_B. The first four random variables take values ± 1 and the last two takes values of 1 or 2.

The first eight outcomes each occur with equal probability x:

$A^{(1)}(\omega)$	$A^{(2)}(\omega)$	$B^{(1)}(\omega)$	$B^{(2)}(\omega)$	$\eta_A(\omega)$	$\eta_B(\omega)$
1	1	1	1	1	1
-1	-1	-1	-1	1	1
1	1	1	1	1	2
-1	-1	-1	-1	1	2
1	1	1	1	2	1
-1	-1	-1	-1	2	1
1	1	1	-1	2	2
-1	-1	-1	1	2	2

The remaining eight outcomes each occur with equal probability y:

$A^{(1)}(\omega)$	$A^{(2)}(\omega)$	$B^{(1)}(\omega)$	$B^{(2)}(\omega)$	$\eta_A(\omega)$	$\eta_B(\omega)$
-1	-1	1	1	1	1
1	1	-1	-1	1	1
-1	-1	1	1	1	2
1	1	-1	-1	1	2
-1	-1	1	1	2	1
1	1	-1	-1	2	1
-1	-1	1	-1	2	2
1	1	-1	1	2	2

The probabilities x and y must be non-negative and $8x + 8y = 1$. One may verify that for $i = 1, 2$ and $\epsilon = \pm 1$:

$$\mathbf{P}(\omega \in \Omega : A^{(i)}(\omega) = \epsilon) \;=\; \frac{1}{2}.$$

Furthermore we can check that for $i, j = 1, 2$ and $\epsilon = \pm 1$:

$$\mathbf{P}(\omega \in \Omega : A^{(i)}(\omega) = \epsilon | \eta_A(\omega) = i)$$

$$= \mathbf{P}(\omega \in \Omega : A^{(i)}(\omega) = \epsilon | \eta_A(\omega) = i, \eta_B(\omega) = j) = \frac{1}{2}$$

and so the *non-signalling* condition holds. A similar set of equations hold for the random variables $B^{(j)}$. We see that

$$< A^{(i)} >=< B^{(j)} >= 0,$$

$$< A^{(1)}, B^{(1)} >=< A^{(2)}, B^{(1)} >= 8x - 8y,$$

and

$$< A^{(1)}, B^{(2)} >=< A^{(2)}, B^{(2)} >= 4x - 4y.$$

The left hand side of inequality (1) becomes $|16x-16y|$, and so (unsurprisingly) (1) holds since $0 \le x, y \le 1/8$.

A further calculation shows that

$$E(A^{(i)}B^{(j)}|\eta_A = i, \eta_B = j) = 8x - 8y, \quad i, j \ne 2 \tag{7}$$

and

$$E(A^{(2)}B^{(2)}|\eta_A = 2, \eta_B = 2) = 8y - 8x. \tag{8}$$

It suffices to set

$$x = \frac{\sqrt{2}+1}{16\sqrt{2}}, \quad y = \frac{\sqrt{2}-1}{16\sqrt{2}}$$

in (7) and (8) to see that equation (4) is indeed satsified for the the expected values given in (5). Again we conclude that there is a probabilistic model consistent with the experimental outcomes given by (5).

Even more striking, perhaps, is the case when $x = 1, y = 0$. From (7) and (8) we have that

$$E(A^{(1)}B^{(1)}|\eta_A = 1, \eta_B = 1) + E(A^{(1)}B^{(2)}|\eta_A = 1, \eta_B = 2) +$$
$$E(A^{(2)}B^{(1)}|\eta_A = 2, \eta_B = 1) - E(A^{(2)}B^{(2)}|\eta_A = 2, \eta_B = 2) = 4$$

and so the left hand side obtains its maximum mathematical value for any distribution of ± 1 valued random variables. Since this is larger than Tsireleson's bound of $2\sqrt{2}$ these outcomes are not obtainable in QM. The above construction gives a perfectly satisfactory probability space consistent with these conditional expectations that satisfies the non-signalling condition.

Remark. The probability space constructed in this section gives values to random variables corresponding to values that are not measured in the EPR-Bohm-Bell experiment. For example, in the probablility space $\omega = (1, 1, 1, -1, 2, 2)$ asserts that $A^{(1)}(\omega) = B^{(1)}(\omega) = 1$ and $\eta(\omega)_A = \eta(\omega)_B = 2$. In an EPR-Bohm-Bell experiment when the PBS's are in their second position there are no readings for $a_{\theta_1}(\omega)$ and $b_{\theta_1'}(\omega)$, and QM gives no predictions about their value. We do not assert that "in reality" for this outcome $a_{\theta_1}(\omega) = b_{\theta_1'}(\omega) = 1$. After all, as pointed out, there may be many consistent ways to assign values to $A^{(1)}(\omega)$ and $B^{(1)}(\omega)$.

The situation of a probabilistic model assigning values to events whose measurement is precluded by QM is by no means specific to our models. Consider, for example, a quantum experiment involving entangled photons for which the outcomes *can* be represented by a joint probability space on four unconditional random variables $A^{(1)}$, $A^{(2)}$, $B^{(1)}$, $B^{(2)}$. Any such joint probability space automatically assigns probabilities to joint events such as $A^{(1)} = 2$, $A^{(2)} = 1$. Since QM does not assign a value to this probability it cannot be verified experimentally, but this does not invalidate the probabilistic model used to describe the experiment.

One interpretation of Bell's theorem is that there does not exist any probability space consistent with (5) for which for all i=1,2 and j=1,2:

$$E(A^{(i)}B^{(j)}|\eta_A = i, \eta_B = j) = E(A^{(i)}B^{(j)}). \tag{9}$$

We merely assert that probability spaces exist that are consistent with all the *available* experimental data. Calculations made within the probability space yielding formulae for which all the parameters can be measured may be tested experimentally.

References

1. A. Yu. Khrennikov, *Interpretations of Probability.* VSP Int. Sc. Publishers, Utrecht/Tokyo, 1999: De Gruyter, Berlin (2009), second edition (completed).
2. A. Yu. Khrennikov, *Contextual approach to quantum probability,* Springer, Berlin-Heidelberg-New York, 2009.
3. B.S. Tsirelson, *Quantum generalizations of Bell's inequality,* Lett. Math. Phys. **2** (1980), 93-100.

Free White Noise Stochastic Equation

Luigi ACCARDI

Centro Vito Volterra

Universita degli studi di Roma Tor Vergata,

via Columbia 2- 00133 Roma, Italy

E-mail: accardi@volterra.uniroma2.it

Wided AYED

Department of mathematic

Institut péparatoire aux études d'ingénieure

El Merezka, Nabeul, 8000, Tunisia

E-mail: wided.ayed@ipein.rnu.tn

Habib OUERDIANE

Department of Mathematics

Faculty of sciences of Tunis

University of Tunis El-Manar

1060 Tunis, Tunisia. E-mail: habib.ouerdiane@fst.rnu.tn

Some estimates of scalar type on the free white noise stochastic integrals are given. These are used to prove an analogue regularity result for solutions of the free white noise stochastic equations with bounded coefficients. Furthermore, normal Form of such equation is given.

1. Introduction

(i) The identification of both classical and quantum stochastic equations with white noise Hamiltonian equations, i.e. the determination of the microscopic structure of these coefficients.
(ii) The explanation of the emergence of the unitarity conditions of Hudson and Parthasarathy as expression of the formal self-adjointness of the associated Hamiltonian equation.
(iii) The explicit expression of the coefficients of the stochastic equation as (nonlinear) functions of the coefficients of the associated Hamiltonian equation.
In this note we begin our program of extending, to the free white noise, the three main achievements of the white noise approach to stochastic analysis, namely:
The present paper is the first step of this program. Here we establish the basic estimates on stochastic integrals which allow to prove the existence and uniqueness theorems for the corresponding white noise Hamiltonian (hence stochastic differential) equations.
As usual the estimates in the free case are simpler than in the q-deformed case with $q \neq 0$ (see Lemma (4.1) below). Finally we use the time consecutive principle of the stochastic limit of quantum theory (see Refs. 1, 4 for its proof) to deduce the normally ordered form of free white noise equations. The unitarity conditions for this equation, as well as the comparison of our results with previous results of Fagnola,[6] Kumerer and Speicher,[5] Skeide[8] will be discussed in the paper.

2. Fock operator valued distributions

Recall that an operator valued distribution is defined by a quadruple $\{\mathcal{E}, \mathcal{H}, B, B^+\}$ where
-$\mathcal{E}$ is a vector space of complex valued (test) functions closed under complex conjugation.
-$\mathcal{H}$ is a Hilbert space (all vector space are complex unless otherwise specified).
- B, B^+ are real linear maps from $\mathcal{E}$ to pre-closed operators on $\mathcal{H}$ satisfying, on the corresponding domains

$$(B_f^+)^* = B_{\bar{f}};\ \forall f \in \mathcal{E} \tag{1}$$

We will use the notation

$$B_f^+ =: \int_{\mathbb{R}} f(s) b_s^+ ds; \ B_f =: \int_{\mathbb{R}} \bar{f}(s) b_s ds$$

and, when the spaces $\mathcal{H}$ and $\mathcal{E}$ are fixed, we will simply speak of the operator valued distribution (b_t) or (b_t^+). with these notations the identity (1) becomes

$$(b_t^+)^* = b_t \tag{2}$$

Definition 2.1. A vector $\Phi \in \mathcal{H}$ is in the domain of the operator valued distribution (b_t) or (b_t^+) if $\forall f \in \mathcal{E}$, $\Phi \in Dom(B_f)$ (or $\Phi \in Dom(B_f^+)$).

In the following we will freely use expression of the form $b_t\Phi$, b_tF_s, $Dom(b_t)$,... without specifying every time that they have to be understood in the distribution sense.

Definition 2.2. An operator valued distribution $\{\mathcal{E}, \mathcal{H}, B, B^+\}$ is called *Fock* if

$$b_t\Phi = 0 \tag{3}$$

and the set of vectors

$$b_{t_n}^+ \cdots b_{t_1}^+ \Phi, \ \ n \in \mathbb{N}, t_1, \cdots, t_n \in \mathbb{R} \tag{4}$$

are total in $\mathcal{H}$ (both identities are meant in the sense that the left hand side is well defined and the identity holds).

Any vector satisfying the above two conditions is called a vacuum vector for (b_t).

3. Free Fock white noise

From now on we fix the space of test functions to be

$$\mathcal{E} = L^2(\mathbb{R})$$

(or much smaller space of test functions is sufficient!) and $\forall f, g \in L^2(\mathbb{R})$ we define

$$\int_{\mathbb{R}}\int_{\mathbb{R}} \bar{f}(s)g(t)\delta(t-s)dsdt := \langle \bar{f}, g\rangle$$

Definition 3.1. A *free white noise* is an operator valued distribution (b_t) with test function space $L^2(\mathbb{R})$ and Hilbert space $\hat{\mathcal{H}}$, satisfying the identity.

$$b_t b_\tau^+ = \delta(t-\tau) \tag{5}$$

weakly in the operator valued distribution sense, i. e. if $f, g \in L^2(\mathbb{R})$ and $\psi \in Dom(b_f^+)$, $\varphi \in Dom(b_g^+)$, then:

$$\langle B_f^+\psi, B_g^+\varphi\rangle = \langle f, g\rangle\langle\psi, \varphi\rangle \tag{6}$$

The following result is well known.

Lemma 3.1. *If (b_t) is a free white noise the operators $B_f, B_f^+; f \in L^2(\mathbb{R})$ are bounded with norm*

$$\|B_f^+\|_\infty = \|B_f\|_\infty = \|f\|$$

Proof: Let $f \in L^2(\mathbb{R})$, $\psi \in Dom(B_f^+)$, then (6) implies that

$$\|B_f^+\psi\|^2 = \langle B_f^+\psi, B_f^+\psi\rangle = \|f\|^2_{L^2(\mathbb{R})}\|\psi\|^2$$

Since, by assumption, the domain of B_f^+ is dense, the thesis follows.

Remark 3.1. The algebra generated by (b_t^+) and (b_s) is the linear span of the identity and products of the form $b_{t_n}^+ \cdots b_{t_1}^+ b_{s_k} \cdots b_{s_1}$; $n, k \in \mathbb{N}$ where if $n = 0$ (resp. $k = 0$) the creators (resp. annihilators) are absent.

Lemma 3.2. *Let (b_t) be a free white noise on a Hilbert space $\hat{\mathcal{H}}$ and let $\Phi \in \hat{\mathcal{H}}$ be a vacuum vector for (b_t). Denote $\mathcal{H}$ the subspace generated by the number vectors (4). Then $\mathcal{H}$ is invariant under the action of the operators*

$B_f, B_f^+; f \in L^2(\mathbb{R})$. *In particular the restriction of* (b_t) *on* $\mathcal{H}$ *is a Fock operator valued distribution.*

4. Free white noise integrals

Let consider the free white noise equation of the form

$$\partial_t U_t = -i(Ab_t^+ + Bb_t + b_t^+ T b_t)U_t \tag{7}$$

where A, B, T are operators on the initial space, b_t, b_t^+ is the free Fock white noise and equation (7) is interpreted as an integral equation

$$U_t = U_0 - i\int_0^t ds(Ab_s^+ + Bb_s + b_s^+ T b_s)U_s \tag{8}$$

Thus to give a meaning to equation (7) we must define the white noise integrals in (8). The first problem is to define products of the form $b_s^+ U_s$, $b_s^+ b_s U_s$, $b_s U_s$.
To this goal we will use the following regularization

$$b_t F_t = \lim_{\varepsilon\to 0}[cb_{t-\varepsilon} + (1-c)b_{t+\varepsilon}]F_t \tag{9}$$

where $c > 0$ is an arbitrary constant and F_t is an arbitrary adapted operator process. We refer to[4] for a description of the meaning of the right hand side of (9).

In particular, if $F_s, s \in \mathbb{R}_+$ is any adapted operator process, we know from[3] that

$$\begin{aligned}
b_t \int_0^t ds b_s^+ F_s &= \lim_{\varepsilon\to 0}[cb_{t-\varepsilon} + (1-c)b_{t+\varepsilon}]\int_0^t ds b_s^+ F_s \\
&= \lim_{\varepsilon\to 0}\int_0^t ds[c\delta(t-\varepsilon-s) + (1-c)\delta(t+\varepsilon-s)]F_s \\
&= c\lim_{\varepsilon\to 0}\int_\varepsilon^{t+\varepsilon} ds\,\delta(t-s)F_s + (1-c)\lim_{\varepsilon\to 0}\int_{-\varepsilon}^{t-\varepsilon} ds\,\delta(t-s)F_s \\
&= cF_{t^+} + (1-c)F_{t^-}
\end{aligned}$$

This gives in particular, for continuous $t \mapsto F_t$:

$$b_t \int_0^t ds b_s^+ F_s = cF_t \tag{10}$$

4.1. *Estimates on the free white noise stochastic integrals*

Lemma 4.1. *Let $F_t, t \geq 0$ be a bounded operator process on an Hilbert space and a bounded subset $I \subseteq \mathbb{R}$. Let $F = \{f_1, \cdots, f_k\}$ be a finite ordered set of locally square integrable test functions, and define the number vector $\psi_F := B^+_{f_k} \cdots B^+_{f_1}\Phi$. Then*

$$\| \int_I ds F_s b_s \psi_F \| \leqslant \|f_k\|_2 \left(\int_I ds \|F_s \psi_{F\setminus\{f_k\}}\|^2 \right)^{\frac{1}{2}} \tag{11}$$

$$\| \int_I ds b_s^+ F_s \psi_F\|^2 = \int_I ds \|F_s \psi_F\|^2 \tag{12}$$

Proof. (11) follows from the relation $b_s B^+_{f_k} = f_k(s)$. In fact

$$\| \int_I ds F_s b_s \psi_F \| = \| \int_I ds F_s b_s B^+_{f_k} \cdots B^+_{f_1}\Phi\| = \| \int_I ds f_k(s) F_s B^+_{f_{k-1}} \cdots B^+_{f_1}\Phi\|$$
$$\leq \|f_k\|_2 \left(\int_I ds \|F_s \psi_{F\setminus\{f_k\}}\|^2 \right)^{\frac{1}{2}}$$

(12) follows from the free relation (5). In fact

$$\begin{aligned}
\| \int_I ds b_s^+ F_s \psi_F\|^2 &= \langle \int_I ds b_s^+ F_s \psi_F, \int_I ds' b_{s'}^+ F_{s'} \psi_F \rangle \\
&= \int_I \int_I ds ds' \langle b_s^+ F_s \psi_F, b_{s'}^+ F_{s'} \psi_F \rangle \\
&= \int_I \int_I ds ds' \langle F_s \psi_F, b_s b_{s'}^+ F_{s'} \psi_F \rangle \\
&= \int_I \int_I ds ds' \langle F_s \psi_F, \delta(s-s') F_{s'} \psi_F \rangle \\
&= \int_I ds \langle F_s \psi_F, F_s \psi_F \rangle \\
&= \int_I ds \|F_s \psi_F\|^2
\end{aligned}$$

Lemma 4.2. *Let $F_t, t \geq 0$ be an adapted continuous bounded operator process on an Hilbert space and a bounded subset $I \subseteq \mathbb{R}$. For each number*

vector $\psi_F = B^+_{f_k} \cdots B^+_{f_1}\Phi$ *where* $f_1, \cdots, f_k$ *are bounded, locally integrable test functions, we have*

$$\| \int_I ds b_s F_s \psi_F \|^2 \leq c_{I,c,\psi_F} \int_I ds \| F_s \psi_{F\setminus\{f_k\}} \|^2 \tag{13}$$

where $c_{I,c,\psi_F} = (|c| \|f_k\|_\infty + |1-c| \|f_k\|_\infty)^2$

$$\| \int_I ds b_s^+ b_s F_s \psi_F \|^2 \leqslant c_{I,c,\psi_F} \int_I ds \| F_s \psi_{F\setminus\{f_k\}} \|^2 \tag{14}$$

Proof: Using the regularization (9), one gets:

$$\begin{aligned}\| \int_I ds b_s F_s \psi_F \| &= \| \int_I ds \lim_{\varepsilon\to 0} [c b_{s-\varepsilon} + (1-c) b_{s+\varepsilon}] F_s \psi_F \| \\ &= \| \lim_{\varepsilon\to 0} \int_I ds [c b_{s-\varepsilon} + (1-c) b_{s+\varepsilon}] F_s \psi_F \|\end{aligned}$$

Since the process F_s is adapted and can be written in a normally ordered strongly convergent series (see Ref. 7). It follows that

$$\begin{aligned}\| \int_I ds b_s F_s \psi_F \|^2 &= \| \lim_{\varepsilon\to 0} \int_I ds [c F_s b_{s-\varepsilon} \psi_F + (1-c) F_s b_{s+\varepsilon} \psi_F] \|^2 \\ &= \| \lim_{\varepsilon\to 0} \int_I ds [c F_s f_k(s-\varepsilon) \psi_{F\setminus\{f_k\}} \\ &+ (1-c) F_s f_k(s+\varepsilon) \psi_{F\setminus\{f_k\}}] \|^2 \\ &= \| \int_I ds [c f_k(s^-) + (1-c) f_k(s^+)] F_s \psi_{F\setminus\{f_k\}} \|^2 \\ &\leq (|c| \|f_k\|_\infty + |1-c| \|f_k\|_\infty)^2 \int_I ds \| F_s \psi_{F\setminus\{f_k\}} \|^2\end{aligned}$$

To prove the estimate (14), we apply first the estimate (12) we get

$$\| \int_I ds b_s^+ b_s F_s \psi_F \|^2 = \int_I ds \| \lim_{\varepsilon\to 0} [c b_{s-\varepsilon} + (1-c) b_{s+\varepsilon}] F_s \psi_F \|^2$$

sine the process F_s is adapted, it follows

$$= \int_I ds \| \lim_{\varepsilon\to 0} [c F_s b_{s-\varepsilon} \psi_F + (1-c) F_s b_{s+\varepsilon} \psi_F] \|^2$$

$$= \int_I ds \| \lim_{\varepsilon\to 0} [c F_s f_k(s-\varepsilon) \psi_{F\setminus\{f_k\}} + (1-c) F_s f_k(s+\varepsilon) \psi_{F\setminus\{f_k\}}] \|^2$$

$$= \int_I ds \| [cf_k(s^-) + (1-c) f_k(s^+)] F_s \psi_{F \setminus \{f_k\}} \|^2$$

$$\leq (|c| \|f_k\|_\infty + |1-c| \|f_k\|_\infty)^2 \int_I ds \| F_s \psi_{F \setminus \{f_k\}} \|^2$$

4.2. *Free White Noise Stochastic Equations*

In this section, using the estimates 14, 12 and 13, we prove the existence of the solution of free white noise equation.

Theorem 4.1. *If A, B, T are bounded operators, the iterated series of equation (7) is normally ordered and strongly convergent on the domain of number vectors.*

Proof The free white noise equation (7) is equivalent to

$$U_t = 1 - i \int_0^t ds (Ab_s^+ + Bb_s + b_s^+ T b_s) U_s \tag{15}$$

The nth term of the expansion for U_t is given by:

$$U_t^n = \int_0^t dt_1 (Ab_{t_1}^+ + Bb_{t_1} + b_{t_1}^+ T b_{t_1}) U_{t_1}^{n-1}$$

after n iterations we get

$$U_t^n = \sum_{j:\text{finite set}} \int_0^t dt_1 \int_0^{t_1} dt_2 \cdots \int_0^{t_{n-1}} dt_n F^{\varepsilon_1} F^{\varepsilon_2} \cdots F^{\varepsilon_n} b_{t_n}^{\varepsilon_n} \cdots b_{t_1}^{\varepsilon_1} \tag{16}$$

where $F^{\varepsilon_i} \in \{A, B, T\}$ and $\varepsilon_i \in \{0, -, +\}, b_t^0 = b_t^+ b_t$. Consider a number vector $\psi_F = b_{f_k}^+ \cdots b_{f_1}^+ \Phi$ where $f_1, \cdots, f_k$ are locally square integrable test functions. To prove the strong convergence of the iterated series, we should calculate $\|U_t^n \psi_F\|^2$ in the different cases, using the lemma 4.1. In fact:

$$\|U_t^n \psi_F\|^2 \leqslant c_{t,c,\psi_F}^{(k)} \int_0^t dt_1 \sum_{\psi_i \in J(\psi_F)} \|U_{t_1}^{n-1} \psi_i\|^2$$

where $c_{\psi_F,c,t}^{(k)} = \max\{c_{\psi_F,c,t}, 1\} \times \max\{\|A\|_\infty, \|B\|_\infty, \|C\|_\infty\}$ and $J(\psi_F) = \{\psi_{F \setminus \{f_k\}}, \psi_F\}$. Similarly:

$$\|U_t^n \psi_F\|^2 \leqslant c_{\psi_F,t}^{(k)} \int_0^t dt_1 \sum_{\psi_i \in J(\psi_F)} c_{\psi_i,t}^{(i)} \int_0^{t_1} dt_2 \sum_{\psi_j \in J(\psi_i)} \|U_{t_2}^{n-2} \psi_j\|^2$$

An n-fold iteration of the same arguments gives us the estimate:

$$\|U_t^n\psi_F\|^2 \leq c_{t,\psi}^n|J(\psi_F)|^{n-1}\sum_{\xi_n\in J(\psi_F)}\int_0^t dt_1\int_0^{t_1} dt_2\cdots\int_0^{t_{n-1}} dt_n\|U_0\psi_h\|^2$$

$$\leq \max_{\eta\in J(\psi_h)}\|\eta\|^2.\|U_0\|^2 c_{t,\psi_f}^n|J(\psi_F)|^n t^n\frac{1}{n!}$$

where $c_{\psi_F,t} = \max_i\{c_{\psi_i,t}^i\}, 1\leq i\leq k$ It follows that

$$\sup_{\|\psi_F\|^2=1}\|U_t^n\psi_F\|^2 \leq \|U_0\|^2 c_{t,\psi}^n|3|^n t^n\frac{1}{n!}$$

where $c_{\psi,t} = \sup_{\|\psi_F\|^2=1}\{c_{\psi_i,t}^i\}$. Therefore the series $\sum_{n=0}^{\infty}U_t^n$ converges in the strong topology on on the number vector uniformly on bounded intervals of $\mathbb{R}_+$.

Corollary 4.1. *If A, B, T are bounded operators, the equation (7) has a unique adapted solution defined on the domain of number vectors.*

Proof The existence of the solution is a consequence of the above theorem. To prove the uniqueness , it will be sufficient to prove that all bounded continuous process $Z_t, t\in\mathbb{R}_+$ satisfying the following free white noise Hamiltonian equation:

$$\partial_t U_t = -i(Ab_t^+ + Bb_t + b_t^+Tb_t)U_t$$

with initial condition be zero. Applying the estimate (11), we obtain:

$$\|Z_t^n\psi_F\|^2 \leq c_{t,\psi}^n|J(\psi_F)|^{n-1}\sum_{\xi_n\in J(\psi_F)}\int_0^t dt_1\int_0^{t_1} dt_2\cdots\int_0^{t_{n-1}} dt_n\|Z_0\psi_h\|^2$$

$$\leq \|Z_0\|^2 c_{t,\psi_F}^n|J(\psi_F)|^n t^n\frac{1}{n!}$$

where $c_{\psi_F,t} = \max_i\{c_{\psi_i,t}^i\}, 1\leq i\leq k$. Since $Z_0 = 0$, therefore $Z_t^n\psi_F = 0, \forall n, \forall t\in[0,T]$, and for any number vector. Then the uniqueness follows. The adapteness is a consequence from the identity (16).

5. Normal Form of the Free White Noise Equations

In the following, the normally ordered form of the free white noise equation will be established. Denote $\mathcal{F}$ the algebra of all operators A which can be represented as series in normally ordered products of the free creators and annihilators. Such algebra is equipped with the topology defined by the weak convergence on number vectors. Consider $P^{(-)}$ the operator defined on the algebra $\mathcal{F}$ by the following prescriptions: $P^{(-)}(F)$ is the operator obtained from the series of F by setting all the terms which contain a creator equal to zero. Equivalently, for any element

$$A = \sum_{t_1,\cdots,t_n,s_1,\cdots,s_p,n,p\in\mathbb{N}} A_{n,p} b_{t_1}^+ \cdots b_{t_n}^+ b_{s_1} \cdots b_{s_p}$$

of $\mathcal{F}$. $P^{(-)}(A)$ is defined by

$$P^{(-)}(A) = \sum_p A_{0,p} b_{s_1} \cdots b_{s_p}$$

Theorem 5.1. *The free white noise equation (7) with the regularization (9) is equivalent to the normally ordered free equation*

$$\partial_t U_t = -i \left[KAb_t^+ U_t + BKb_t P^{(-)}(U_t) + b_t^+ TKb_t P^{(-)}(U_t) - iBKcA\, U_t \right] \quad (17)$$

where K and F satisfy the relation

$$K = (1 + icT)^{-1} \quad (18)$$

Proof The free white noise equation (7) is equivalent to the integral equation

$$U_t = 1 - i \int_0^t ds (Ab_s^+ + Bb_s + b_s^+ Tb_s) U_s \quad (19)$$

weakly on the number vectors.

By multiplication from the left by b_t, one gets

$$b_t U_t = b_t - iAb_t \int_0^t ds b_s^+ U_s - iBb_t \int_0^t ds b_s U_s - iTb_t \int_0^t ds b_s^+ b_s U_s \quad (20)$$

Using the regularization (9) and the conditions (5) and (3), we get

$$\begin{aligned} b_t \int_0^t ds b_s^+ b_s &= \lim_{\varepsilon \to 0} [cb_{t-\varepsilon} + (1-c) b_{t+\varepsilon}] \int_0^t ds b_s^+ b_s \\ &= \lim_{\varepsilon \to 0} \int_0^t ds [c\delta(t - \varepsilon - s) b_s + (1-c)\delta(t + \varepsilon - s) b_s] \\ &= cb_t \end{aligned}$$

Also by applying (10), equation (10) becomes

$$b_tU_t = b_t - icAU_t - iBb_t\int_0^t dsb_sU_s - iTcb_tU_t$$

Since under the assumption that $K := (1+icT)^{-1}$ is invertible, it follows that

$$b_tU_t = K(b_t - iBb_t\int_0^t dsb_sU_s) - icKAU_t \tag{21}$$

Applying $P^{(-)}$ to both sides of (19) we obtain

$$P^{(-)}(U_t) = 1 - i\int_0^t dsBP^{(-)}(b_sU_s)$$

Multiplying this on the left by b_t, we find

$$b_tP^{(-)}(U_t) = b_t - iBb_t\int_0^t dsP^{(-)}(b_sU_s) \tag{22}$$

To finish the proof, we need to justify that

$$b_t\int_0^t dsP^{(-)}(b_sU_s) = b_t\int_0^t dsb_sU_s \tag{23}$$

To this goal we prove our result by induction on the n-th term of the expansion of U_s in the normal form. In fact we have $U^0 = 1$ so

$$b_t\int_0^t dsP^{(-)}(b_sU^0) = b_t\int_0^t dsP^{(-)}(b_s) = b_t\int_0^t dsb_s = b_t\int_0^t dsb_sU^0$$

Suppose now that for any $0 \le k \le n$ one has

$$b_t\int_0^t dsP^{(-)}(b_sU_s^k) = b_t\int_0^t dsb_sU_s^k \tag{24}$$

and let prove the result for the $(n+1)$-term:

$$b_t\int_0^t dsP^{(-)}(b_sU_s^{n+1}) = b_t\int_0^t dsb_sU_s^{n+1} \tag{25}$$

In fact the right hand side of identity (25) gives

$$b_t\int_0^t dsb_sU_s^{n+1} = b_t\int_0^t dsb_s[U^0 - i\int_0^s ds_1(Ab_{s_1}^+ + Bb_{s_1} + b_s^+Tb_{s_1})U_{s_1}^n] \tag{26}$$

$$= b_t\int_0^t dsb_sU^0 - ib_t\int_0^t dsb_s\int_0^s ds_1Ab_{s_1}^+U_{s_1}^n - ib_t\int_0^t dsb_s\int_0^s ds_1Bb_{s_1}U_{s_1}^n$$

$$-ib_t \int_0^t dsb_s \int_0^s ds_1 b_{s_1}^+ Tb_{s_1} U_{s_1}^n$$

Let now study separately

$$b_t \int_0^t dsb_s U^0 = b_t \int_0^t dsP^{(-)}(b_s U^0)$$

$$b_t \int_0^t dsb_s \int_0^s ds_1 Ab_{s_1}^+ U_{s_1}^n = b_t \int_0^t ds \int_0^s ds_1 A\delta(s-s_1) U_{s_1}^n = b_t \int_0^t dsAU_s^n$$

$$= b_t \int_0^t dsA[U^0 - i\int_0^s ds_1 (Ab_{s_1}^+ + Bb_{s_1} + b_{s_1}^+ Tb_{s_1}) U_{s_1}^{n-1}]$$

$= b_t \int_0^t dsAU^0 - iA^2 \int_0^t ds \int_0^s ds_1 b_t b_{s_1}^+ U_{s_1}^{n-1} - iAB \int_0^t dsb_t \int_0^s ds_1 b_{s_1} U_{s_1}^{n-1} - iTA \int_0^t ds \int_0^s ds_1 b_t b_{s_1}^+ b_{s_1} U_{s_1}^{n-1}$

$$= b_t \int_0^t dsAU^0 - iA^2 \int_0^t ds \int_0^s ds_1 \delta(t-s_1) U_{s_1}^{n-1}$$

$$-iAB \int_0^t dsb_t \int_0^s ds_1 P^{(-)}(b_{s_1} U_{s_1}^{n-1})$$

$$-iTA \int_0^t ds \int_0^s ds_1 \delta(t-s_1) b_{s_1} U_{s_1}^{n-1}$$

$$= b_t \int_0^t dsAU^0 - iAB \int_0^t dsb_t \int_0^s ds_1 P^{(-)}(b_{s_1} U_{s_1}^{n-1})$$

because of the induction assumption and that terms contain $(\delta(t-s_1))$ which is not time consecutive, so it gives zero contribution. For the third term of identity (26), one has

$$b_t \int_0^t dsb_s \int_0^s ds_1 Bb_{s_1} U_{s_1}^n = Bb_t \int_0^t dsb_s \int_0^s ds_1 P^{(-)}(b_{s_1} U_{s_1}^n)$$

by the induction assumption.
The fourth term of identity (26) gives

$$Tb_t \int_0^t ds \int_0^s ds_1 b_s b_{s_1}^+ b_{s_1} U_{s_1}^n = Tb_t \int_0^t ds \int_0^s ds_1 \delta(s-s_1) b_{s_1} U_{s_1}^n$$

$$= Tb_t \int_0^t dsb_s U_s^n$$

$$= Tb_t \int_0^t dsP^{(-)}(b_s U_s^n)$$

by the induction assumption.
Let now calculate the left hand side of identity (25). We have

$$b_t \int_0^t dsP^{(-)}(b_s U_s^{n+1})$$

$$= b_t \int_0^t dsP^{(-)}(b_s[U^0 - i \int_0^s ds_1(Ab_{s_1}^+ + Bb_{s_1} + b_s^+ Tb_{s_1})U_{s_1}^n])$$

$$= b_t \int_0^t dsP^{(-)}(b_s U^0) - ib_t \int_0^t dsP^{(-)}(b_s \int_0^s ds_1 Ab_{s_1}^+ U_{s_1}^n)$$

$$-ib_t \int_0^t dsP^{(-)}(b_s \int_0^s ds_1 Bb_{s_1} U_{s_1}^n) - ib_t \int_0^t dsP^{(-)}(b_s \int_0^s ds_1 b_{s_1}^+ Tb_{s_1} U_{s_1}^n)$$

$$= b_t \int_0^t dsb_s U^0 - iAb_t \int_0^t dsP^{(-)}(\int_0^s ds_1 \delta(s - s_1) U_{s_1}^n)$$

$$-ib_t \int_0^t dsP^{(-)}(b_s \int_0^s ds_1 Bb_{s_1} U_{s_1}^n) - iTb_t \int_0^t dsP^{(-)}(\int_0^s ds_1 \delta(s-s_1) b_{s_1} U_{s_1}^n)$$

$$= b_t \int_0^t dsb_s U^0 - iAb_t \int_0^t dsP^{(-)}(U_s^n) - iBb_t \int_0^t dsb_s P^{(-)}(\int_0^s ds_1 b_{s_1} U_{s_1}^n)$$

$$-iTb_t \int_0^t dsP^{(-)}(b_s U_s^n)$$

Since

$$Ab_t \int_0^t dsP^{(-)}(U_s^n) = Ab_t \int_0^t dsP^{(-)}([U^0 - i \int_0^s ds_1(Ab_{s_1}^+ + Bb_{s_1} + b_{s_1}^+ Tb_{s_1})U_{s_1}^{n-1}])$$

$$= Ab_t \int_0^t dsP^{(-)}(U^0) - iAb_t \int_0^t dsP^{(-)}(\int_0^s ds_1 Ab_{s_1}^+ U_{s_1}^{n-1})$$

$$-iAb_t \int_0^t dsP^{(-)}(\int_0^s ds_1 Bb_{s_1} U_{s_1}^{n-1}) - iAb_t \int_0^t dsP^{(-)}(\int_0^s ds_1 b_{s_1}^+ Tb_{s_1} U_{s_1}^{n-1})$$

$$= Ab_t \int_0^t dsP^{(-)}(U^0) - iA \int_0^t dsP^{(-)}(\int_0^s ds_1 A\delta(t - s_1) U_{s_1}^{n-1})$$

$$-iABb_t \int_0^t dsP^{(-)}(\int_0^s ds_1 b_{s_1} U_{s_1}^{n-1}) - iAT \int_0^t dsP^{(-)}(\int_0^s ds_1 \delta(t-s_1) b_{s_1} U_{s_1}^{n-1})$$

$$= Ab_t \int_0^t dsP^{(-)}(U^0) - iABb_t \int_0^t dsP^{(-)}(\int_0^s ds_1 b_{s_1} U_{s_1}^{n-1})$$

because that terms contain $(\delta(t-s_1))$ which is not time consecutive, so it gives zero contribution. It follows that

$$b_t \int_0^t dsP^{(-)}(b_s U_s^{n+1})$$

$$= b_t \int_0^t dsb_s U^0 - i[Ab_t \int_0^t dsP^{(-)}(U^0) - iABb_t \int_0^t dsP^{(-)}(\int_0^s ds_1 b_{s_1} U_{s_1}^{n-1})]$$

$$-iBb_t \int_0^t dsb_s P^{(-)}(\int_0^s ds_1 b_{s_1} U_{s_1}^n) - iTb_t \int_0^t dsP^{(-)}(b_s U_s^n) = b_t \int_0^t dsb_s U_s^{n+1}$$

this finish the induction. Then (22) and (25) leads to

$$b_t P^{(-)}(U_t) = b_t - iBb_t \int_0^t dsb_s U_s \tag{27}$$

Then (27) becomes

$$b_t U_t = Kb_t P^{(-)}(U_t) - icKAU_t \tag{28}$$

replacing (28) in (7) we get

$$\partial_t U_t = -i\left[(A - iTKcF^-)b_t^+ U_t + BKb_t P^{(-)}(U_t) + b_t^+ TKb_t P^{(-)}(U_t) - iBKcAU_t\right]$$

and then the result.

References

1. L. Accardi, I. G. Lu and I. Volovich: *Quantum Theory and Its Stochastic Limit*, Springer (2002).
2. Accardi L., Fagnola F., Quaegebeur: *A representation free Quantum Stochastic Calculus,* Journ. Funct. Anal. 104 (1) (1992) 149–197 Volterra preprint N. 18 (1990) submitted Dec. (1989).
3. Accardi L., Lu Y.G., Volovich I.: Nonlinear extensions of classical and quantum stochastic calculus and essentially infinite dimensional analysis, in: Probability Towards 2000; L. Accardi, Chris Heyde (eds.) Springer LN in Statistics 128 (1998) 1–33 Proceedings of the Symposium: Probability towards two thousand, Columbia University, New York, 2–6 October (1995)

4. L.Accardi, I.V.Vlovich and Y.G.Lu: A White Noise Approach to Classical and Quantum Stochastic Calculus, Volterra Preprint 375, Rome, July 1999.
5. B. Kummerer, R. Speicher: *Stochastic Integration on the Cuntz Algebra* O_∞, Jou. Func. Anal. **103**,2, (1992), 372–408.
6. F. Fagnola: *On Quantum Stochastic Integration With Respect to Free Noise*, Infin. Dimens. Anal. Quantum Probab. Relat. Top. **7** (2004), 183–194.
7. N. Obata: *White noise calculus and Fock space*, Lecture Notes in Math., No. 1577, Springer-Verlag, 1994.
8. M. Skeide: *On Quantum Stochastic Calculus on Full Fock Modules*, Preprint Volterre, N. 374, July (1999).

Lévy Models Robustness and Sensitivity

Fred Espen Benth

Centre of Mathematics for Applications and Department of Mathematics. University of Oslo. P.O. Box 1053 Blindern, N-0316 Oslo, Norway.

E-mail: fredb@math.uio.no

Giulia Di Nunno

Centre of Mathematics for Applications and Department of Mathematics. University of Oslo. P.O. Box 1053 Blindern, N-0316 Oslo, Norway.

E-mail: g.d.nunno@cma.uio.no

Asma Khedher

Centre of Mathematics for Applications and Department of Mathematics. University of Oslo. P.O. Box 1053 Blindern, N-0316 Oslo, Norway.

E-mail: asma.khedher@cma.uio.no

We study the robustness of the sensitivity with respect to parameters in expectation functionals with respect to various approximations of a Lévy process. As sensitivity parameter, we focus on the *delta* of an European option as the derivative of the option price with respect to the current value of the underlying asset. We prove that the delta is stable with respect to natural approximations of a Lévy process, including approximating the small jumps by a Brownian motion. Our methods are based on the density method, and we propose a new *conditional density method* appropriate for our purposes. Several examples are given, including numerical examples demonstrating our results in practical situations.

1. Introduction

The dynamics of asset prices seems to be well modeled by Lévy processes and most of current research in mathematical finance is focused around this class (see e.g. Cont and Tankov[6]). When modeling the asset dynamics, it can be difficult to determine the most appropriate process among the wide variety in the Lévy family. For example, some authors promote the generalized hyperbolic class (see for instance Eberlein[8]) and Barndorff-Nielsen[2]), while others prefer the CGMY model (see Carr *et al.*[4]). Furthermore, it is a philosophical question whether asset prices are driven by pure-jump processes, or if there is a diffusion in the non-Gaussian dynamics (see e.g. Eberlein and Keller[9] for a discussion). From a statistical point of view it may be very hard to determine whether a model should have a diffusion term or not.

In this paper we are dealing with the robustness of expectation functionals of models chosen within the Lévy family. In particular, we study the sensitivity with respect to parameters in the functionals with respect to various approximations of a Lévy process. As sensitivity parameter, we choose to focus on the *delta* of an European option as the derivative of the option price with respect to the current value of the underlying asset.

There are several methods of computation of the delta, but in this presentation we consider the so-called *density method* which is proper of Lévy models admitting a probability density absolutely continuous with respect to the Lebesgue measure. This method was introduced in Broadie and Glasserman[3] for Brownian models, but it is not restricted to this case. Following the density method, the computation of the delta leads to a formula of the type:

$$\Delta := \frac{\partial}{\partial s_o}\mathbb{E}\left[g(S^{s_o}(T))\right] = \mathbb{E}\left[g(S^{s_o}(T)) \times \pi\right] , \tag{1}$$

where $(S^{s_o}(t))_{t\geq 0}$, $(S(0) = s_o > 0)$, is the discounted price of the asset and $g(S^{s_o}(T))$ is the payoff of a European option with maturity $T > 0$. The random variable π is called *weight*. The underlying idea is to move the differentiation with respect to the parameter s_o to the density function, thus the weight π is given by the log-derivative of the density. Since most asset price models are defined as the exponential of a Lévy process (see for instance Eberlein[8]), we choose to work in this paper with expectation functionals of the form

$$F(x) = \mathbb{E}\left[f(x + L(T))\right] .$$

We easily recover the above by letting $x = \ln s_o$ and $f(y) = g(\mathrm{e}^y)$. The application of the density method will yield a weight π which is the log-derivative of the density of $L(T)$.

As a variation to this method we introduce the *conditional density method* which allows some flexibility in the computation when dealing with Lévy models not of Brownian nature. The conditional density method relies on the observation that we may use conditioning in order to separate out differentiable density in the expectation function. More precisely, if we have a random variable which may be represented as a sum of two independent random variable, where one possesses a differentiable density, we may use conditional expectation and the "classical" density approach to move the differentiation to this density. We recall from the Lévy-Kintchine representation of Lévy processes that any Lévy process can be represented as a pure-jump process and an independent drifted Brownian component. The application of the conditional density method provides *different* weights than the density method. The fact that the weights are not unique is well-known, as this appears also by application to other methods of computations, e.g. the so-called Malliavin methods. We stress that the delta is in any case the same, only the computation method is different. It is well-known that the density method provides an expression for the delta which has minimal variance. This is the meaning of *optimality* for weights. The weights derived by the conditional density method are not optimal.

From the point of view of robustness to model choice, our point of departure is the paper of Asmussen and Rosinski,[1] where it is proven that the small jumps of a Lévy process $(L(t))_{t\geq 0}$ can be approximated by a Brownian motion scaled with the standard deviation of the small jumps, that is,

$$L(t) \approx \sigma(\varepsilon)B(t) + N^{\varepsilon}(t),$$

where N^{ε} is a Lévy process with jumps bigger than ε and $(B_t)_{t\geq 0}$ is an independent Brownian motion. Note that if the Lévy process has a continuous martingale part, N^{ε} includes it. The function $\sigma(\varepsilon)$ is the standard deviation of the jumps smaller than ε of the Lévy process, which can be computed as the integral of z^2 with respect to the Lévy measure in a ball of radius ε. Obviously, $\sigma(\varepsilon)$ tends to zero with ε. In fact, this approximative Lévy process converges in distribution to the original one. Based on this approximation, which is popular when simulating the paths of different Lévy processes like the normal inverse Gaussian (see Rydberg[13]), we investigate the relationship of the deltas derived from the two respective models. From the results of Asmussen and Rosinski,[1] we know that the respective option prices converge when ε goes to zero. The question is if the same holds true for the deltas, and in this paper we show that this is indeed the case.

In itself it is maybe not a priori surprising that the deltas are robust with respect to Lévy models which are approximately equal, but it turns out that for pure-jump Lévy processes one obtain weights for the approximating model which explode when ε tends to zero. Hence, the random variable inside the expectation diverges. However, due to an independence property in the limit which is not found in the classical setting of the density method, the delta converges anyhow. However, the variance of the expression explodes, which in turn implies that the weights are highly inefficient from a Monte Carlo point of view. The same problem does not occur for Lévy processes having a continuous martingale part. Hence, we conclude that even though the delta is robust towards these approximations, the resulting expressions for the deltas may become inefficient for practical simulation, at least in the pure-jump case. We study numerical examples discussing this problem. Also, we provide convergence rates for the approximative deltas.

Our presentation is organized as follows. After a short introduction on Lévy processes, we discuss the computational methods for the delta based on the existence of a probability density function: we revise the density method and we introduce the conditional density method. Then we discuss the problems related to model robustness and we present our results in connection to the analysis of sensitivity. Several examples are provided, including different classes of Lévy process and relevant functions f. A numerical study investigates our findings in a practical setting based on Monte Carlo simulations. Comments on our results and future research are given as conclusion.

2. Density based methods for the computation of derivatives

In this Section we introduce and analyze the so-called density method for the calculation of derivatives with respect to parameters in expectation functionals. The density method is a classical approach based on the representation of the derivative of the expectation functionals as an expectation involving the logarithmic derivative of the density function. We recall basic results in this area, and propose a *conditional density* approach which may be useful in certain contexts.

2.1. *Some mathematical preliminaries*

Let $(\Omega, \mathcal{F}, \mathbb{P})$ be a complete probability space equipped with a filtration $\{\mathcal{F}_t\}_{t\in[0,T]}$ satisfying the *usual conditions.* We introduce the generic notation $L(t)$ for a Lévy process on the probability space, and denote by $B(t)$ a Brownian motion, with $t \in [0,T]$ and $L(0) = B(0) = 0$ by convention. In the sequel of the paper, we fix $T = 1$ for simplicity in notation. We work with the RCLL[a] version of the Lévy process, and let $\triangle L(t) = L(t) - L(t-)$. Denote the Lévy measure of $L(t)$ by $Q(dz)$. This is a σ-finite Borel measure on $\mathbb{R}_0 \triangleq \mathbb{R} - \{0\}$.

We recall the Lévy-Itô decomposition for a Lévy process (See e.g. Sato[14]).

Theorem 2.1. ***Lévy-Itô decomposition.***

Let $L(t)$ be a Lévy process and Q its Lévy measure. Then we have:

- *Q verifies*

$$\int_{\mathbb{R}_0} \min(1, z^2)\, Q(dz) < \infty.$$

- *The jump measure of $L(t)$, denoted by $N(dt, dz)$, is a Poisson random measure on $[0, \infty[\times\mathbb{R}$ with intensity measure $Q(dz)\, dt$.*
- *There exists a Brownian motion $W(t)$ such that*

$$L(t) = at + bW(t) + Z(t) + \lim_{\varepsilon\downarrow 0} \widetilde{Z}_\varepsilon(t), \tag{2}$$

where

$$Z(t) \triangleq \sum_{s\in[0,t]} \triangle L(s)\mathbf{1}_{\{|\triangle L(s)|\geq 1\}} = \int_0^t \int_{|z|\geq 1} z\, N(ds, dz)$$

[a]Right-continuous with left limits, also called *càdlàg*.

and

$$\widetilde{Z}_\varepsilon(t) \triangleq \sum_{s\in[0,t]} \triangle L(s)\mathbf{1}_{\{\varepsilon\leq|\triangle L(s)|<1\}} - t\int_{\varepsilon\leq|z|<1} z\,Q(dz)$$

$$= \int_0^t \int_{\varepsilon\leq|z|<1} z\,\widetilde{N}(ds,dz)\,,$$

where $\widetilde{N}(dt,dz) := N(dt,dz) - Q(dz)dt$ *is the compensated Poisson random measure of* $L(t)$ *and* $a, b \in \mathbb{R}$ *are two constants. The limit in* (2) *is meant with convergence almost sure and uniform in* $t \in [0,1]$, *The components* W, Z *and* $\widetilde{Z}_\varepsilon$ *are independent.*

We introduce the following notation for the variation of the Lévy process $L(t)$ close to the origin. For $0 < \varepsilon \leq 1$,

$$\sigma^2(\varepsilon) \triangleq \int_{|z|<\varepsilon} z^2\,Q(dz)\,. \tag{3}$$

Since any Lévy measure $Q(dz)$ integrates z^2 in an open interval around zero, we have that $\sigma^2(\varepsilon)$ is finite for any $\varepsilon > 0$. The $\sigma^2(\varepsilon)$ represents the variance of the jumps smaller than ε of $L(t)$ in the case it is symmetric and has mean zero. We will frequently make use of $\sigma^2(\varepsilon)$ for our studies. But first, we recall a result of Orey[12] which relates the asymptotic behavior of the Lévy measure at zero (that is, the asymptotic behavior of $\sigma^2(\varepsilon)$ as ε tends to zero) to the smoothness of the probability density of $L(t)$.

Theorem 2.2. *Let* $L(t)$ *be a Lévy process, then it follows:*

- *If* $b > 0$ *or* $Q(\mathbb{R}_0) = \infty$, *then* $L(t)$ *has a continuous probability density* $p_t(.)$ *on* $\mathbb{R}$.
- *If there exists* $\gamma \in]0,2[$ *such that* $Q(dz)$ *satisfies*

$$\liminf_{\varepsilon\to 0} \frac{\sigma^2(\varepsilon)}{\varepsilon^\gamma} > 0\,, \tag{4}$$

then the probability density p_t of $L(t)$ is infinitely continuously differentiable and for all $n \geq 1$,

$$\lim_{|x|\to\infty} \frac{\partial^n p_t}{\partial x^n}(x) = 0\,.$$

We observe that both the α-stable and the normal inverse Gaussian (NIG) Lévy processes satisfy condition (4) ensuring the existence of a smooth density. Indeed, the Lévy measure of an α-stable process with $\alpha \in]0,2[$ is (see for instance Sato[14])

$$Q(dz) = c_1|z|^{-1-\alpha}\mathbf{1}_{\{z<0\}}\,dz + c_2 z^{-1-\alpha}\mathbf{1}_{\{z>0\}}\,dz\,,$$

with $c_1, c_2 \geq 0$ and $c_1 + c_2 > 0$. Therefore,

$$\sigma^2(\varepsilon) = \frac{c_1+c_2}{2-\alpha}\varepsilon^{2-\alpha}\,.$$

Hence, choose $\gamma = 2-\alpha$ to verify condition (4). The NIG Lévy process has Lévy measure (see Barndorff-Nielsen[2]),

$$Q(dz) = \frac{\alpha\delta}{\pi|z|}K_1(\alpha|z|)e^{\beta z}\,dz\,,$$

where α, β, δ are parameters satisfying $0 \leq \beta \leq \alpha$ and $\delta > 0$, and $K_1(z)$ is the modified Bessel function of the third kind with index 1. Using properties of the Bessel functions (see Assmussen and Rosinski[1]), one finds

$$\sigma^2(\varepsilon) = \frac{2\delta}{\pi}\varepsilon\,.$$

Hence, letting $\gamma = 1$ we readily verify condition (4) also for the NIG Lévy process.

Thm. 2.2 is useful in our analysis since it ensures that the density function of a Lévy process is differentiable, which is the basic requirement for the applicability of the so-called density method which we study next.

2.2. *The density method*

In this paper we are concerned with the derivative of the expectation of functionals of the form

$$F(x) \triangleq \mathbb{E}\left[f(x+Y)\right]\,, \tag{5}$$

for a random variable Y and a measurable function f such that $f(x+Y) \in L^1(\mathbb{P})$ for each $x \in \mathbb{R}$ (or in some subset of $\mathbb{R}$). Here, we denote by $L^1(\mathbb{P})$

the space of all random variables which are integrable with respect to $\mathbb{P}$. In most of our forthcoming analysis, Y will be a Lévy process $L(t)$ or some approximation of such including Brownian motion. We call a random variable π a *weight* if $f(x+Y)\pi \in L^1(\mathbb{P})$ for $x \in \mathbb{R}$ and

$$F'(x) \triangleq \frac{dF(x)}{dx} = \mathbb{E}\left[f(x+Y)\pi\right] . \tag{6}$$

A straightforward derivation inside the expectation operator would lead to $F'(x) = \mathbb{E}[f'(x+Y)]$, so a sensitivity weight can be viewed as the result after a kind of "integration-by-parts" operation. The advantage with an expression of the form (6) is that we can consider the derivative of expectation functionals where the function f is not differentiable. Examples where this is relevant include the calculation of delta-hedge ratios in option pricing for "payoff-functions" f being non-differentiable (digital options, say). Other examples are the sensitivity of risk measures with respect to a parameter, where the risk measure may be a non-differentiable function of the risk (Value-at-Risk, say, which is a quantile measure).

There exist by now at least two methods to derive sensitivity weights for functionals like $F(x)$. The classical approach is the density method, which transfers the dependency of x to the density function of Y, and then differentiate. An alternative method is the Malliavin approach, applying the tools from Malliavin calculus to perform an integration-by-parts utilizing the Malliavin derivative rather than classical differentiation. We refer to Fournié *et al.*[11] for more information on this approach.

Let us discuss the density method (see Broadie and Glasserman[3] for applications to finance). Suppose Y has a density p_Y with respect to the Lebesgue measure dt. Then, from classical probability theory, we have that

$$F(x) = \int_{\mathbb{R}} f(x+y)p_Y(y)\,dy = \int_{\mathbb{R}} f(y)p_Y(y-x)\,dy . \tag{7}$$

Hence, the expectation functional $F(x)$ can be expressed as a convolution between f and p_Y. Recalling Thm. 8.10 in Folland,[10] as long as $f \in L^1(\mathbb{R})$ and $p_Y \in C_0^n(\mathbb{R})$, F is n times continuously differentiable and its derivatives can be expressed as

$$F^{(k)}(x) = \int_{\mathbb{R}} f(y)(-1)^k \frac{d^k}{dy^k} p_Y(y-x)\,dy ,$$

for $k \leq n$ and $F^{(k)}$ denoting the k'th derivative of F. Here we have denoted the space of Lebesgue integrable functions on $\mathbb{R}$ by $L^1(\mathbb{R})$ and the space of differentiable (up to order n) functions on $\mathbb{R}$ vanishing at indfinity by $C_0^n(\mathbb{R})$.

Restricting our attention to $n = 1$, and assuming that $p_Y(y) > 0$ for $y \in \mathbb{R}$, we find that

$$F'(x) = \int_{\mathbb{R}} f(x+y)(-\frac{d}{dy}\ln p_Y(y))p_Y(y)\,dy = \mathbb{E}\left[f(x+Y)(-\partial \ln p_Y(Y))\right].$$

Thus, the density method yields a weight $\pi = -\partial \ln p_Y(Y)$, the logarithmic derivative of the density. As we see from the above, under very mild assumptions on the density of Y and the function f, we can find a weight π for calculating the derivative of F without having to differentiate f.

Assuming that $f \in L^1(\mathbb{R})$ is rather strict in many applications. We can relax the conditions on f considerably as follows. Suppose that p_Y is differentiable and strictly positive, and $f(\cdot)p'_Y(\cdot - x)$ is bounded uniformly in x by an integrable function on $\mathbb{R}$. Then, according to Thm. 2.27 in Folland,[10] we have

$$\begin{aligned}
F'(x) &= \frac{d}{dx}\int_{\mathbb{R}} f(y)p_Y(y-x)\,dy \\
&= \int_{\mathbb{R}} f(y)(-1)p'_Y(y-x)\,dy \\
&= \int_{\mathbb{R}} f(x+y)(-\frac{d}{dy}\ln p_Y(y))p_Y(y)\,dy \\
&= \mathbb{E}\left[f(x+Y)(-\partial \ln p_Y(Y))\right].
\end{aligned}$$

We obtain the same weight $\pi = -\partial \ln p_Y(Y)$ as above, naturally. However, we can include functions f which can grow at infinity as long as the density (and its derivative) dampens this growth sufficiently. This ensures that we can apply the density method in financial contexts like calculating the delta of a call option.

2.3. *The conditional density method*

Another case which is important in our analysis will be the situation where we have two independent strictly positive random variables Y and Z with densities p_Y and p_Z, respectively. In some situations that we will encounter in the sequel, only one of the two densities may be known, or one of the two may be simpler to be used for computational purposes. We propose a *conditional density method* for such cases.

Obviously, if the density of $Y + Z$ is known, we are in the situation described in the previous subsection. Under the hypotheses stated there, we may apply the standard density method in order to find the derivative of the functional

$$F(x) = \mathbb{E}\left[f(x+Y+Z)\right] .$$

In this case we find

$$F'(x) = F'_{Y+Z}(x) = \mathbb{E}\left[f(x+Y+Z)(-\partial \ln p_{Y+Z}(Y+Z))\right] .$$

We use the notation $F'_{Y+Z}(x)$ to emphasize that we apply the density method to the sum $Y+Z$.

On the onter hand, if only one of the two densities p_Y or p_Z is known or better fitting computations, we can apply the *conditional density method* as follows. Since by conditioning we have

$$F(x) = \mathbb{E}\left[\mathbb{E}\left[f(x+Y+Z)\,|\,Y\right]\right] = \mathbb{E}\left[\mathbb{E}\left[f(x+Y+Z)\,|\,Z\right]\right] ,$$

we find (see Sato, Prop. 1.16)

$$F(x) = \int_{\mathbb{R}} \mathbb{E}\left[f(y+Z)\right] p_Y(y-x)\,dy = \int_{\mathbb{R}} \mathbb{E}\left[f(z+Y)\right] p_Z(z-x)\,dz .$$

This holds as long as $\mathbb{E}\left[f(\cdot+Z)\right] p_Y(\cdot-x)$ is integrable (or, symmetrically, $\mathbb{E}\left[f(\cdot+Y)\right] p_Z(\cdot-x)$ is integrable). Strictly speaking, the Proposition 1.16 in Sato is only valid under boundedness conditions, however, these can be relaxed by standard limiting arguments. The expressions of $F(x)$ can be used in two ways to derive the derivative $F'(x)$: First, we find

$$F'_Y(x) = \mathbb{E}\left[f(x+Y+Z)(-\partial \ln p_Y(Y))\right] ,$$

as long as p_Y is differentiable and $\mathbb{E}\left[f(\cdot-Z)\right] p'_Y(\cdot-x)$ is bounded by an integrable function uniformly in x, say. Symmetrically, we obtain

$$F'_Z(x) = \mathbb{E}\left[f(x+Y+Z)(-\partial \ln p_Z(Z))\right] ,$$

whenever $\mathbb{E}\left[f(\cdot-Y)\right] p'_Z(\cdot-x)$ is bounded by an integrable function uniformly in x. Obviously, $F'_Y(x) = F'_Z(x) = F'_{Y+Z}(x) = F'(x)$, however, the three different calculations lead to three different weights, being, respectively,

$$\begin{aligned} \pi_{Y+Z} &\triangleq -\partial \ln p_{Y+Z}(Y+Z) \\ \pi_Y &\triangleq -\partial \ln p_Y(Y) \\ \pi_Z &\triangleq -\partial \ln p_Z(Z) . \end{aligned}$$

The two last are resulting from the conditional density method, while the first one is from the density method. These three weights are genuinely different.

3. Robustness of the delta to model choice

In this Section we will analyze the sensitivity of expectation functionals with respect to Lévy processes and their approximations. Our main focus will be on cases where a Lévy process $L(t)$ and its approximation $L_\varepsilon(t)$ are indistinguishable in practical contexts for small ε. Hence, in a concrete application, we may think of two models $L(t)$ and $L_\varepsilon(t)$ for the same random phenomenon which we cannot in practical terms separate. For instance, we may think of two speculators in a financial market who want to price an option. The first investor believes in a model given by $L(t)$, while the other chooses a model $L_\varepsilon(t)$, being slightly different than the former. The distributions of the two models will be very close, and thus also the derived option price. However, the main question we want to analyze in this paper is whether the same holds true for the sensitivities (or the Greeks in financial terminology). We refer to this question as a problem of robustness of sensitivities to model choice.

We analyze a particular class of approximations of $L(t)$, namely the one introduced by Asmussen and Rosinski[1] where the small jumps of $L(t)$ are substituted by an appropriately scaled Brownian motion. Before analyzing the sensitivity parameter delta for such approximations, we include for the convenience of the reader some details on small jump approximations and convergence.

3.1. *Small jump approximations of Lévy processes*

In applications of Lévy processes, it is often useful to approximate the small jumps by a Brownian motion. This approximation was advocated in Rydberg[13] as a way to simulate the path of a Lévy process with NIG distributed increments, and later studied in detail by Asmussen and Rosinski.[1]

Recall the Lévy-Itô decomposition of a Lévy process $L(t)$ as given in (2) and introduce now an approximating Lévy process

$$L_\varepsilon(t) \triangleq at + bW(t) + \sigma(\varepsilon)B(t) + Z(t) + \widetilde{Z}_\varepsilon(t)\,, \tag{8}$$

with $\sigma^2(\varepsilon)$ defined in (3) and $B(t)$ being a Brownian motion independent of $L(t)$ (which in fact also means independent of $W(t)$). From the definition of $\widetilde{Z}_\varepsilon$, we see that we have substituted the small jumps (compensated by their expectation) in $L(t)$ by a Brownian motion scaled with $\sigma(\varepsilon)$. Hence, we approximate the small jumps by a Brownian motion with the same variance as the compensated small jumps. We have the following result:

Proposition 3.1. *Let the processes $L(t)$ and $L_\varepsilon(t)$ be defined as in equation (2) and (8), respectively. Then, for every t, we have:*

$$\lim_{\varepsilon \longrightarrow 0} L_\varepsilon(t) = L(t) \quad \textit{the convergence is in } L^1 \textit{ and } \mathbb{P}-a.s.$$

Proof. Whenever $\varepsilon \to 0$, it follows that

$$\begin{aligned}
\mathbb{E}\left[|L_\varepsilon(t) - L(t)|\right] &= \mathbb{E}\left[|\sigma(\varepsilon)B(t) - \int_0^t \int_{0\le|z|\le\varepsilon} z\widetilde{N}(ds,dz)|\right] \\
&\le \sigma(\varepsilon)\mathbb{E}\left[|B(t)|\right] + \mathbb{E}\left[|\int_0^t \int_{0\le|z|\le\varepsilon} z\widetilde{N}(ds,dz)|\right] \\
&\le \sigma(\varepsilon)\mathbb{E}\left[B^2(t)\right]^{1/2} + \mathbb{E}\left[\Big(\int_0^t \int_{0\le|z|\le\varepsilon} z\widetilde{N}(ds,dz)\Big)^2\right]^{1/2} \\
&\le 2\sigma(\varepsilon)\sqrt{t} \longrightarrow 0\,,
\end{aligned}$$

from the triangle and Cauchy-Schwarz inequalities. This proves the convergence in L^1. As for the $\mathbb{P}$-a.s. convergence, this follows directly from the proof of the Lévy-Kintchine formula (See Thm. 19.2 in Sato[14]). □

The study in Asmussen and Rosinski[1] gives a central limit type of result for the approximation of the small jumps. It says that the small jumps are, after scaling by $\sigma(\varepsilon)$, indeed close to be standard normally distributed. We note that the above result only says that, for every t, the two random variables $L(t)$ and $L_\varepsilon(t)$ are close in distribution, but nothing about the asymptotic distribution of the small jumps in the limit. Indeed, under an asymptotic condition on $\sigma(\varepsilon)$, the result in Ref. 1 is:

Theorem 3.1. *If*

$$\lim_{\varepsilon \to 0} \frac{\sigma(\varepsilon)}{\varepsilon} = \infty, \tag{9}$$

then

$$\lim_{\varepsilon \to 0} \sigma^{-1}(\varepsilon)\widetilde{Z}_\varepsilon = B\,,$$

where B is a Brownian motion and the convergence is in distribution.

This results supports the choice of using a Brownian motion and the scale $\sigma(\varepsilon)$ for the small jumps of a Lévy process.

3.2. *Robustness to model choice*

Assume that $f : \mathbb{R} \mapsto \mathbb{R}$ is a measurable function and that for each x belonging to a compact set of $\mathbb{R}$, there exists a random variable $U \in L^1(\mathbb{P})$ such that $|f(x + L_\varepsilon(1))| \leq U$ for all ε. Without loss of generality, we can consider $x \in [x_1, x_2]$, for some $x_1, x_2 \in \mathbb{R}$. Since $f(x + L_\varepsilon(1))$ converges almost surely to $f(x + L(1))$, by dominated convergence it holds that

$$\lim_{\varepsilon \downarrow 0} \mathbb{E}\left[f(x + L_\varepsilon(1))\right] = \mathbb{E}\left[f(x + L(1))\right] = F(x)\,. \tag{10}$$

Such expectation functionals arise in pricing of options, where f is the payoff function from the option and $x + L(1)$ is the state of the underlying asset at exercise time 1. If $f(x) = \mathbf{1}_{\{x<q\}}$, we may view the expectation as coming from a simple quantile risk measure on the random variable $x + L(1)$, where the x is the initial state of the system under consideration. For notational simplification, we introduce

$$F_\varepsilon(x) \triangleq \mathbb{E}\left[f(x + L_\varepsilon(1))\right] \tag{11}$$

and set $F_0(x) = F(x)$. We analyze $F'_\varepsilon(x)$ and its convergence to $F'(x)$.

To differentiate $F_\varepsilon(x)$, we have in fact a multiple of different approaches. Motivated from the Malliavin method of Davis and Johansson[7] for jump diffusions, it is natural to use the conditional density method with respect to the Brownian motion. However, this leads to three possibilities. Either we can differentiate with respect to the original Brownian motion $W(t)$, or with respect to $B(t)$, or finally with respect to a *new* Brownian motion defined as the sum of the two. This will lead to three new expectation operators, which we would like to converge to $F'(x)$ when $\varepsilon \downarrow 0$. As we will see, this is indeed the case, however, the three choices have different properties. Obviously, if the distribution of $L_\varepsilon(t)$ is available, one would prefer to use the density method directly on this. However, due to the truncation of the jumps at ε, this will not, in most practical applications, be available.

To get some intuition on the problem we are facing, let us consider a trivial case of a Brownian motion

$$L(t) = bW(t)\,.$$

Of course, in this case we have by a straightforward application of the density method

$$F'(x) = \mathbb{E}\left[f(x + bW(1))\frac{W(1)}{b}\right].$$

Now, introduce

$$L_\varepsilon(t) = \sqrt{b^2 - \varepsilon^2}W(t) + \varepsilon B(t)$$

where $B(t)$ is an independent Brownian motion. Note that $L_\varepsilon(t)$ in distribution is identical to $L(t)$, so it is in this sense not an approximation of $L(t)$ as above for the Lévy case. However, we are mimicking the approximation procedure above by representing the "small jumps" of $W(t)$ by a new Brownian motion $B(t)$. If we first apply the conditional density method on $W(t)$, we find

$$\begin{aligned} F'_\varepsilon(x) &= \frac{d}{dx}\int_{\mathbb{R}} \mathbb{E}\left[f(u + \varepsilon B(1))\right] p_{\sqrt{b^2-\varepsilon^2}W(1)}(u - x)\, dx \\ &= \mathbb{E}\left[f(x + L_\varepsilon(1))\frac{W(1)}{\sqrt{b^2 - \varepsilon^2}}\right], \end{aligned}$$

where we recall that p_X denotes the probability density of the random variable X. We see easily that $F'_\varepsilon(x)$ converges nicely to $F'(x)$. Next, differentiating using the distribution of $\sqrt{b^2 - \varepsilon^2}W(1) + \varepsilon B(1)$ leads similarly to

$$F'_\varepsilon(x) = \mathbb{E}\left[f(x + L_\varepsilon(1))\frac{\sqrt{b^2 - \varepsilon^2}W(1) + \varepsilon B(1)}{b^2}\right].$$

This will again converge nicely to $F'(x)$. Finally, apply the procedure with respect to $B(t)$ to find

$$F'_\varepsilon(x) = \mathbb{E}\left[f(x + L_\varepsilon(1))\frac{B(1)}{\varepsilon}\right].$$

This results in a sensitivity weight $B(1)/\varepsilon$ which explodes when $\varepsilon \downarrow 0$, and it is not immediately clear by direct inspection of the functional if it is nicely behaving when taking the limit. However, since all the three approaches above lead to the same derivative $F'_\varepsilon(x)$, we are ensured that the limit also in this case is equal to $F'(x)$. However, from a practical perspective the weight will have a very high variance compared to the two first approaches, and therefore it is not useful in numerical simulations. This illustrates that the approach of Davis and Johannson[7] is not necessarily leading to sensitivity weights which are "good". In particular we should notice that in the case of pure-jump Lévy processes we will not have any $W(t)$-term, and we are somehow "forced" to use the density method with respect to $B(t)$, the approximating Brownian motion. This may lead to problems when understanding the limit since we will not have any comparison.

Let us go back to the general case, where we first suppose that $b > 0$ in (2) and apply the density method on the combination of $W(t)$ and $B(t)$. To distinguish between the different sensitivity weights, we introduce some notation. Let the derivative of $F_\varepsilon(x)$ with respect to x resulting from applying the density method on $W(\mathrm{t})$, which is the Brownian motion in the Lévy-Kintchine representation of $L(t)$, be denoted by $F'_{\varepsilon,W}(x)$. Further, we use the notation $F'_{\varepsilon,B}(x)$ and $F'_{\varepsilon,B,W}(x)$ for the derivative when we use the density method with respect to the small-jump approximating process B or $bW(t) + \sigma(\varepsilon)B(t)$, respectively. Note that even though we may have the density of $L(t)$, it may be very hard to find the density of $L_\varepsilon(t)$, and thus to apply the density method on the approximating process directly.
We denote by C_b^k the space of k-times continuously differentiable functions with all derivatives bounded, C_b^0 will be denoted by C_b, the space of bounded and continuous functions. It is simple to derive the following result:

Proposition 3.2. *Suppose $f \in C_b$. For every $\varepsilon > 0$, we have that*

$$F'_{\varepsilon,B}(x) = \mathbb{E}\left[f(x + L_\varepsilon(1))\frac{B(1)}{\sigma(\varepsilon)}\right]$$
$$F'_{\varepsilon,B,W}(x) = \mathbb{E}\left[f(x + L_\varepsilon(1))\frac{bW(1) + \sigma(\varepsilon)B(1)}{b^2 + \sigma^2(\varepsilon)}\right].$$

If $b > 0$, we have in addition that

$$F'_{\varepsilon,W}(x) = \mathbb{E}\left[f(x + L_\varepsilon(1))\frac{W(1)}{b}\right].$$

Proof. Using the conditional density method applied to $B(t)$, we get

$$F'_\varepsilon(x) = \frac{\partial}{\partial x}\int_{\mathbb{R}} \mathbb{E}[f(u + a + bW(1) + Z(1) + \widetilde{Z}_\varepsilon(1))]p_{\sigma(\varepsilon)B(1)}(u - x)dx.$$

Here we can dominate the density $P_{\sigma(\varepsilon)B(1)}(u - x)$ uniformly in x by an integrable function which is a sufficient condition to take the derivative inside the integral if f is bounded. Applying the conditional density method to $bW(t) + \sigma(\varepsilon)B(t)$ and $W(t)$, respectively, and using the same arguments above to take the derivative inside the integral, we get the result. □

Note that $F'_{\varepsilon,B}(x) = F'_{\varepsilon,B,W}(x) = F'_{\varepsilon,W}(x)$ for all $\varepsilon > 0$. Moreover, we have the following robustness result when $b > 0$, that is, when the Lévy process $L(t)$ has a continuous martingale term.

Proposition 3.3. *Suppose that the diffusion coefficient $b > 0$ and that $f \in C_b$. Then we have*

$$\lim_{\varepsilon \downarrow 0} F'_{\varepsilon,W}(x) = \lim_{\varepsilon \downarrow 0} F'_{\varepsilon,B,W}(x) = \lim_{\varepsilon \downarrow 0} F'_{\varepsilon,B}(x) = \mathbb{E}\left[f(x + L(1))\frac{W(1)}{b}\right] = F'(x)\,.$$

Proof. This hinges on the fact that,

$$F'(x) = \mathbb{E}\left[f(x + L(1))\frac{W(1)}{b}\right]\,.$$

Now, by the assumption on $f(x + L_\varepsilon(1))W(1)$ and the dominated convergence theorem, we find that

$$\lim_{\varepsilon \downarrow 0} F'_{\varepsilon,W}(x) = F'(x)\,.$$

Furthermore, since $F'_{\varepsilon,B}(x) = F'_{\varepsilon,B,W}(x) = F'_{\varepsilon,W}(x)$, we have that the limit of $F'_{\varepsilon,B,W}(x)$ and $F'_{\varepsilon,B}(x)$ also exist and are equal to $F'(x)$. This proves the result. □

Remark that although we cannot bound $B(1)/\sigma(\varepsilon)$ by some integrable random variable, we still obtain the convergence. This depends on the fact that the derivative $F'_{\varepsilon,B}(x)$ is equal to $F'_{\varepsilon,W}(x)$ when $b > 0$. When $b = 0$, we can not use this argument anymore, however, we have the following simple result when f is smooth.

Proposition 3.4. *Suppose $f \in C_b^1$ and that there exists a random variable $U \in L^1(\mathbb{P})$ such that $|f'(x + L_\varepsilon(1)| \leq U$ uniformly in x and ε. Then*

$$\lim_{\varepsilon \to 0} F'_{\varepsilon,B}(x) = F'(x) = \mathbb{E}\left[f'(x + L(1))\right]\,.$$

Proof. First, observe that $|f'(x + L(1))| \leq U$ uniformly in x by choosing $\varepsilon = 0$ in the assumption. Hence, by Thm. 2.27 in Folland,[10] $F(x)$ is differentiable, and we can move the differentiation inside the expectation operator to obtain

$$F'(x) = \mathbb{E}\left[f'(x + L(1))\right]\,.$$

This proves the second equality. Next, by the same argument, we have that

$$F'_\varepsilon(x) = \frac{d}{dx}\mathbb{E}\left[f(x+L_\varepsilon(1))\right] = \mathbb{E}\left[f'(x+L_\varepsilon(1))\right] .$$

From the conditional density method, we know that $F'_{\varepsilon,B}(x) = F'_\varepsilon(x)$. By dominated convergence, it holds that

$$\lim_{\varepsilon\to 0} F'_\varepsilon(x) = F'(x)$$

and the proof is complete. □

Note that the result holds for all $b \geq 0$, and we could have used it to prove the limits for $b > 0$ as well in the smooth case of f.

In many applications, like for instance in finance, the assumption that f should be continuous and bounded is too restrictive. For example, a call option will lead to an unbounded function, whereas a digital option gives a discontinuous f. Hence, it is natural to look for extensions of the above results to classes of functions where the conditions on f are weakened. One natural approach is to look at classes of functions f which can be approximated by functions in C_b. Another path, which we shall take here, is to apply Fourier methods.

Let now $f \in L^1(\mathbb{R})$. The Fourier transform of f is defined by

$$\widehat{f}(u) = \int_{\mathbb{R}} f(y)\mathrm{e}^{\mathrm{i}uy}\,dy\,. \tag{12}$$

Suppose in addition that $\widehat{f} \in L^1(\mathbb{R})$. Then the inverse Fourier transform is well-defined, and we have

$$f(y) = \frac{1}{2\pi}\int_{\mathbb{R}} \mathrm{e}^{-\mathrm{i}uy}\widehat{f}(u)\,du\,. \tag{13}$$

With these two definitions at hand, we can do the following calculation taken from Carr and Madan.[5] Assume for every x that $f(x+\cdot)$ is integrable with respect to the distribution $p_{L(1)}(dy)$ of $L(1)$. Then

$$\mathbb{E}[f(x+L(1))] = \int_{\mathbb{R}} f(x+y)p_{L(1)}(dy)\,.$$

Invoking the inverse Fourier transformed representation of f in (13), and applying Fubini-Tonelli to commute the integration, we find

$$\begin{aligned}\mathbb{E}[f(x+L(1))] &= \int_{\mathbb{R}}\Big\{\frac{1}{2\pi}\int_{\mathbb{R}} e^{-i(x+y)u}\widehat{f}(u)du\Big\}p_{L(1)}(dy)\\ &= \frac{1}{2\pi}\int_{\mathbb{R}} e^{-iux}\Big\{\int_{\mathbb{R}} e^{-iuy}p_{L(1)}(dy)\Big\}\widehat{f}(u)du\,.\end{aligned}$$

Thus, it follows that

$$\mathbb{E}[f(x+L(1))] = \frac{1}{2\pi}\int_{\mathbb{R}} e^{-iux}\varphi_{L(1)}(u)\widehat{f}(u)du\,, \tag{14}$$

where $\varphi_{L(1)}$ is the characteristic function of $L(1)$ defined from the Lévy-Kintchine formula as

$$\varphi_{L(1)}(u) = \exp\left(iau - \frac{1}{2}b^2u^2 + \int_{\mathbb{R}_0} \mathrm{e}^{\mathrm{i}uz} - 1 - \mathrm{i}uz1_{|z|<1}\, Q(dz)\right). \tag{15}$$

We have the following Lemma for the delta.

Lemma 3.1. *Under the condition $u\widehat{f}(u) \in L^1(\mathbb{R})$ we have*

$$F'(x) = \frac{\partial}{\partial x}\mathbb{E}[f(x+L(1))] = \frac{1}{2\pi}\int_{\mathbb{R}} -iue^{-iux}\varphi_{L(1)}(u)\widehat{f}(u)\, du\,.$$

Proof. We differentiate the integrand in (14) and dominate it uniformly in x:

$$|\frac{\partial}{\partial x}\mathrm{e}^{-iux}\varphi_{L(1)}(u)\widehat{f}(u)| = |-iu\frac{\mathrm{e}^{-iux}}{2\pi}\varphi_{L(1)}(u)\widehat{f}(u)| \le |u\widehat{f}(u)|\,.$$

The result follows by appealing to Prop. 2.27 in Folland.[10] □

Note that the condition $u\widehat{f}(u) \in L^1(\mathbb{R})$ is related to the derivative of f, since as long as f is differentiable we have $\widehat{f'}(u) = u\widehat{f}(u)$ whenever $f' \in L^1(\mathbb{R})$.

We finally reach the desired stability result for non-smooth f's.

Proposition 3.5. *Suppose that*

$$u\widehat{f}(u) \in L^1(\mathbb{R}).$$

Then we have

$$\lim_{\varepsilon\to 0}\frac{\partial}{\partial x}\mathbb{E}[f(x+L_\varepsilon(1))] = \frac{\partial}{\partial x}\mathbb{E}[f(x+L(1))]\,.$$

Proof. From Lemma 3.1 applied to $L_\varepsilon(1)$ we have

$$\frac{\partial}{\partial x}\mathbb{E}[f(x+L_\varepsilon(1))] = \frac{1}{2\pi}\int_{\mathbb{R}} -iue^{-iux}\varphi_{L_\varepsilon(1)}(u)\widehat{f}(u)\, du\,.$$

But,

$$|-iue^{-iux}\varphi_{L_\varepsilon(1)}(u)\widehat{f}(u)| \le |u\widehat{f}(u)|$$

which, from the assumption, permits us to take the limit inside the integral and the result follows by Prop. 2.24 in Folland.[10] □

Observe that in the Proposition above we handle $b \geq 0$, and there is no need to differentiate between the cases $b = 0$ and $b > 0$. There is no requirement of continuity of f in the above arguments. However, the integrability restriction excludes unbounded functions f, like for instance those coming from option pricing. However, we can easily deal with such by introducing a damped function f in the following manner. Define for $\alpha > 0$ the function

$$g_\alpha(y) = \mathrm{e}^{-\alpha y} f(y)\,. \tag{16}$$

Assuming that $g_\alpha \in L^1(\mathbb{R})$ and $\widehat{g}_\alpha \in L^1(\mathbb{R})$ for some $\alpha > 0$, we can apply the above results for g_α. To translate to f, observe that

$$f(y) = \frac{1}{2\pi}\int_{\mathbb{R}} \mathrm{e}^{(\alpha - \mathrm{i}u)y}\widehat{g}_\alpha(u)\,du\,,$$

and

$$\widehat{g}_\alpha(u) = \widehat{f}(u + \mathrm{i}\alpha)\,.$$

Hence, Prop. 3.5 holds for any f such that there exists $\alpha > 0$ for which we have the following assumptions

$$(\alpha - \mathrm{i}u)\widehat{f}(u + \mathrm{i}\alpha) \in L^1(\mathbb{R}) \quad \text{and} \quad e^{\alpha y} p_{L(1)}(dy) \in L^1(\mathbb{R}).$$

As illustration we consider two examples. First, let f be the payoff from a call option written on an asset with price defined as $S(t) = S(0)\exp(L(t))$ $(S(0) > 0)$. Then, with $x = \ln S(0)$, we have

$$f(y) = \max(\mathrm{e}^y - K, 0)$$

for $K > 0$ being the strike price. For $\alpha > 1$, we have that $g_\alpha \in L^1(\mathbb{R})$. Moreover,

$$\widehat{g}_\alpha(u) = \frac{K\mathrm{e}^{(\mathrm{i}u-\alpha)\ln K}}{(\mathrm{i}u - \alpha)(\mathrm{i}u - \alpha + 1)}\,,$$

which is in $L^1(\mathbb{R})$. By a direct calculation, we find that

$$(\alpha - \mathrm{i}u)\widehat{f}(u + \mathrm{i}\alpha) = \frac{K^{1+\mathrm{i}u-\alpha}}{1 + \mathrm{i}u - \alpha},$$

which belongs to $L^1(\mathbb{R})$. Hence, Prop. 3.5 ensures that the approximation $L_\varepsilon(1)$ gives a delta which converges to the delta resulting from the model with $L(1)$.

We consider now a digital option written on an asset with price defined as $S(t) = S(0)\exp(L(t))$ $(S(0) > 0)$. Then, with $x = \ln S(0)$, we have

$$f(y) = 1_{\{e^y > B\}}, \quad B \in \mathbb{R}_+ .$$

For $\alpha > 0$, we have that $g_\alpha \in L^1(\mathbb{R})$. Moreover,

$$\widehat{g}_\alpha(u) = \frac{-B^{iu-\alpha}}{iu-\alpha},$$

which is in $L^1(\mathbb{R})$. By a direct calculation, we find that

$$(\alpha - iu)\widehat{f}(u + i\alpha) = B^{iu-\alpha},$$

which belongs to $L^1(\mathbb{R})$.

3.3. *Robustness to smoothing of a Lévy process*

In the above analysis we have focused entirely on the approximation of the small jumps in a Lévy process. However, we can apply our analysis also in a slightly different situation. Suppose that we are dealing with is a Lévy process for which the density of $L(t)$ (and its log-derivative) may be hard to compute, or may not even be existent analytically. In this case one may approximate the derivative of $F(x)$ by considering the following Lévy process:

$$\widehat{L}_\varepsilon(t) \triangleq L(t) + \widehat{\sigma}(\varepsilon)B(t)\,, \tag{17}$$

where $B(t)$ is a Brownian motion independent of $L(t)$ and

$$\lim_{\varepsilon \downarrow 0} \widehat{\sigma}(\varepsilon) = 0\,.$$

We call $\widehat{L}_\varepsilon(t)$ a *smoothing* of $L(t)$, since we add an independent Brownian motion which has a smooth density, and thus $\widehat{L}_\varepsilon(t)$ will possess a smooth density as well.

Using the same proof as in Prop. 3.1, we have that $\widehat{L}_\varepsilon(t)$ converges in $L^1(\mathbb{P})$ to $L(t)$. Furthermore, since obviously $\widehat{\sigma}(\varepsilon)B(1)$ converges *a.s.* to zero, $\widehat{L}_\varepsilon(1)$ converges *a.s.* to $L(t)$. We have

$$\widehat{F}'_\varepsilon(x) = \mathbb{E}\left[f(x + \widehat{L}_\varepsilon(1)) \frac{B(1)}{\widehat{\sigma}(\varepsilon)} \right] . \tag{18}$$

Tracing through the arguments in the preceeding subsection and assuming the right conditions on f, we find that

$$\lim_{\varepsilon \to 0} \widehat{F}'_\varepsilon(x) = F'(x)\,. \tag{19}$$

This provides us with another stability result. The derivative of $F(x)$ is continuous with respect to perturbation of $L(t)$ by $\widehat{\sigma}(\varepsilon)B(t)$. As a curiosity, we can do the following: By independence of $L(1)$ and $B(1)$, we have

$$\begin{aligned}\mathbb{E}\left[f(x+\widehat{L}_\varepsilon(1))\frac{B(1)}{\widehat{\sigma}(\varepsilon)}\right] &= \mathbb{E}\left[\left(f(x+\widehat{L}_\varepsilon(1)) - f(x+L(1))\right)\frac{B(1)}{\widehat{\sigma}(\varepsilon)}\right] \\ &\quad + \mathbb{E}\left[f(x+L(1))\frac{B(1)}{\widehat{\sigma}(\varepsilon)}\right] \\ &= \mathbb{E}\left[B^2(1)\,\frac{f(x+\widehat{\sigma}(\varepsilon)B(1)+L(1)) - f(x+L(1))}{\widehat{\sigma}(\varepsilon)B(1)}\right].\end{aligned}$$

Notice, that the fraction on the right is close to a Malliavin derivative, since we in fact are looking at a derivative of f along B. Loosely speaking, when taking the limit we are looking at a derivative of $f(x + L(1))$ in the direction of $B(1)$, which resembles the idea of Malliavin differentiation. Furthermore, it is to be expected that this limit will be independent of $B(1)$, which has variance equal to 1. Informally, we have therefore given a link between the Malliavin derivative based on Brownian motion and the delta for Lévy processes. Hence, this motivates that the approach by Davis and Johansson[7] may be extended to more general Lévy processes than merely Brownian motion and Poisson processes as is the case in their paper. The formalization of this procedure is left to a future study.

Let us consider an example where the smoothing of $L(t)$ may be an attractive procedure. The so-called CGMY distribution was suggested in Carr *et al.*[4] to model asset price returns. It does not have any explicit density function, but is defined through its cumulant function[b]

$$\psi_{\mathrm{CGMY}}(\theta) = C\Gamma(-Y)\left\{(M-\mathrm{i}\theta)^Y + (G-\mathrm{i}\theta)^Y - G^Y\right\}, \tag{20}$$

with $\Gamma(x)$ being the Gamma-function and constants C, G, M and Y. We suppose that C, G and M are positive, and $Y \in [0, 2)$. The CGMY distribution is infinitely divisible, and we can define a CGMY-Lévy process $L(t)$ with Lévy measure

$$Q(dz) = C|z|^{-1-Y}\exp\left(-(G1(z<0) + M1(z>0))|z|\right)dz\,. \tag{21}$$

Since we do not have explicitly the density function of $L(t)$, the density method can not be used for deriving sensitivity estimates $F'(x)$. Instead we can apply the density method on the Brownian motion after smoothing $L(t)$. We first verify the condition of Thm. 2.2 for the CGMY-Lévy process, showing that a smooth density indeed exists.

[b]The cumulant function is here the logarithm of the characteristic function

We have

$$\sigma^2(\varepsilon) = C \int_{|z|<\varepsilon} |z|^{1-Y} \exp\left(-(G1(z<0) + M1(z>0))|z|\right) dz\,,$$

and thus it is sufficient to verify the condition in Thm. 2.2 for $z > 0$. Note that $\mathrm{e}^{-Mz} \geq M\varepsilon$ for $0 \leq z \leq \varepsilon$, and it follows that

$$\int_0^\varepsilon z^{1-Y} \mathrm{e}^{-Mz}\, dz \geq \frac{1}{2-Y} \varepsilon^{2-Y} \mathrm{e}^{-M\varepsilon}\,.$$

Let $\gamma = 2 - Y$ in Thm. 2.2, and we see that

$$\liminf_{\varepsilon \downarrow 0} \frac{\sigma^2(\varepsilon)}{\varepsilon^\gamma} \geq \frac{1}{2-Y} > 0\,,$$

as long as $Y > 0$. Hence, there exists a smooth density for the CGMY-distribution when $Y \in (0,2)$ and, if we would have this available, we could calculate $F'(x)$ via its logarithmic derivative. By smoothing we can approximate $F'(x)$ by

$$\widehat{F}'_\varepsilon(x) = \mathbb{E}\left[f(x + \widehat{L}_\varepsilon(1)) \frac{B(1)}{\sigma(\varepsilon)}\right]\,.$$

The sensitivity weight will have a large variance for small ε, but it provides us with an expression that can be calculated using Monte Carlo simulations based on sampling of the CGMY-distribution and an independent normal distribution.

As an application, we consider an example from insurance. Let the loss of an insurance company be described by $L(t)$, and x being the premium charged by the company to accept this risk. The question for the insurance company is to find a level x such that the net loss $x + L(1)$ is acceptable. A simple measure could be that the insurance company can only bear losses which are above a certain threshold, K say. Given a premium x, they want to calculate the probability of falling below the threshold K, which can be expressed by $P(x + L(1) < K)$. We find

$$P(x + L(1) < K) = \mathbb{E}\left[1_{\{x+L(1)<K\}}\right]\,,$$

which therefore is an expectation functional on the form we have analyzed in this paper with $f(z) = 1_{\{z<K\}}$. Consider the derivative of this probability with respect to x, which we call the *marginal premium rate*:

$$F'(x) = \frac{d}{dx}\mathbb{E}\left[1_{\{x+L(1)<K\}}\right]\,.$$

The marginal premium rate tells us how sensitive the loss probability is with respect to the premium. Of course, if we know the density of $L(1)$, $p_{L(1)}$, and this is differentiable, the marginal premium rate is straightforwardly calculated to be

$$F'(x) = -p_{L(1)}(K - x)\,.$$

Thus, changing the premium by dx leads to a change in the loss probability of $-p_{L(1)}(K - x)\,dx$. However, if now the density of $L(1)$ is not known as is the case for the CGMY-distribution, we can not perform this simple calculation. By smoothing $L(t)$, we find the approximation

$$\widehat{F}'_\varepsilon(x) = \mathbb{E}\left[1_{\{x+\widehat{L}_\varepsilon(1)<K\}}\frac{B(1)}{\widehat{\sigma}(\varepsilon)}\right].$$

Computations using conditional expectation lead to

$$\begin{aligned}\widehat{F}'_\varepsilon(x) &= \mathbb{E}\left[\mathbb{E}\left[1_{\{x+\widehat{\sigma}(\varepsilon)B(1)+L(1)<K\}}\frac{B(1)}{\widehat{\sigma}(\varepsilon)} \mid L(1)\right]\right] \\ &= -\mathbb{E}\left[p_{\widehat{\sigma}(\varepsilon)B(1)}(K - L(1) - x)\right],\end{aligned}$$

with $p_{\widehat{\sigma}(\varepsilon)B(1)}$ being the density function of $\widehat{\sigma}(\varepsilon)B(1)$. Thus, also the approximation can be expressed as a density evaluated in $K - x$, however, in this case we need to take the expectation over $L(1)$. Furthermore, the density is singular when going to the limit.

Using the theory of distribution functions, we give a direct argument for the convergence of $\widehat{F}'_\varepsilon(x)$ to $F'(x)$. In fact by integration-by-parts, we have

$$\widehat{F}'_{\varepsilon,B}(x) = -\left(p_{\widehat{\sigma}(\varepsilon)B(1)}(K - x - \cdot), p_{L(1)}\right)_2 = -\left(p_{\widehat{\sigma}(\varepsilon)B(1)}, p_{L(1)}(K - x - \cdot)\right)_2$$

where $(\cdot,\cdot)_2$ is the inner product in $L^2(\mathbb{R})$, the space of square-integrable functions on $\mathbb{R}$. Since $p_{\widehat{\sigma}(\varepsilon)B(1)} \to \delta_0$ when $\widehat{\sigma}(\varepsilon) \to 0$, we find

$$\lim_{\varepsilon\to 0}\widehat{F}'_{\varepsilon,B}(x) = -\left(\delta_0, p_{L(1)}(K - x - \cdot)\right)_2 = p_{L(1)}(K - x) = F'(x)\,.$$

This procedure may be carried through rigorously by using Schwartz distribution theory.

3.4. *Some numerical issues*

In the last part of this Section we turn our attention to some numerical issues concerning the use of the conditional density method for the above approximations.

For numerical purposes, it is of interest to know the rate of convergence of $F'_\varepsilon(x)$ to $F'(x)$. We have the following convergence speed for $F'_{\varepsilon,B,W}(x)$:

Proposition 3.6. *Suppose $b > 0$, $L(t)$ having finite variance and f being a Lipschitz continuous function. Then there exists a constant C depending on x, b, the Lipschitz constant of f, and the variance of $L(1)$ such that*

$$|F'_{\varepsilon,B,W}(x) - F'(x)| \leq C\sigma(\varepsilon)\,.$$

Proof. From the triangle and Cauchy-Schwarz inequalities, we have

$$\begin{aligned}
&|F'_{\varepsilon,B,W}(x) - F'(x)| \\
&\quad \leq \mathbb{E}\left[|f(x+L_\varepsilon(1)) - f(x+L(1))|\frac{bW(1)+\sigma(\varepsilon)B(1)}{b^2+\sigma^2(\varepsilon)}\right] \\
&\qquad + \mathbb{E}\left[|f(x+L(1))||\frac{bW(1)+\sigma(\varepsilon)B(1)}{b^2+\sigma^2(\varepsilon)} - \frac{W(1)}{b}|\right] \\
&\quad \leq \mathbb{E}\left[|f(x+L_\varepsilon(1)) - f(x+L(1))|^2\right]^{1/2} \frac{\mathbb{E}\left[(bW(1)+\sigma(\varepsilon)B(1))^2\right]^{1/2}}{b^2+\sigma^2(\varepsilon)} \\
&\qquad + \mathbb{E}\left[f^2(x+L(1))\right]^{1/2} \mathbb{E}\left[|\frac{bW(1)+\sigma(\varepsilon)B(1)}{b^2+\sigma^2(\varepsilon)} - \frac{W(1)}{b}|^2\right]^{1/2}.
\end{aligned}$$

Letting K being the Lipschitz constant (which we assume being equal to the growth constant of f for convenience), we get

$$\begin{aligned}
|F'_{\varepsilon,B,W}(x) - F'(x)| \leq{}& \frac{K}{\sqrt{b^2+\sigma^2(\varepsilon)}}\mathbb{E}\left[|L_\varepsilon(1) - L(1)|^2\right]^{1/2} + K\mathbb{E}\left[(1+x+L(1))^2\right]^{1/2} \\
&\times \mathbb{E}\left[|\left(\tfrac{b}{b^2+\sigma^2(\varepsilon)} - \tfrac{1}{b}\right)W(1) + \tfrac{\sigma(\varepsilon)}{b^2+\sigma^2(\varepsilon)}B(1)|^2\right]^{1/2}.
\end{aligned}$$

Since $W(1)$ and $B(1)$ are independent, we find the last expectation to be (after taking the square-root)

$$\sigma(\varepsilon)/b\sqrt{(b^2+\sigma^2(\varepsilon))}\,.$$

Moreover,

$$L_\varepsilon(1) - L(1) = \sigma(\varepsilon)B(T) + \widetilde{Z}_\varepsilon(1) - \lim_{\varepsilon\downarrow 0}\widetilde{Z}_\varepsilon(1)\,.$$

Note that the difference between $\widetilde{Z}_\varepsilon(1)$ and $\lim_{\varepsilon\downarrow 0}\widetilde{Z}_\varepsilon(1)$ is the jumps between 0 and ε. Due to independence of the jumps and the Brownian motion B, we get

$$\mathbb{E}\left[|L_\varepsilon(1) - L(1)|^2\right] = 2\sigma^2(\varepsilon)\,.$$

Hence, the result follows. □

We note that with minor modifications of the above proof we can show that

$$|F'_{\varepsilon,B}(x) - F'(x)| \leq C\sigma(\varepsilon)\,,$$

where C is a positive constant (not necessarily equal to the constant in the Proposition above). To show this result, we can simply let $b = 0$ in the proof and modify accordingly. Finally, it holds true for $\widehat{F}'_\varepsilon(x)$ as well by similar arguments.

In practice, one uses Monte Carlo methods in order to calculate $F'_\varepsilon(x)$. We consider the case $F'_{\varepsilon,B}(x)$, and recall that the estimated value of this based on N Monte Carlo simulations is

$$F'_{\varepsilon,B}(x) \approx \sum_{n=1}^{N} f\left(x + l_{\varepsilon,n}\right) \frac{b_n}{\sigma(\varepsilon)}$$

where b_n and $l_{\varepsilon,n}$ are independent random draws of $B(1)$ and $L_\varepsilon(1)$, respectively. Note that in order to draw from $L_\varepsilon(1)$, we use the draw from $B(1)$. The Monte Carlo error (or rather the standard deviation of the error) is given by

$$\operatorname{std}\left(f(x + L_\varepsilon(1))B(1)\right)/(\sqrt{N}\sigma(\varepsilon))\,.$$

Assume now for technical simplicity that f is bounded. Then, from dominated convergence and independence of L and B, we find

$$\begin{aligned}\lim_{\varepsilon\to 0} \operatorname{Var}\Big[f(x + L_\varepsilon(1))B(1)\Big] &= \lim_{\varepsilon\to 0}\Big\{\mathbb{E}\Big[f^2(x + L_\varepsilon(1))B^2(1)\Big] \\ &\quad - \mathbb{E}\Big[f(x + L_\varepsilon(1))B(1)\Big]^2\Big\} \\ &= \mathbb{E}\Big[f^2(x + L(1))\Big].\end{aligned}$$

Hence,

$$\lim_{\varepsilon\to 0} \operatorname{Var}\left(f(x + L_\varepsilon(1))\frac{B(1)}{\sigma(\varepsilon)}\right) = \infty.$$

From this we can conclude the following: If we decide to use the conditional density method on pure-jump Lévy processes after first doing an approximation, the expression to simulate will have a large variance for small ε. Indeed, when ε tends to zero the variance explodes. This means that for close approximations of $x + L(1)$ we will have an expression to simulate which has a very high variance, and therefore we need a very high number of samples to get a confident estimate of the delta. In conclusion, the method may become very inefficient and unstable, and variance-reducing techniques are called for in order to get reliable estimates.

4. Numerical examples

We consider some examples to illustrate the conditional density method and our findings on approximations.

Let us assume that $L(t)$ is an NIG-Lévy process, that is, a Lévy process with NIG-distributed increments. Supposing $L(1)$ being NIG distributed with parameters α, β, δ and μ, the density is (see Barndorff-Nielsen[2])

$$p_{\mathrm{NIG}}(x;\alpha,\beta,\delta,\mu) = \frac{\alpha\delta}{\pi} e^{\delta\sqrt{\alpha^2-\beta^2}+\beta(x-\mu)} \frac{K_1\left(\alpha\sqrt{\delta^2+(x-\mu)^2}\right)}{\sqrt{\delta^2+(x-\mu)^2}}. \tag{22}$$

Here, K_λ is the modified Bessel function of the second order with parameter λ, which can be represented by the integral

$$K_\nu(z) = \frac{\sqrt{\pi} z^\nu}{2^\nu \Gamma(\nu+\frac{1}{2})} \int_1^\infty e^{-zt}(t^2-1)^{\nu-\frac{1}{2}}\,dt\,,$$

for $\nu > -\frac{1}{2}$ and $z > 0$. We apply the density method to find a sensitivity weight π: A direct differentiation gives

$$-\partial \ln P_{\mathrm{NIG}}(x) = -\beta + \tfrac{x-\mu}{\delta^2+(x-\mu)^2} \times \left\{ 1 - \frac{\alpha\sqrt{\delta^2+(x-\mu)^2} K_1'\left(\alpha\sqrt{\delta^2+(x-\mu)^2}\right)}{K_1\left(\alpha\sqrt{\delta^2+(x-\mu)^2}\right)} \right\}.$$

We can now use the recursive relation for the derivative of the Bessel function K_λ, saying (see e.g. Rydberg[13])

$$K_1'(x) = -\frac{1}{2}K_0(x) - \frac{1}{2}K_2(x)\,.$$

Using the recursion $K_2(z) = K_0(z) + (2/z)K_1(z)$ we reach

$$K_1'(z) = -\frac{1}{z}K_1(z) - K_0(z)\,.$$

Inserting this into the expression of $-\partial \ln p_{\mathrm{NIG}}$ yields,

$$\begin{aligned}\pi &= -\partial \ln p_{\mathrm{NIG}}(L(1)) \\ &= -\beta + \frac{L(1)-\mu}{\delta^2+(L(1)-\mu)^2} \qquad (23)\\ &\times \left\{ 2 + \alpha\sqrt{\delta^2+(L(1)-\mu)^2}\frac{K_0(\alpha\sqrt{\delta^2+(L(1)-\mu)^2})}{K_1(\alpha\sqrt{\delta^2+(L(1)-\mu)^2})} \right\}. \qquad (24)\end{aligned}$$

Since this is a function of $L(1)$, it will be a variance optimal weight.

In applying the Monte Carlo simulation technique, it may be rather cumbersome to calculate the two modified Bessel functions K_1 and K_2 in order to calculate an outcome of the sensitivity weight π. In fact, for each draw we must perform such a calculation, which makes the method very inefficient due to the heavy computational burden involved in calculating Bessel functions. An alternative will then be to use an approximation, like for instance considering the smoothed random variable $\widehat{L}_\varepsilon(1)$ defined in (17). Using the conditional density argument, we find that the delta can be calculated by the expectation operator

$$\widehat{F}'_\varepsilon(x) = \mathbb{E}\left[f(x + L(1) + \varepsilon B(1))\frac{B(1)}{\varepsilon}\right].$$

Hence, rather than doing numerical calculation of Bessel functions, we simulate from a normal distribution. From the analysis in this paper, letting $\varepsilon \to 0$ brings us back to the derivative we are interested in. Hence, for small ε's, $\widehat{F}'_\varepsilon(x)$ should be reasonably close to $F'(x)$. We have tested this numerically in the following examples.

Let $\alpha = 50$, $\beta = \mu = 0$ and $\delta = 0.015$. These figures are not unreasonable estimates for the logreturns of a stock price on a daily scale, see Rydberg.[13] Further, we consider a function f being the payoff from a call option with strike $K = 100$, that is,

$$f(x) = \max(0, \exp(x) - 100).$$

We implemented to density method in Matlab by sampling a NIG-distribution using the technique in Rydberg[13] and calculating the Bessel functions $K_0(z)$ and $K_1(z)$ using the built-in Matlab function `besselk`. The approximation $\widehat{F}'_\varepsilon(x)$ was calculated by drawing samples from a standard normal distribution.

In Figure 1 we show the resulting derivatives for $x = \ln(S_0)$ with $S_0 = 100$ and $\varepsilon = 0.01$. Along the horizontal axis we have the number of samples (in 10^5) used in the estimation of the expectation operator, and the two expressions are calculated using common random numbers. The density method is depicted with a broken line, and we see that it has slightly less variance than the approximated derivative. But looking at the scale on the vertical axis, the approximation is pretty good, although it seems that it is slightly overestimating the true derivative. By reducing ε, we observe a convergence towards $F'(x)$, however, at the expense of a higher variance in the estimation of the expectation in $\widehat{F}'_\epsilon(x)$. This is shown in Figure 2, where we plot the estimates as function of samples for three different values of ε,

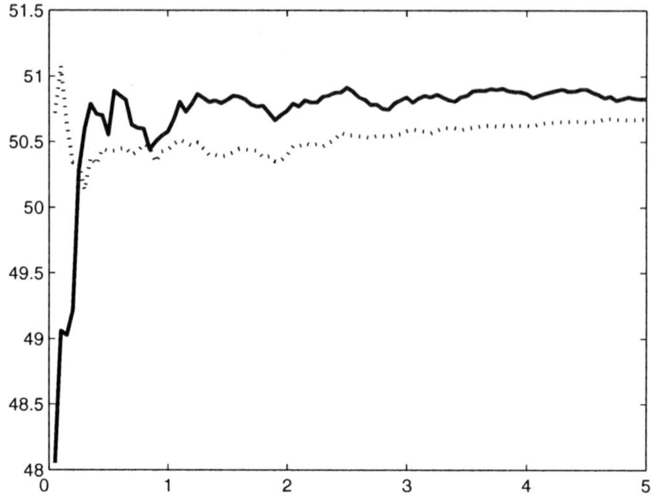

Fig. 1. The estimated derivative based on the density method (broken line) versus an approximation using an added Brownian motion (solid line) as a function of the number of samples (in 10^5).

$\varepsilon = 0.01, 0.005$ and $\varepsilon = 0.001$. The smaller ϵ, the higher variance, which leads to a higher number of samples for ensuring accuracy of the estimate.

We note that our numerical example covers the delta of an at-the-money call option on a stock, where the delta is calculated one time step (one day, say) prior to the exercise date of the option. We get the delta by dividing the derivatives $F'(x)$ and $\widehat{F}'_\varepsilon(x)$ by $S_0 = 100$, resulting from an application of the chain rule.

To test our method on discontinuous functions f, we considered the above set-up for a digital option, that is, a payoff function $f(x) = 1(\mathrm{e}^x > K)$ for some positive threshold K. The simulations showed that the derivative $\widehat{F}'_\varepsilon(x)$ had a significantly higher variance when estimated by Monte Carlo simulations. In fact, one needed to choose number of samples several scales above what was required for the call option in order to get reasonable estimates for the approximated derivative. Hence, from a numerical perspective, discontinuous functions f seem to behave badly under approximations by Brownian motions when one applies it to pure-jump Lévy processes. Variance reducing techniques like *quasi*-Monte Carlo simulations may be

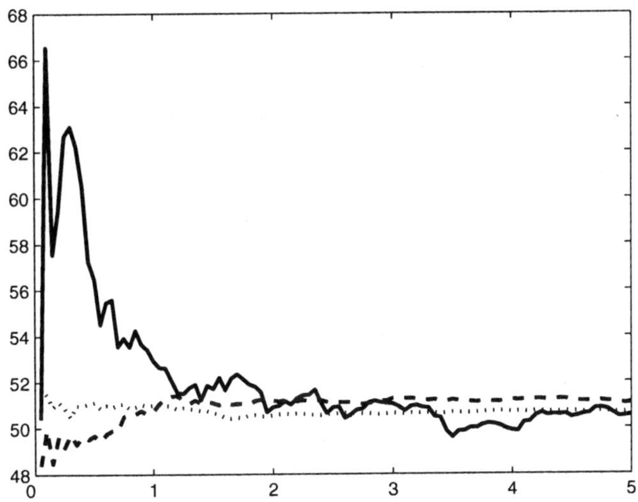

Fig. 2. The estimated derivative based on an approximation using an added Brownian motion as a function of the number of samples (in 10^5). $\varepsilon = 0.01$ in dotted line, $\varepsilon = 0.005$ in broken line and $\varepsilon = 0.001$ in solid line.

fruitful and speed up the convergence in such situations.

As a final note in this numerical subsection, let us briefly discuss the issues concerning approximating the small jumps of a Lévy process by a Brownian motion. Following the idea in this paper, the small jumps are approximated by $\sigma(\varepsilon)B(t)$ for a suitable scaling $\sigma(\varepsilon)$. The sensitivity weight is of the form

$$\pi = \frac{B(1)}{\sigma(\varepsilon)},$$

if we have no continuous martingale part in the Lévy process and decide to use the density method with respect to the Brownian motion B. In order to simulate $F'_{\varepsilon,B}(x)$, we must sample from $B(1)$ and $L_\varepsilon(1)$. The latter is equivalent to sample from a compound Poisson process since we have only jumps of size bigger than ε. Indeed, we must sample from a compound Poisson process with jump size distribution given by

$$\mathbf{1}_{|z|\geq\varepsilon}\,Q(dz)/c$$

where the normalizing constant c is defined as

$$c = Q(|z| \geq \varepsilon)\,.$$

The jump intensity will be c. This is in principle simple to simulate as long as one has a routine to sample for the truncated Lévy measure *and* knows the constant c. However, using for instance Markov Chain Monte Carlo methods, one can sample from the jump distribution without knowing the constant c.

5. Conclusion

In this study we have considered the problem of robustness of the sensitivity parameter delta to model choice. Our models are selected within the Lévy family, but they differ according to how the presence of small jumps is taken into account.

First, following the study in Asmussen and Rosinski,[1] we have considered models with small jumps, see $L(t)$ in (2) and their approximations given by models with of type $L_\varepsilon(t)$ (8), where a continuous martingale part with controlled standard deviation is replacing the small jumps. In this case both models have the same total variance. In this case $L_\varepsilon(t) \longrightarrow L(t)$, for $\varepsilon \downarrow 0$. Secondly, we have considered a smoothing $\widehat{L}_\varepsilon(t)$ of the Lévy process $L(t)$. Aslo in this case we have $\widetilde{L}_\varepsilon(t) \longrightarrow L(t)$, for $\varepsilon \downarrow 0$, but there is no control on the variances between the two models. The two situations can be usefully applied in different contexts.

In both cases we have addressed the question of the robustness of the parameter delta

$$\begin{aligned} F'(x) &= \frac{d}{dx} E[f(x + L(t)] \\ F'_\varepsilon(x) &= \frac{d}{dx} E[f(x + L_\varepsilon(t)] \\ \widehat{F}'_\varepsilon(x) &= \frac{d}{dx} E[f(x + \widehat{L}_\varepsilon(t)]. \end{aligned}$$

We have applied different methods of computation: the classical density method and the newly introduced conditional density method. The different computational techniques for the delta lead to different weights. However the values of the parameter is the same. Qualitatively, this last one is an application of computations similar to the ones in the density method, but applied after having performed some conditioning (this inspired by the Malliavin method á la Davis and Johansson[7]). In our analysis we have considered functions f with different degrees of regularity, always keeping in mind the needs coming from applications to finance and insurance. Our examples include, for example, also the digital option.

Indeed a robustness result is proved, i.e.

$$F'_\varepsilon(x) \longrightarrow F'(x), \quad \varepsilon \downarrow 0$$
$$\widehat{F}'_\varepsilon(x) \longrightarrow F'(x), \quad \varepsilon \downarrow 0.$$

If this is reassuring when coming to applications, we also remark that we experience some curious situations important from the numerical point of view. In fact, according to the different methods applied some representations of the deltas turn out to be highly inefficient. This is evident when we consider models $L(t)$ with no original continuous martingale component (i.e. $b = 0$ in (2)) and we take the corresponding $L_\varepsilon(t)$ as approximating model. In this case the conditional density method shows an exploding variance of the random variable that must be simulated. This yielding to the need of a large number of samples to get some confident estimate of the delta.

Our future studies will direct in two directions. On one side we will consider the problem of robustness of the delta to model choice for path dependent options, i.e. for functional f. This will require techniques of Malliavin calculus. On the other side we will consider the problem of robustness of prices to model choice under change of measure. To explain, when dealing with no-arbitrage pricing an equivalent martingale measure (EMM) is required. This is a probability measure under which the discounted price processes are martingales. In incomplete market models, such as those driven by $L(t)$, the EMMs available are infinite. However, they do not correspond to the set of EMMs for $L_\varepsilon(t)$ or $\widehat{L}_\varepsilon(t)$. Our goal is to understand the robustness of pricing when different models are considered.

References

1. Asmussen, S., and Rosinski, J. *Approximations of small jump Lévy processes with a view towards simulation*, J. Appl. Prob., **38**, (2001), 482–493.
2. Barndorff-Nielsen, O. E. *Processes of normal inverse Gaussian type*, Finance Stoch., **2** no. 1 (1998), 41–68.
3. Broadie, M., and Glasserman, P. *Estimating security price derivatives using simulation*, Manag. Science, **42** (1996), 169–285.

4. Carr, P., Geman, H., Madan, D., and Yor, M. *The fine structure of asset returns: an empirical investigation*, J. Business, **75** no.2 (2002), 305–332.
5. Carr, P., and Madan D. B. *Option valuation using fast Fourier transform*, J. Comp. Finance, **2** (1998), 61–73.
6. Cont, R., and Tankov, P. *Financial Modelling with Jump Processes.* Chapman Hall (2004).
7. Davis, M. H. A., and Johansson, M. P. *Malliavin Monte Carlo Greeks for jump diffusions*, Stoch. Processes. Appl., **116** no. 1 (2006), 101–129.
8. Eberlein, E. (2001). Applications of generalized hyperbolic Lévy motion to finance. In *Lévy Processes – Theory and Applications*, Barndorff-Nielsen, O. E., Mikosch, T., and Resnick, S (eds.), Birkhäuser: Boston, (2001), 319–336.
9. Eberlein, E., and Keller, U. *Hyperbolic distributions in finance*, Bernoulli, **1** (1995), 281–299.
10. Folland, G. B. . *Real Analysis – Modern Techniques and their Applications.* John Wiley & Sons (1984).
11. Fournié, E., Lasry, J. M., Lébucheux, J., Lions, P. L., and Touzi, N., Applications of Malliavin calculus to Monte Carlo methods in finance, Finance Stoch., **3** (1999), 391–412.
12. Orey, S. *On continuity properties of infinitely divisible distribution functions*, Annals Math. Statistics, **39** (1968), 936–937.
13. Rydberg, T. H. *The normal inverse Gaussian Lévy process: simulation and approximation*, Comm. Stat. Stoch. Models, **13** (1997), 887–910.
14. Sato, K.-I., *Lévy Processes and Infinitely Divisible Distributions*, Cambridge University Press (1999).

Quantum Heat Equation with Quantum K-Gross Laplacian: Solutions and Integral Representation

Samah Horrigue* and Habib Ouerdiane

Department of Mathematics, Faculty of Sciences of Tunis, University of Tunis El-Manar, 1060 Tunis, Tunisia

** E-mail: samah.horrigue@fst.rnu.tn*

habib.ouerdiane@fst.rnu.tn

In this paper, we first introduce and study the quantum (K_1, K_2)-Gross Laplacian denoted $\Delta_{QG}(K_1, K_2)$. Then, we prove that $\Delta_{QG}(K_1, K_2)$ is a well defined and linear continuous operator acting on the space of continuous operators and has a quantum stochastic integral. Finally, we give an explicit solution of the quantum heat equation associated with $\Delta_{QG}(K_1, K_2)$. Then, under some positive conditions, we give an integral representation of this solution.

Keywords: Space of entire functions with growth condition, symbols and kernels of operators, convolution product, K-quantum Gross Laplacian, quantum heat equations, integral representation.

1. Introduction

Gross initiated in Ref. 6, the study of the Laplacian operator on an infinite dimensional space called Gross Laplacian Δ_G. Based on white noise analysis, Kuo in Ref. 12 formulated the Gross Laplacian Δ_G in terms of the Hida differentiation ∂_t and its adjoint ∂_t^*. In Ref. 3, Chung and Ji studied the generalized Laplacian $\Delta_G(K)$, called the $K-$Gross Laplacian. To establish some properties of $\Delta_G(K)$, they employ the formal integral expression given by:

$$\Delta_G(K) = \int_{\mathbb{R}^2} \tau_K(s,t)\partial_s\partial_t ds dt,$$

where K is an operator from a complex nuclear Fréchet space E into his strong topological dual E' and $\tau_K \in (E \otimes E)'$ verifying

$$\langle \tau_K, \xi \otimes \eta \rangle := \langle K\xi, \eta \rangle = \int_{\mathbb{R}^2} \tau_K(s,t)\xi(t)\eta(s) dt ds, \qquad \xi, \eta \in E.$$

It is clear that in the particular case where K is the identity, we obtain the usual Gross Laplacian studied in Refs. 12, 13.

We would like to mention the quantum extension of the Gross Laplacian Δ_{QG} (see Refs. 8, 9) which allows us to solve the quantum Cauchy problem. Such extension is based entirely on the combination of two powerful ideas: the Gross Laplacian in two infinite dimensional variables and the Schwartz-Grothendieck theorem.

In this paper, we first introduce a generalized quantum Laplacian $\Delta_{QG}(K_1, K_2)$, called the quantum (K_1, K_2)-Gross Laplacian, which is a continuous linear operator acting on the space $L(\mathcal{F}_{\theta_1}(N_1'), \mathcal{F}^*_{\theta_2}(N_2'))$ of continuous operators from the space of test functions with some growth conditions $\mathcal{F}_{\theta_1}(N_1')$ into the space of the distributions $\mathcal{F}^*_{\theta_2}(N_2')$. Then, we give a quantum stochastic integral of $\Delta_{QG}(K_1, K_2)$. Finally, we present an explicit solution of the quantum Cauchy problem for the heat equation associated with $\Delta_{QG}(K_1, K_2)$. Moreover, under some conditions, we find an integral representation of this solution.

The paper is organized as follows. In section 2, we give a brief background and known results in white noise analysis. In section 3, we define and study the quantum $(K_1, K_2)-$Gross Laplacian $\Delta_{QG}(K_1, K_2)$ given by

$$\Delta_{QG}(K_1, K_2)(\Xi) = \mathcal{T}_{(K_1,K_2)} * \Xi^K, \quad \forall \Xi \in L(\mathcal{F}_{\theta_1}(N_1'), \mathcal{F}_{\theta_2}(N_2')),$$

where Ξ^K is the kernel of the operator Ξ. In section 4, using the quantum white noise derivatives defined in Ref. 10, we present a quantum stochastic integral of $\Delta_{QG}(K_1, K_2)$:

$$\Delta_{QG}(K_1, K_2) = \int_{\mathbb{R}^2} K_1(s,t) D_s^+ D_t^+ ds dt + \int_{\mathbb{R}^2} K_2(s,t) D_s^- D_t^- ds dt,$$

where for $t \in \mathbb{R}$, D_t^+ and D_t^- are respectively the creation derivative and annihilation derivative defined by:

$$D_t^+\Xi = \partial_t\Xi - \Xi\partial_t, \quad D_t^-\Xi = \Xi\partial_t^* - \partial_t^*\Xi, \forall \Xi \in L(\mathcal{F}_{\theta_1}(N_1'), \mathcal{F}^*_{\theta_2}(N_2')).$$

In section 5, we find in Theorem 5.1, the explicit solution of the following quantum stochastic differential equation

$$(\mathcal{P})\begin{cases}\frac{d}{dt}\Xi(t)=\Delta_{QG}(K_1,K_2)\Xi(t)+\Theta(t)\\ \Xi(0)=\Xi_0,\end{cases} \tag{1}$$

where $t\mapsto\Theta(t)$ is a quantum stochastic process defined on an interval containing the origin $I\subset\mathbb{R}$ and the initial condition $\Xi_0\in L(\mathcal{F}_{\theta_1}(N_1'),\mathcal{F}^*_{\theta_2}(N_2'))$. As a direct application of Theorem 5.1 and Theorem 3.2 in Ref. 1, we give under some conditions, an integral representation of the solution.

2. Preliminaries

In this section, we introduce the framework needed later on. Let $\mathbb{N}=\{0,1,2,\ldots\}$. For $j=1,2$, the starting point is the complex nuclear Fréchet space N_j with topology given by an increasing family $\{|\cdot|_{j,p},p\in\mathbb{N}\}$ of Hilbertian norms. Then

$$N_j=\operatorname*{proj\,lim}_{p\to\infty}(N_j)_p\,,\quad\text{and } N_j'=\operatorname*{ind\,lim}_{p\to\infty}(N_j)_{-p}\,,$$

where $(N_j)_p$ is the completion of N_j with respect to the norm $|\cdot|_{j,p}$, $(N_j)_{-p}$ is the topological dual space of the Hilbert space $(N_j)_p$ and N_j' is the strong dual of N_j. For all $p\in\mathbb{N}$, we denote by $|\cdot|_{j,-p}$ the norm on $(N_j)_{-p}$ and by $\langle\cdot,\cdot\rangle_j$ the $\mathbb{C}$-bilinear form on $N_j'\times N_j$.

In the following, H denote the direct Hilbertian sum of $(N_1)_0$ and $(N_2)_0$, i. e., $H=(N_1)_0\oplus(N_2)_0$.

Let $\theta_j:[0,\infty[\to[0,\infty[$ be a Young function (i. e., θ_j is continuous, convex and increasing verifying $\theta_j(0)=0$ and $\lim_{x\to+\infty}\frac{\theta_j(x)}{x}=+\infty$). Obviously, the conjugate function θ_j^* of θ_j defined by

$$\forall x\geq 0,\quad \theta_j^*(x):=\sup_{t\geq 0}(tx-\theta_j(t)), \tag{2}$$

is also a Young function. For every $n\in\mathbb{N}$, let

$$(\theta_j)_n=\inf_{r>0}\frac{e^{\theta_j(r)}}{r^n}. \tag{3}$$

2.1. *Space of entire functions with growth condition*

In this subsection, we present a collection of definitions and results from Ref. 11, that are necessary in this paper. Let $\mathcal{H}(N_1\times N_2)$ be the space of all entire functions on $N_1 \times N_2$. For $p, q \in \mathbb{N}$, $a_1, a_2 > 0$, the norm $\|.\|_{(\theta_1,\theta_2),(p,q),(a_1,a_2)}$ on $\mathcal{H}(N_1 \times N_2)$ is defined by

$$\|f\|_{(\theta_1,\theta_2),(p,q),(a_1,a_2)}$$
$$= \sup\{|f(z_1, z_2)|e^{-\theta_1(a_1|z_1|_{1,p})-\theta_2(a_2|z_2|_{2,q})}, (z_1, z_2) \in (N_1)_p \times (N_2)_q\}.$$

Then, the space of entire functions on $N_1' \times N_2'$ with (θ_1, θ_2)-exponential growth of minimal type is naturally defined by

$$\mathcal{F}_{(\theta_1,\theta_2)}(N_1' \times N_2')$$
$$= \operatorname*{proj\,lim}_{\substack{p,q\to\infty \\ (a_1,a_1)\downarrow(0,0)}} \{f \in \mathcal{H}(N_1 \times N_2); \|f\|_{(\theta_1,\theta_2),(-p,-q),(a_1,a_2)} < \infty\}. \tag{4}$$

Similarly, the space of entire functions on $N_1 \times N_2$ with (θ_1, θ_2)-exponential growth of finite type is naturally defined by

$$\mathcal{G}_{(\theta_1,\theta_2)}(N_1 \times N_2) = \operatorname*{ind\,lim}_{\substack{p,q\to\infty \\ (a_1,a_2)\to 0}} \{f \in \mathcal{H}(N_1 \times N_2); \|f\|_{(\theta_1,\theta_2),(p,q),(a_1,a_2)} < \infty\}. \tag{5}$$

Denote by $\mathcal{F}^*_{(\theta_1,\theta_2)}(N_1' \times N_2')$ the topological strong dual of $\mathcal{F}_{(\theta_1,\theta_2)}(N_1' \times N_2')$, called the space of distributions on $N_1' \times N_2'$.

For any $n \in \mathbb{N}$ and $j = 1, 2$, we denote by $N_j^{\odot n}$ the n-fold symmetric tensor product of N_j equipped with the π-topology. Let $a_1, a_2 > 0$ and $p, q \in \mathbb{N}$ be given. For $\overrightarrow{\varphi} = (\varphi_{n,m})_{n,m\in\mathbb{N}}$ and $\overrightarrow{\Phi} = (\Phi_{n,m})_{n,m\in\mathbb{N}}$ with $\varphi_{n,m} \in (N_1)_p^{\odot n} \otimes (N_2)_q^{\odot m}$ and $\Phi_{n,m} \in (N_1)_{-p}^{\odot n} \otimes (N_2)_{-q}^{\odot m}$, we put

$$|||\overrightarrow{\varphi}|||^2_{(\theta_1,\theta_2);(p,q);(a_1,a_2)} = \sum_{n,m\in\mathbb{N}} (\theta_1)_n^{-2} (\theta_2)_m^{-2} a_1^{-n} a_2^{-m} |\varphi_{n,m}|^2_{p,q}, \tag{6}$$

$$|||\overrightarrow{\Phi}|||^2_{(\theta_1,\theta_2);(-p,-q);(a_1,a_2)} = \sum_{n,m\in\mathbb{N}} [n!m!\,(\theta_1)_n\,(\theta_2)_m]^2\, a_1^n a_2^m |\Phi_{n,m}|^2_{-p,-q}\quad , \tag{7}$$

where for $j = 1, 2$, $(\theta_j)_{n\in\mathbb{N}}$ is the sequence defined in (10).

Then, we have

$$F_{(\theta_1,\theta_2)}(N_1 \times N_2)$$
$$= \operatorname*{proj\,lim}_{\substack{p,q\to\infty \\ a_1,a_2\downarrow 0}}\{\overrightarrow{\varphi} = (\varphi_{n,m})_{n,m\in\mathbb{N}}; |||\overrightarrow{\varphi}|||_{(\theta_1,\theta_2);(p,q);(a_1,a_2)} < \infty\} \tag{8}$$

and

$$G_{(\theta_1,\theta_2)}(N_1' \times N_2')$$
$$= \operatorname*{ind\,lim}_{\substack{p,q\to\infty \\ a_1,a_2\to 0}}\{\overrightarrow{\Phi} = (\Phi_{n,m})_{n,m\in\mathbb{N}}; |||\overrightarrow{\Phi}|||_{(\theta_1,\theta_2);(-p,-q);(a_1,a_2)} < \infty\}. \tag{9}$$

By general duality theory, the topological dual of $F_{(\theta_1,\theta_2)}(N_1 \times N_2)$ is identified with the space $G_{(\theta_1,\theta_2)}(N_1' \times N_2')$ with the dual pairing:

$$\langle\langle \Phi, \varphi \rangle\rangle = \sum_{n,m\in\mathbb{N}} n!m!\langle \Phi_{n,m}, \varphi_{n,m}\rangle, \tag{10}$$

where $\overrightarrow{\varphi} = (\varphi_{n,m})_{n,m\in\mathbb{N}} \in F_{(\theta_1,\theta_2)}(N_1 \times N_2)$ and $\overrightarrow{\Phi} = (\Phi_{n,m})_{n,m\in\mathbb{N}} \in G_{(\theta_1,\theta_2)}(N_1' \times N_2')$. The Taylor series map $S.T$ (at zero) associates to any entire function the sequence of coefficients of its Taylors series expansion at the origin. Then, we have the following two topological isomorphisms:

$$\text{S.T} \ : \ \mathcal{F}_{(\theta_1,\theta_2)}(N_1' \times N_2') \ni \phi \longmapsto \overrightarrow{\phi} = (\phi_{n,m})_{n,m\in\mathbb{N}} \in F_{(\theta_1,\theta_2)}(N_1 \times N_2),$$

$$\text{S.T} \ : \ \mathcal{G}_{(\theta_1^*,\theta_2^*)}(N_1 \times N_2) \ni \Phi \longmapsto \overrightarrow{\Phi} = (\Phi_{n,m})_{n,m\in\mathbb{N}} \in G_{(\theta_1,\theta_2)}(N_1' \times N_2').$$

In the sequel, we identify every test function $\phi \in \mathcal{F}_{(\theta_1,\theta_2)}(N_1' \times N_2')$ (resp. every generalized function $\Phi \in \mathcal{G}_{(\theta_1,\theta_2)}(N_1 \times N_2)$) with its formal power series $\overrightarrow{\phi} = (\phi_{n,m})_{n,m\in\mathbb{N}} \in F_{(\theta_1,\theta_2)}(N_1 \times N_2)$ (resp. $\overrightarrow{\Phi} = (\Phi_{n,m})_{n,m\in\mathbb{N}} \in G_{(\theta_1,\theta_2)}(N_1' \times N_2')$).

Let $(\xi_1, \xi_2) \in N_1 \times N_2$. The exponential function is defined by

$$e_{(\xi_1,\xi_2)}(x_1, x_2) = \sum_{n,m\in\mathbb{N}} \langle x_1^{\otimes n} \otimes x_2^{\otimes m}, \frac{\xi_1^{\otimes n} \otimes \xi_2^{\otimes m}}{n!m!}\rangle, \ (x_1, x_2) \in N_1' \times N_2'. \tag{11}$$

So, the Laplace transform is given, for $\Phi \in \mathcal{F}^*_{(\theta_1,\theta_2)}(N_1' \times N_2')$, by

$$\mathcal{L}(\Phi)(\xi_1, \xi_2) = \langle\langle \Phi, e_{(\xi_1,\xi_2)} \rangle\rangle = \widehat{\Phi}(\xi_1, \xi_2), \quad \forall (\xi_1, \xi_2) \in N_1 \times N_2. \tag{12}$$

Then, we have the following topological isomorphism

$$\mathcal{L} : \mathcal{F}^*_{(\theta_1,\theta_2)}(N_1' \times N_2') \to \mathcal{G}_{(\theta_1^*,\theta_2^*)}(N_1 \times N_2). \tag{13}$$

2.2. *Operator theory*

We denote by $\mathrm{L}(\mathcal{F}_{\theta_1}(N_1'), \mathcal{F}^*_{\theta_2}(N_2'))$ the space of continuous linear operators from $\mathcal{F}_{\theta_1}(N_1')$ into $\mathcal{F}^*_{\theta_2}(N_2')$ endowed with the bounded convergence topology.

The symbol map σ is defined for all $\Xi \in \mathrm{L}(\mathcal{F}_{\theta_1}(N_1'), \mathcal{F}^*_{\theta_2}(N_2'))$ by

$$\sigma(\Xi)(\xi_1, \xi_2) = \langle\langle \Xi e_{\xi_1}, e_{\xi_2} \rangle\rangle = \widehat{\Xi^\kappa}(\xi_1, \xi_2), \quad \forall (\xi_1, \xi_2) \in N_1 \times N_2, \tag{14}$$

where Ξ^κ is an element of $\mathcal{F}^*_{(\theta_1,\theta_2)}(N_1' \times N_2')$ called the kernel of the operator Ξ. In view of the kernel theorem, there is an isomorphism

$$\mathrm{L}(\mathcal{F}_{\theta_1}(N_1'), \mathcal{F}^*_{\theta_2}(N_2')) \to \mathcal{F}^*_{\theta_1}(N_1') \otimes \mathcal{F}^*_{\theta_2}(N_2') \simeq \mathcal{F}^*_{(\theta_1,\theta_2)}(N_1' \times N_2') \to \mathcal{G}_{(\theta_1^*,\theta_2^*)}$$
$$\Xi \mapsto \sigma(\Xi)(\xi,\eta) = \sum_{n,m\in\mathbb{N}} \langle \kappa_{n,m}, \xi^{\otimes n} \otimes \eta^{\otimes m}\rangle \mapsto \mathcal{L}(\sigma(\Xi)) = K = (\kappa_{n,m})_{n,m\in\mathbb{N}}. \tag{15}$$

We define the convolution product $\Phi * \varphi$ of a distribution $\Phi \in \mathcal{F}^*_{(\theta_1,\theta_2)}(N_1' \times N_2')$ and a test function $\varphi \in \mathcal{F}_{(\theta_1,\theta_2)}(N_1' \times N_2')$ as follows

$$\Phi * \varphi(z,t) = \langle\langle \Phi, \tau_{(z,t)}\varphi\rangle\rangle = \langle\langle \Phi, \varphi(z+.,t+.)\rangle\rangle,\ (z,t) \in N_1' \times N_2'. \tag{16}$$

Moreover, it is proved in Ref. 8 that the Taylor expansion of the convolution product of the distribution $\Phi = (\Phi_{n,m})_{n,m\in\mathbb{N}}$ and the test function $\varphi = (\varphi_{n,m})_{n,m\in\mathbb{N}}$ is given by

$$\Phi * \varphi = \left(\sum_{k,l\in\mathbb{N}} \left\langle \frac{(n+k)!}{n!} \frac{(m+l)!}{m!} \Phi_{k,l}, \varphi_{n+k,m+l} \right\rangle_{k,l} \right)_{n,m\in\mathbb{N}}. \tag{17}$$

So, for any $\Psi_1, \Psi_2 \in \mathcal{F}^*_{(\theta_1,\theta_2)}(N_1' \times N_2')$, we define the convolution product $\Psi_1 * \Psi_2$ by

$$\langle\langle \Psi_1 * \Psi_2, \varphi\rangle\rangle = \langle\langle \Psi_1, \Psi_2 * \varphi\rangle\rangle, \ \text{ for any } \varphi \in \mathcal{F}_{(\theta_1,\theta_2)}(N_1' \times N_2').$$

Since $\mathcal{G}_{(\theta_1^*,\theta_2^*)}(N_1 \times N_2)$ is closed under point-wise multiplication (see Refs. 11, 14), the convolution product of two operators $\Xi_1, \Xi_2 \in L(\mathcal{F}_{\theta_1}(N_1'), \mathcal{F}_{\theta_2}(N_2'))$, denoted by $\Xi_1 * \Xi_2$ is defined as the unique operator such that

$$\sigma(\Xi_1 * \Xi_2) = \sigma(\Xi_1)\sigma(\Xi_2). \tag{18}$$

Moreover, it is proved in Ref. 2 that

$$\exp^*(\Xi) := \sum_{n\in\mathbb{N}} \frac{1}{n!} \Xi^{*n} \in L(\mathcal{F}_{(e^{\theta_1^*}-1)^*}(N_1'), \mathcal{F}^*_{(e^{\theta_2^*}-1)^*}(N_2')). \tag{19}$$

2.3. *Integral Kernel Operators and Expansion Theorem*

Let $S(\mathbb{R})$ be the Schwartz space and $S'(\mathbb{R})$ be the Schwartz distributions space. In this subsection, we consider in the particular case, $N = N_1 = N_2 = S_{\mathbb{C}}(\mathbb{R}) = S(\mathbb{R}) + iS(\mathbb{R})$, $\theta_1 = \theta_2 = \theta$. Now we recall the most fundamental white noise operators. To each $\xi \in N$, we associate the annihilation operator $a(x)$ defined by

$$a(\xi) : \phi = (\phi_n)_{n\in\mathbb{N}} \mapsto ((n+1)\xi \otimes_1 \phi_{n+1})_{n\in\mathbb{N}},$$

where $\xi \otimes_1 \phi_{n+1}$ stands for the contraction. It is known that $a(\xi) \in \mathcal{L}(\mathcal{F}_\theta(N'), \mathcal{F}_\theta(N'))$. Its adjoint operator $a^*(\xi) \in \mathcal{L}(\mathcal{F}^*_\theta(N'), \mathcal{F}^*_\theta(N'))$ is called the creation operator and satisfies

$$a^*(\xi) : \Phi = (\Phi_n)_{n\in\mathbb{N}} \mapsto (\xi \otimes \Phi_{n-1})_{n\in\mathbb{N}},$$

understanding that $\Phi_{-1} = 0$.

For $t \in \mathbb{R}$ and $\varphi \in \mathcal{F}_\theta(N')$, we define the annihilation operator at a point $t \in \mathbb{R}$ or Hida's differential operator by

$$\partial_t\varphi(x) := a(\delta_t) = \lim_{\varepsilon\to 0} \frac{\varphi(x+\varepsilon\delta_t) - \varphi(x)}{\varepsilon}, \quad x \in N', \tag{20}$$

and the creation operator as its adjoint operator denoted by $\partial_t^+ := a^*(\delta_t)$. Then $\partial_t \in L(\mathcal{F}_\theta(N'), \mathcal{F}_\theta(N'))$ and $\partial_t^* \in L(\mathcal{F}^*_\theta(N'), \mathcal{F}^*_\theta(N'))$. By a straightforward computation, we have

$$\partial_t e_\xi = \xi(t) e_\xi, \quad \forall \xi \in N. \tag{21}$$

It is proved in Refs. 12, 13 via the isomorphism (15), that for $\Xi = (\Xi_{n,m})_{n,m\in\mathbb{N}}$, there exits $\kappa \in N'^{\otimes n+m}$ such that we have the following formal integral expression:

$$\Xi_{n,m} = \int_{\mathbb{R}^{n+m}} \kappa(s_1,\ldots,s_n,t_1,\ldots,t_m)\partial^*_{s_1}\ldots\partial^*_{s_n}\partial_{t_1}\ldots\partial_{t_m} ds_1\ldots ds_n dt_1\ldots dt_m. \tag{22}$$

3. Quantum K-Gross Laplacian

For $j = 1, 2$, let $K_j \in \mathrm{L}(N_j, N_j')$. We denote by τ_{K_j} the kernel associated to K_j in $(N_j^{\otimes 2})'$ which is defined by

$$\langle \tau_{K_j}, \xi_j \otimes \eta_j \rangle = \langle K_j\xi_j, \eta_j \rangle, \qquad \forall(\xi_j, \eta_j) \in N_j^{\otimes 2}. \tag{23}$$

For simplicity, we denote the pair of operators (K_1, K_2), throughout this paper, by K.

We define the distribution $\mathcal{T}_K = \mathcal{T}_{(K_1,K_2)} = \left(\left(\mathcal{T}_{(K_1,K_2)}\right)_{n,m}\right)_{n,m\in\mathbb{N}} \in \mathcal{F}^*_{(\theta_1,\theta_2)}(N_1' \times N_2')$ by

$$(\mathcal{T}_K)_{n,m} = \left(\mathcal{T}_{(K_1,K_2)}\right)_{n,m} = \begin{cases} \tau_{K_1},\ n = 2,\ m = 0, \\ \tau_{K_2},\ n = 0,\ m = 2, \\ 0, \text{ otherwise.} \end{cases} \tag{24}$$

Definition 3.1. The K-Gross Laplacian on the space of test functions in two infinite dimensional variables, denoted $\Delta_G(K)$, is a well defined operator from $\mathcal{F}_{(\theta_1,\theta_2)}(N_1' \times N_2')$ into itself, given by

$$\Delta_G(K)\varphi = \mathcal{T}_K * \varphi, \quad \forall \varphi \in \mathcal{F}_{(\theta_1,\theta_2)}(N_1' \times N_2'). \tag{25}$$

Using the formula (17), it is easy to see that, for all $\varphi = (\varphi_{n,m})_{n,m\in\mathbb{N}} \in \mathcal{F}_{(\theta_1,\theta_2)}(N_1' \times N_2')$, the quantum K-Gross Laplacian $\Delta_G(K)$ is given by

$$\begin{aligned}
&\Delta_G(K)\varphi(x,y) \\
&= \sum_{n,m\in\mathbb{N}} \langle x^{\otimes n} \otimes y^{\otimes m}, (n+2)(n+1)\langle \tau_{K_1}, \varphi_{n+2,m}\rangle^{1}_{2,0}\rangle \\
&+ \sum_{n,m\in\mathbb{N}} \langle x^{\otimes n} \otimes y^{\otimes m}, (m+2)(m+1)\langle \tau_{K_2}, \varphi_{n,m+2}\rangle^{2}_{0,2}\rangle,
\end{aligned} \tag{26}$$

for all $(x,y) \in N_1' \times N_2'$.

Proposition 3.1. *In the particular case where $N_1 = N_2 = S_{\mathbb{C}}(\mathbb{R}) = S(\mathbb{R}) + iS(\mathbb{R})$ and $\theta_1 = \theta_2 = \theta$, we have the stochastic integral:*

$$\Delta_G(K) = \int_{T^2} (\tau_{K_1}(s,t)(\partial_s \otimes I)(\partial_t \otimes I) + \tau_{K_2}(s,t)(I \otimes \partial_s)(I \otimes \partial_t))\, dsdt \tag{27}$$

where $\partial_t \otimes I$ and $I \otimes \partial_t$ are defined by

$$\partial_t \otimes I(f \otimes g) = \partial_t(f) \otimes g, \quad I \otimes \partial_t(f \otimes g) = f \otimes \partial_t(g), \quad \forall f, g \in \mathcal{F}_\theta(N').$$

Proof. Let $N_1 = N_2 = S_{\mathbb{C}}(\mathbb{R}) = S(\mathbb{R}) + iS(\mathbb{R})$, $\theta_1 = \theta_2 = \theta$ and $\xi_1, \xi_2 \in N$. Using (21), it is easy to see that for all $t \in T$

$$\partial_t \otimes I(e_{(\xi_1,\xi_2)}) = \partial_t \otimes I(e_{\xi_1} \otimes e_{\xi_2}) = \xi_1(t) e_{\xi_1} \otimes e_{\xi_2} = \xi_1(t) e_{(\xi_1,\xi_2)},$$

and

$$I \otimes \partial_t(e_{(\xi_1,\xi_2)}) = I \otimes \partial_t(e_{\xi_1} \otimes e_{\xi_2}) = \xi_2(t) e_{\xi_1} \otimes e_{\xi_2} = \xi_2(t) e_{(\xi_1,\xi_2)}.$$

Then, we obtain

$$\int_{T^2} (\tau_{K_1}(s,t)(\partial_s \otimes I)(\partial_t \otimes I) + \tau_{K_2}(s,t)(I \otimes \partial_s)(I \otimes \partial_t))\, dsdt e_{(\xi_1,\xi_2)}$$
$$= \int_{T^2} \tau_{K_1}(s,t)(\partial_s \otimes I)(\partial_t \otimes I)e_{(\xi_1,\xi_2)} + \tau_{K_2}(s,t)(I \otimes \partial_s)(I \otimes \partial_t)e_{(\xi_1,\xi_2)} dsdt$$
$$= \int_{T^2} \tau_{K_1}(s,t)\xi_1(s)\xi_1(t)e_{(\xi_1,\xi_2)} dsdt + \int_{T^2} \tau_{K_2}(s,t)\xi_2(s)\xi_2(t)e_{(\xi_1,\xi_2)} dsdt$$
$$= (\langle K_1\xi_1, \xi_1\rangle + \langle K_2\xi_2, \xi_2\rangle)\, e_{(\xi_1,\xi_2)}.$$

On the other hand, we have

$$\Delta_G(K)e(\xi_1,\xi_2)(x,y)$$
$$= \sum_{n,m\in\mathbb{N}} \langle x^{\otimes n} \otimes y^{\otimes m}, \langle (n+2)(n+1)\mathcal{T}_{K_1}, \frac{\xi_1^{\otimes n+2}}{(n+2)!} \otimes \frac{\xi_2^{\otimes n}}{n!}\rangle_{2,0}$$
$$+ \sum_{n,m\in\mathbb{N}} \langle x^{\otimes n} \otimes y^{\otimes m}, \langle (m+2)(m+1)\langle \mathcal{T}_{K_2}, \frac{\xi_1^{\otimes n}}{n!} \otimes \frac{\xi_2^{\otimes n+2}}{(n+2)!}\rangle_{0,2}\rangle$$
$$= (\langle K_1\xi_1, \xi_1\rangle + \langle K_2\xi_2, \xi_2\rangle)e_{(\xi_1,\xi_2)}(x,y), \quad \forall x, y \in N'.$$

As $\{e(\xi_1,\xi_2),\ \xi_1,\xi_2 \in N\}$ spans a dense subspace of $\mathcal{F}_{(\theta_1,\theta_2)}(N_1' \times N_2')$, we obtain the stochastic integral representation (27). □

For $j = 1, 2$, assuming that the Young function θ_j satisfies the condition

$$\limsup_{x\to\infty} \frac{\theta_j(x)}{x^2} < +\infty, \tag{28}$$

we obtain the following two Gel'fand triplets (see Ref. 4)

$$\mathcal{F}_{\theta_j}(N_j') \to L^2(X_j', \gamma_j) \to \mathcal{F}^*_{\theta_j}(N_j'), \tag{29}$$

and

$$\mathcal{F}_{(\theta_1,\theta_2)}(N_1' \times N_2') \to L^2(X_1' \times X_2', \gamma_1 \otimes \gamma_2) \to \mathcal{F}^*_{(\theta_1,\theta_2)}(N_1' \times N_2'), \tag{30}$$

where γ_j is the Gaussian measure on the real Fréchet nuclear space X_j' (i.e. $N_j' = X_j' + iX_j'$), defined via the Bochner-Minlos theorem [7], by its characteristic function:

$$\int_{X_j'} e^{i\langle x,\xi_j\rangle_j} d\gamma(x) = e^{-\frac{1}{2}|\xi_j|_0^2}, \xi_j \in X_j. \tag{31}$$

Using the Gel'fand triplet (30) and the K-Gross Laplacian defined by (25), we extend the K-Gross Laplacian to the generalized functions space $\mathcal{F}^*_{(\theta_1,\theta_2)}(N_1' \times N_2')$ as follows.

Definition 3.2. The extended K-Gross Laplacian, denoted also $\Delta_G(K)$, acting on generalized functions space $\mathcal{F}^*_{(\theta_1,\theta_2)}(N'_1 \times N'_2)$ is defined by

$$\Delta_G(K)\Psi = \mathcal{T}_K * \Psi, \ \Psi \in \mathcal{F}^*_{(\theta_1,\theta_2)}(N'_1 \times N'_2). \tag{32}$$

Using the topological isomorphism

$$L(\mathcal{F}_{\theta_1}(N'_1), \mathcal{F}^*_{\theta_2}(N'_2)) \ni \Xi \mapsto \Xi^\kappa \in \mathcal{F}^*_{(\theta_1,\theta_2)}(N'_1 \times N'_2)$$

and the extended K-Gross Lapalcian given in (32), we can define the quantum K-Gross Laplacian as follows:

Definition 3.3. For all $\Xi \in L(\mathcal{F}_{\theta_1}(N'_1), \mathcal{F}^*_{\theta_2}(N'_2))$, the quantum K-Gross Laplacian $\Delta_{QG}(K)$ is defined by:

$$\Delta_{QG}(K)(\Xi) = \mathcal{T}_K * \Xi^\kappa. \tag{33}$$

Proposition 3.2. *For all $\Xi \in L(\mathcal{F}_{\theta_1}(N'_1), \mathcal{F}^*_{\theta_2}(N'_2))$, the symbol of the quantum K-Gross Laplacian is given for all $(\xi_1, \xi_2) \in N'_1 \times N'_2$, by:*

$$\sigma(\Delta_{QG}(K)(\Xi))(\xi_1, \xi_2) = (\langle K_1\xi_1, \xi_1\rangle_1 + \langle K_2\xi_2, \xi_2\rangle_2)\sigma(\Xi)(\xi_1, \xi_2). \tag{34}$$

Proof. Using the property (18) of the symbol of the convolution product of two operators and the definition (14), we have

$$\sigma(\Delta_{QG}(K)(\Xi)) = \mathcal{L}(\mathcal{T}_K * \Xi^k) = \widehat{\mathcal{T}_K}\widehat{\Xi^k} = \widehat{\mathcal{T}_K}\sigma(\Xi), \ \Xi \in L(\mathcal{F}_{\theta_1}(N'_1), \mathcal{F}^*_{\theta_2}(N'_2)).$$

On the other hand, using the Taylor expansion of $\mathcal{T}_K$, we obtain

$$\begin{aligned}
\widehat{\mathcal{T}_K}(\xi_1, \xi_2) &= \langle\langle \mathcal{T}_K, e_{(\xi_1,\xi_2)}\rangle\rangle \\
&= \langle (\mathcal{T}_K)_{2,0}, \xi_1^{\otimes 2}\rangle_1 + \langle (\mathcal{T}_K)_{0,2}, \xi_2^{\otimes 2}\rangle_2 \\
&= \langle \tau_{K_1}, \xi_1^{\otimes 2}\rangle_1 + \langle \tau_{K_2}, \xi_2^{\otimes 2}\rangle_2 \\
&= \langle K_1\xi_1, \xi_1\rangle_1 + \langle K_2\xi_2, \xi_2\rangle_2, \quad (\xi_1, \xi_2) \in N'_1 \times N'_2,
\end{aligned}$$

which proves the above proposition. □

4. Quantum white noise derivatives of quantum stochastic integrals

In this section, we recall some definitions from Ref. 10. We take $X_1 = X_2 = S(\mathbb{R})$ and $\theta_1 = \theta_2 = \theta$.

4.1. *Gross derivative*

Definition 4.1. We say that $\Phi \in \mathcal{F}_\theta^*(N')$ is Gross differentiable if for any $\zeta \in H$ the translation $T_{\varepsilon\zeta}\Phi$ is defined for small $|\varepsilon| < \varepsilon_0$ and if

$$\langle\langle D_\zeta \Phi, \varphi\rangle\rangle = \langle\langle \Phi, D_{-\zeta}\varphi\rangle\rangle, \qquad \forall \varphi \in \mathcal{F}_\theta(N') \tag{35}$$

where for all $\varphi \in \mathcal{F}_\theta(N')$, the operator $D_{-\zeta}\varphi$ is given by

$$D_{-\zeta}\varphi \equiv \lim_{\epsilon\mapsto 0} \frac{T_{\varepsilon\zeta}\varphi - \Phi}{\varepsilon} = \lim_{\epsilon\mapsto 0} \frac{\varphi(\varepsilon\zeta + \cdot) - \varphi}{\varepsilon} \tag{36}$$

converges in $\mathcal{F}_\theta(N')$ with respect to the weak topology. $D_\zeta\Phi$ is called the Gross derivative of Φ in direction ζ.

Proposition 4.1. *Let $\zeta \in H$. Then, D_ζ is a continuous operator on $\mathcal{F}_\theta^*(N')$ and for all $\Phi = (\Phi_n)_{n\in\mathbb{N}} \in \mathcal{F}_\theta^*(N')$, we have:*

$$D_\zeta\Phi = ((n+1)\langle \zeta, \Phi_{n+1}\rangle_1)_{n\in\mathbb{N}}.$$

Proof. Let $\zeta \in H$ and $\epsilon > 0$. For all $\varphi = (\varphi_n)_{n\in\mathbb{N}} \in \mathcal{F}_\theta(N')$, it is easy to see that:

$$T_\zeta\varphi(x) = \varphi(x+\zeta) = \sum_{n,m\in\mathbb{N}} \binom{n+m}{m} \langle x^{\otimes n} \otimes \zeta^{\otimes m}, \varphi_{n+m}\rangle.$$

Then, we obtain

$$\frac{T_{\epsilon\zeta}\varphi(x)-\varphi(x)}{\epsilon} = \frac{\sum_{n,m\in\mathbb{N}}\binom{n+m}{m}\epsilon^m\langle x^{\otimes n}\otimes\zeta^{\otimes m},\varphi_{n+m}\rangle - \sum_{n\in\mathbb{N}}\langle x^{\otimes n},\varphi_n\rangle}{\epsilon}$$

$$= \sum_{n\in\mathbb{N}}\langle x^{\otimes n}, \sum_{m\geq 1}\binom{n+m}{m}\epsilon^{m-1}\langle\zeta^{\otimes m},\varphi_{n+m}\rangle_m\rangle, \tag{37}$$

for all $x \in N'$. Then, we have

$$D_\zeta\varphi(x) = \lim_{\epsilon\to 0}\frac{T_{\epsilon\zeta}\varphi(x)-\varphi(x)}{\epsilon}$$

$$= \lim_{\epsilon\to 0}\frac{\sum_{n,m\in\mathbb{N}}\binom{n+m}{m}\epsilon^m\langle x^{\otimes n}\otimes\zeta^{\otimes m},\varphi_{n+m}\rangle - \sum_{n\in\mathbb{N}}\langle x^{\otimes n},\varphi_n\rangle}{\epsilon}$$

$$= \lim_{\epsilon\to 0}\sum_{n\in\mathbb{N}}\langle x^{\otimes n}, \sum_{m\geq 1}\binom{n+m}{m}\epsilon^{m-1}\langle\zeta^{\otimes m},\varphi_{n+m}\rangle_m\rangle$$

$$= \sum_{n\in\mathbb{N}}\langle x^{\otimes n},(n+1)\langle\zeta,\varphi_{n+1}\rangle_1\rangle, \quad \forall x\in N'. \tag{38}$$

On the other hand, for all $x \in N'$, it is easy to see that

$$\left(\frac{T_{\epsilon\zeta}\varphi-\varphi}{\epsilon} - D_\zeta\varphi\right)(x) = \sum_{n\in\mathbb{N}}\langle x^{\otimes n}, \sum_{m=2}^{\infty}\binom{n+m}{n}\epsilon^{m-1}\langle\zeta^{\otimes m},\varphi_{n+m}\rangle_m\rangle.$$

Then, we have for all $p \in \mathbb{N}$ and $a, a' > 0$:

$$\left\|\frac{T_{\epsilon\zeta}\varphi-\varphi}{\epsilon} - D_\zeta\varphi\right\|^2_{(\theta,-p,a)} = \sum_{n\in\mathbb{N}}\theta_n^{-2}a^{-n}\left|\sum_{m=2}^{\infty}\binom{n+m}{n}\epsilon^{m-1}\langle\zeta^{\otimes m},\varphi_{n+m}\rangle_m\right|^2_p$$

$$\leq \sum_{n\in\mathbb{N}}\theta_n^{-2}a^{-n}\left\{\sum_{m=2}^{\infty}\binom{n+m}{m}\epsilon^{m-1}|\zeta^{\otimes m}|_p^2|\varphi_{n+m}|_{-p}^2\right\}^2$$

$$\leq \|\varphi\|^2_{(\theta,-p,a')}\sum_{n\in\mathbb{N}}\theta_n^{-2}a^{-n}\left\{\sum_{m=2}^{\infty}\binom{n+m}{m}\epsilon^{m-1}|\zeta|_p^{2m}\theta_{n+m}a'^{n+m}\right\}^2$$

$$\leq \|\varphi\|^2_{(\theta,-p,a')}\left\{\sum_{\substack{n\in\mathbb{N}\\ m\geq 2}}\theta_n^{-2}a^{-n}\binom{n+m}{m}\epsilon^{m-1}|\zeta^{\otimes m}|_p^2\theta_{n+m}a'^{n+m}\right\}^2.$$

To conclude, we recall from Ref. 4, that

$$2^{-(n+m)}\theta_{n+m} \leq \theta_n\theta_m \leq 2^{n+m}\theta_{n+m}, \quad \forall n,m\in\mathbb{N}, \tag{39}$$

and

$$\theta_m |\zeta|_p^{2m} \leq e^{\theta(|\zeta|_p)}, \quad \forall \xi \in H, \, p \in \mathbb{N}.$$

Using the fact $\binom{n+m}{m} \leq 2^{n+m}$, it follows for $\varepsilon a' < 1$ and $\frac{a'}{a} < 1$, that

$$\left\| \frac{T_{\epsilon\zeta}\varphi - \varphi}{\epsilon} - D_\zeta \varphi \right\|_{(\theta,-p,a)} \leq \frac{e^{\theta(|\xi|_p)}}{(1-\frac{a}{a'})(1-a'\varepsilon)} \|\varphi\|_{(\theta,-p,a')}.$$

Thus we have shown that (38) converges in norm and the desired assertion follows. Using the duality paring, it is easy to see that

$$D_\zeta \Phi = ((n+1)\langle \zeta, \Phi_{n+1}\rangle_1)_{n\in\mathbb{N}}, \quad \forall \Phi \in \mathcal{F}_\theta^*(N'), \, \zeta \in H.$$

Finally, the distribution $\Psi = D_\zeta \Phi = ((n+1)\langle \zeta, \Phi_{n+1}\rangle_1)_{n\in\mathbb{N}}$ is an element of $\mathcal{F}_\theta^*(N')$. In fact, using the right estimation in (39), we have:

$$\begin{aligned}
\|\Psi\|_{(\theta,p,a)}^2 &= \sum_{n\in\mathbb{N}} [n!\theta_n]^2 a^n |\Psi_n|_{-p}^2 \\
&\leq |\zeta|_p \sum_{n\in\mathbb{N}} [n!\theta_n]^2 (n+1)^2 a^{-n} |\Phi_{n+1}|_{-p}^2 \\
&= \frac{|\zeta|_p}{a} \sum_{n\in\mathbb{N}} [(n+1)!\theta_{n+1}]^2 a^{n+1} |\Phi_{n+1}|_{-p}^2 \left(\frac{\theta_n}{\theta_{n+1}}\right)^2 \\
&\leq \frac{2|\zeta|_p}{a\theta_1} \|\Phi\|_{(\theta,p,a)}^2,
\end{aligned}$$

for all $p \in \mathbb{N}$ and $a > 0$. □

4.2. *Quantum stochastic integral of the quantum Gross Laplacian*

It is easy to prove the following proposition.

Proposition 4.2. *Let $\zeta \in H$. The maps D_ζ^+ and D_ζ^- given for all $\Xi \in \mathrm{L}(\mathcal{F}_\theta(N'), \mathcal{F}_\theta^*(N'))$ by*

$$D_\zeta^- \Xi = a(\zeta)\Xi - \Xi a(\zeta), \qquad D_\zeta^+ \Xi = \Xi a^+(\zeta) - a^+(\zeta)\Xi, \tag{40}$$

are well defined from $\mathrm{L}(\mathcal{F}_\theta(N'), \mathcal{F}_\theta^*(N'))$ into itself.

The maps D_ζ^+ and D_ζ^- defined in Proposition 4.2 are called respectively the creation derivative and annihilation derivative. In the particular case where $\zeta = \delta_t$ for any $t \in T$, D_ζ^+ and D_ζ^- are denoted respectively D_t^+ and D_t^- and called pointwise creation derivative and pointwise annihilation derivative.

Definition 4.2. Let $\Xi \in \mathrm{L}(\mathcal{F}_\theta(N'), \mathcal{F}_\theta^*(N'))$. Ξ is said to be pointwisely differentiable if there exists a measurable map $t \mapsto D_t^\pm \Xi \in \mathrm{L}(\mathcal{F}_\theta(N'), \mathcal{F}_\theta(N'))$ such that

$$\langle\langle (D_\zeta^\pm \Xi)e_\xi, e_\eta\rangle\rangle e^{-\langle \xi,\eta\rangle} = \int_{\mathbb{R}_+} \langle\langle (D_t^\pm \Xi)e_\xi, e_\eta\rangle\rangle \zeta(t)dt, \;\; (\xi,\eta) \in N_1 \times N_2. \quad (41)$$

Then, $D_t^+\Xi$(resp. $D_t^-\Xi$) is called the pointwise creation (resp. annihilation) derivative of Ξ.

Proposition 4.3. *The Quantum Gross Laplacian has the following stochastic integral:*

$$\Delta_{QG}(K) = \int K_1(s,t) D_s^+ D_t^+ dsdt + \int K_2(s,t) D_s^- D_t^- dsdt. \qquad (42)$$

Proof. For all $\xi, \eta \in N$, we have

$$\langle\langle \int K_1(s,t) D_s^+ D_t^+ \Xi e_\xi dsdt, e_\eta\rangle\rangle = \int K_1(s,t) \langle\langle (\partial_s D_t^+ \Xi - D_t^+ \Xi \partial_s)e_\xi, e_\eta\rangle\rangle dsdt$$

$$= \int K_1(s,t) \langle\langle D_t^+ \Xi e_\xi, \partial_s^+ e_\eta\rangle\rangle dsdt - \int K_1(s,t)\langle\langle D_t^+\Xi\partial_s e_\xi, e_\eta\rangle\rangle dsdt$$

$$= \int K_1(s,t)\xi(s)\langle\langle D_t^+\Xi e_\xi, \partial_s^+ e_\eta\rangle\rangle dsdt - \int \xi(s) K_1(s,t) \langle\langle D_t^+\Xi e_\eta, e_\xi\rangle\rangle dsdt$$

$$= \widehat{\Xi}(\eta,\xi) \int K_1(s,t)\eta(s)\eta(t)dsdt + \int K_1(s,t)\langle\langle (\partial_t\partial_s - \xi(t)\partial_s - \xi(s)\partial_t)\Xi e_\eta, e_\xi\rangle\rangle ds$$

$$= \widehat{\Xi}(\eta,\xi) \int K_1(s,t)\eta(s)\eta(t)dsdt$$

$$= \widehat{\Xi}(\eta,\xi)\langle K_1\xi,\xi\rangle,$$

and by the same computation, we obtain

$$\langle\langle \int K_2(s,t) D_s^- D_t^- \Xi e_\xi dsdt, e_\eta\rangle\rangle = \widehat{\Xi}(\eta,\xi)\langle K_2\eta,\eta\rangle.$$

Hence the assertion follows from Proposition 3.2. □

5. Quantum Cauchy problem associated to the K-Gross Laplacian

5.1. *Solution of the Quantum Cauchy problem associated to the K-Gross Laplacian*

We consider two Young functions θ_1 and θ_2, satisfying

$$\limsup_{x\to\infty}\frac{\theta_j(x)}{x^2}<\infty,\ j=1,2.$$

Let $\{\Theta(t),\, t\in I\}$ be a quantum stochastic processes defined at each point t of an interval $0\in I\subset\mathbb{R}$ and taking values in $L(\mathcal{F}_{\theta_1}(N_1'),\mathcal{F}^*_{\theta_2}(N_2'))$ and $\Xi_0\in L(\mathcal{F}_{\theta_1}(N_1'),\mathcal{F}^*_{\theta_2}(N_2'))$. Consider now the following initial value problem:

$$(\mathcal{P})\begin{cases}\frac{d}{dt}\Xi(t)=\Delta_{QG}(K)\Xi(t)+\Theta(t)\\ \Xi(0)=\Xi_0.\end{cases}\tag{43}$$

Theorem 5.1. *The Cauchy problem* $(\mathcal{P})$ *has a unique solution in* $L(\mathcal{F}_{(e^{\theta_1^*})^*}(N_1'),\mathcal{F}^*_{(e^{\theta_2^*})^*}(N_2'))$ *given by*

$$\Xi(t)=\Xi_0*e^{*\frac{t}{2}\mathcal{T}_{\mathbf{K}}}+\int_0^t e^{*\frac{t-s}{2}\mathcal{T}_{\mathbf{K}}}*\Theta(s)ds.\tag{44}$$

5.2. *Integral representation of the solution*

The set of positive test functions is defined by

$$\{f\in\mathcal{F}_{(\theta_1,\theta_2)}(N_1'\times N_2'),f((x+i0),(y+i0))\geq 0,\forall(x,y)\in N_1'\times N_2'\}.\tag{45}$$

A generalized function $\Phi\in\mathcal{F}^*_{(\theta_1,\theta_2)}(N_1'\times N_2')$ is positive in the usual sense, if it satisfies the following condition

$$\langle\langle\Phi,f\rangle\rangle\geq 0,\ \forall f\in\mathcal{F}_{(\theta_1,\theta_2)}(N_1'\times N_2')_+.\tag{46}$$

An operator $\Xi\in L(\mathcal{F}_{\theta_1}(N_1'),\mathcal{F}^*_{\theta_2}(N_2'))$ is positive if its kernel Ξ^K is an element of $\mathcal{F}^*_{(\theta_1,\theta_2)}(N_1'\times N_2')_+$.

Theorem 5.2. $\Delta_{QG}(K)$ *is a positive operator if and only if* K_i *is a positive operator, for* $i=1,2$.

We denote by X_j the real Fréchet space such that $N_j = X_j + iX_j$, for $j = 1, 2$. The next result, which is an application of Theorem 3.2 in Ref. 1, gives a sufficient condition to have a positive solution of the stochastic quantum differential equation $(\mathcal{P})$.

Theorem 5.3. *If the initial condition* Ξ_0 *and the process* Θ_t *are positive operators, then there exists a unique positive Borel measure* $\mu_{\Xi_t^\kappa}$ *on* $X_1 \times X_2$ *such that*

$$\langle\langle \Xi_t^K, \phi \rangle\rangle = \int_{X_1 \times X_2} \phi((x + i0, y + i0)) d\mu_{\Xi_t^\kappa}(x \oplus y), \quad \forall \phi \in \mathcal{F}_{(\theta_1, \theta_2)}(N_1' \times N_2') \tag{47}$$

where the Laplace transform of $\mu_{\Xi_t^\kappa}$ *is given for all* $(\xi_1, \xi_2) \in N_1 \times N_2$, *by*

$$\mathrm{L}(\mu_{\Xi_t^\kappa})(\xi_1, \xi_2) = \widehat{\Xi_0} e^{t \frac{\langle K_1 \xi_1, \xi_1 \rangle_1 + \langle K_2 \xi_2, \xi_2 \rangle_2}{2}} + \int_0^t e^{(t-s) \frac{\langle K_1 \xi_1, \xi_1 \rangle_1 + \langle K_2 \xi_2, \xi_2 \rangle_2}{2}} \widehat{\Theta_s} ds. \tag{48}$$

6. Remarks

(1) In the particular case where $\theta_1 = \theta_2 = \theta$, $N_1 = N$ and $N_2 = \{0\}$, all results on the space of entire functions in two infinite dimensional variables developed in this paper are a generalization of those studied in Ref. 3.

(2) For $j = 1, 2$, in the particular case where $K_j = I_j$, τ_{K_j} is the trace operator τ_j defined by

$$\langle \tau_j, \xi_j \otimes \eta_j \rangle_j = \langle \xi_j, \eta_j \rangle_j, \quad \forall (\xi_j, \eta_j) \in N_j \times N_j,$$

and of course $\Delta_{QG}(I)$ is the Quantum Gross Laplacian defined in Refs. 8 and 9.

References

1. W. Ayed and H. Ouerdiane, *Positive operators and integral representation*, Taiwanese J. of Math. 11, No. 5, (2007), pp. 1456-1475.

2. M. Ben Chrouda, M. El Oued, and H. Ouerdiane, *Convolution calculus and applications to stochastic differentential equations*, Soochow J. of Math., 28, (2002), pp. 375-388.
3. D. M. Chung and U. C. Ji, *Transforms on white noise functionals with their applications to Cauchy problems*, Nagoya Math. J., 147, (1997), pp. 1-23.
4. R. Gannoun, R. Hachaichi, H. Ouerdiane and A. Rezgui, *Un théorème de dualité entre espaces de fonctions holomorphes à croissance exponentielle,* J. Func. Anal, Vol. 171, No. 1, (2000), pp. 1-14.
5. S. Gheryani, S. Horrigue and H. Ouerdiane, *Evolution equation associated to the powers of the quantum Gross Laplacian*,Stochastics An International Journal of Probability and Stochastic Processes, Vol. 81, Issue 3 and 4 June 2009 , pp. 355-366.
6. L. Gross, *Potential theory on Hilbert space*, J. Funct. Anal. 1, (1967), pp. 123-181.
7. T. Hida, H. Kuo, J. Potthof and L. Streit, *White Noise: An Infinite Dimensional Calculus*, Kluwer Academic Publishers, (1993).
8. S. Horrigue and H. Ouerdiane, *Gross Laplacian Acting on Operators*, Acta Applicanda Mathematicae 105: (2008), pp. 227-239.
9. S. Horrigue and H. Ouerdiane, *Quantum Laplacian and Applications*, accepted to publication in Acta Applicandae Mathematicae.
10. U. C. Ji and N. Obata, *Annihilation-derivative, creation-derivative and representation of quantum martingales*, to appear in Commun. Math. Phys.
11. U. C. Ji, N. Obata and H. Ouerdiane, *Analytic Characterization of Generalized Fock Space Operators as Two-variable Entire Functions with Growth Condition*, World Scientific, Vol. 5, No. 3, (2002), pp. 395-407.
12. H. H. Kuo, *White noise distribution theory*, CRC Press, (1996).
13. N. Obata, *White noise calculus and Fock space*, Lect. Notes in Math. Vol.

1577, Springer- Verlag, (1994).

14. N. Obata, *Quantum white noise calculus based on nuclear algebras of entire functions*, Trends in Infinite Dimensional Analysis and Quantum Probability (Kyoto 2001), RIMS No. 1278, pp. 130-157.

ON MARGINAL MARKOV PROCESSES OF QUANTUM QUADRATIC STOCHASTIC PROCESSES

Farrukh MUKHAMEDOV*

Department of Computational & Theoretical Sciences

Faculty of Science, International Islamic University Malaysia

P.O. Box, 141, 25710, Kuantan

Pahang, Malaysia *

E-mail: far75m@yandex.ru; farrukh_m@iiu.edu.my

www.iium.edu.my

In this paper quantum quadratic stochastic process (q.q.s.p.) on von Meumann algebra $\mathcal{M}$ is considered. It is defined two marginal Markov processes on von Neumann algebras $\mathcal{M}$ and $\mathcal{M} \otimes \mathcal{M}$, respectively, to given q.q.s.o. We prove that such marginal processes uniquely determine the q.q.s.p. Moreover, certain ergodic relations between marginal processes and q.q.s.p. are established.

Keywords: quantum quadratic stochastic process; marginal Markov process; ergodic principle

1. Introduction

It is known that Markov processes, are well-developed field of mathematics, which have various applications in physics, biology and so on. But there are some physical models which cannot be described by such processes. One of such models is a model related to population genetics. Namely, this model is described by quadratic stochastic processes (see Refs. 2, 8, 11). To define it, we denote

$$\ell^1 = \{x = (x_n) : \|x\|_1 = \sum_{n=1}^{\infty} |x_n| < \infty;\ x_n \in \mathbb{R}\},$$
$$S^{\infty} = \{x \in \ell^1 : x_n \geq 0;\ \|x\|_1 = 1\}.$$

Hence this process is defined as follows (see Refs. 2, 11): Consider a family of functions $\{p_{ij,k}^{[s,t]} : i, j, k \in \mathbb{N},\ s,\ t \in \mathbb{R}_+,\ t - s \geq 1\}$. Such a family is said to be *quadratic stochastic process (q.s.p.)* if for fixed $s, t \in \mathbb{R}_+$ it satisfies the following conditions:

(i) $p_{ij,k}^{[s,t]} = p_{ji,k}^{[s,t]}$ for any $i, j, k \in \mathbb{N}$.

(ii) $p_{ij,k}^{[s,t]} \geq 0$ and $\sum_{k=1}^{\infty} p_{ij,k}^{[s,t]} = 1$ for any $i, j, k \in \mathbb{N}$.

(iii) An analogue of Kolmogorov-Chapman equation; here there are two variants: for the initial point $x^{(0)} \in S^{\infty}$, $x^{(0)} = (x_1^{(0)}, x_2^{(0)}, \cdots)$ and $s < r < t$ with $t - r \geq 1, r - s \geq 1$ one has either

(iii_A)

$$p_{ij,k}^{[s,t]} = \sum_{m,l=1}^{\infty} p_{ij,m}^{[s,r]} p_{ml,k}^{[r,t]} x_l^{(r)},$$

where $x_k^{(r)}$ is given by

$$x_k^{(r)} = \sum_{i,j=1}^{\infty} p_{ij,k}^{[0,r]} x_i^{(0)} x_j^{(0)};$$

or

(iii_B)

$$p_{ij,k}^{[s,t]} = \sum_{m,l,g,h=1}^{\infty} p_{im,l}^{[s,r]} p_{jg,h}^{[s,r]} p_{lh,k}^{[r,t]} x_m^{(s)} x_g^{(s)}.$$

It is said that the q.s.p. $\{p_{ij,k}^{s,t}\}$ is of *type (A) or (B)* if it satisfies the fundamental equations either (iii_A) or (iii_B), respectively. In this definition the function $p_{ij,k}^{[s,t]}$ denotes the probability that under the interaction of the elements i and j at time s the element k comes into effect at time t. Since for physical, chemical and biological phenomena a certain time is necessary for the realization of an interaction, we shall take the greatest such a time to be equal to 1 (see the Boltzmann model[5] or the biological model[8]). Thus the probability $p_{ij,k}^{[s,t]}$ is defined for $t - s \geq 1$.

It should be noted that such quadratic stochastic processes are related to the notion of quadratic stochastic operator introduced in Ref. 1, in the same way as Markov processes are related to linear transformations (i.e. Markov operators). A problem of studying the behaviour of trajectories of quadratic stochastic operators was stated in Ref. 14. A lot of papers (see for example Refs. 6, 8, 9, 15) were devoted to study limit behaviour and ergodic properties of trajectories of such operators.

We note that quadratic stochastic processes describe physical systems defined above, but they do not occupate the cases in a quantum level. So, it is natural to define a concept of quantum quadratic processes. In Refs. 3, 10 quantum (noncommutative) quadratic stochastic processes (q.q.s.p.) were defined on a von Neumann algebra and studied certain ergodic properties ones. In Ref. 3 it is obtained necessary and sufficient conditions for the validity of the ergodic principle for q.q.s.p. From the physical point of view, such a principle means that for sufficiently large values of time a system described by q.q.s.p. does not depend on the initial state of the system. Moreover, it has been found relations between quantum quadratic stochastic processes and non-commutative Markov processes. In Ref. 10 an expansion of q.q.s.p. into a so-called fibrewise Markov process is given, and it is proved that such an expansion uniquely determines the q.q.s.p. As an application, it is established a criterion (in terms of this expansion) for the q.q.s.p. to satisfy the ergodic principle. By means of such a result, it is proved that a q.q.s.p. satisfies the ergodic principle if and only if the associated Markov process satisfies that principle. It is natural to ask: is the defined Markov process determines the given q.q.s.p. uniquely, or how many Markov processes are needed to uniquely determine the q.q.s.p.? In this paper we are going to solve this problem. Namely, we shall show that there two non-stationary Markov processes defined on different von Neumann algebras $\mathcal{M}$ and $\mathcal{M} \otimes \mathcal{M}$, respectively, called marginal Markov processes, which uniquely determine the given to quantum quadratic stochastic process. Such a description allows us to investigate other properties of q.q.s.p. by means of Markov processes. Moreover, certain ergodic relations between them are established.

2. Preliminaries

Let $\mathcal{M}$ be a von Neumann algebra acting on a Hilbert space H. The set of all continuous (resp. ultra-weak continuous) functionals on $\mathcal{M}$ is denoted by $\mathcal{M}^*$ (resp. $\mathcal{M}_*$), and put $\mathcal{M}_{*,+} = \mathcal{M}_* \cap \mathcal{M}^*_+$, here $\mathcal{M}^*_+$ denotes the set of all positive linear functionals. By $\mathcal{M} \otimes \mathcal{M}$ we denote tensor product of $\mathcal{M}$ into itself. The sets S and S^2 denote the set of all normal states on $\mathcal{M}$ and $\mathcal{M}\otimes\mathcal{M}$, respectively. By U we denote a linear operator $U : \mathcal{M}\otimes\mathcal{M} \to \mathcal{M}\otimes\mathcal{M}$ such that $U(x \otimes y) = y \otimes x$ for all $x, y \in \mathcal{M}$. Given a state $\varphi \in S$, define the conditional expectation operator $E_\varphi : \mathcal{M}\otimes\mathcal{M} \to \mathcal{M}$ on elements $a \otimes b$, $a, b \in \mathcal{M}$ by

$$E_\varphi(a \otimes b) = \varphi(a)b \tag{1}$$

and extend it by linearity and continuity to $\mathcal{M} \otimes \mathcal{M}$. Clearly, such an operator is completely positive and $E_\varphi \mathbb{1}_{\mathcal{M}\otimes\mathcal{M}}=\mathbb{1}_{\mathcal{M}}$, here $\mathbb{1}_{\mathcal{M}}$ and $\mathbb{1}_{\mathcal{M}\otimes\mathcal{M}}$ are the identity operators in $\mathcal{M}$ and $\mathcal{M} \otimes \mathcal{M}$, respectively. We refer the reader to Ref. 13, for more details on von Neumann algebras.

Now consider a family of linear operators $\{P^{s,t} : \mathcal{M} \to \mathcal{M} \otimes \mathcal{M}, s, t \in \mathbb{R}_+, t - s \geq 1\}$.

Definition 2.1. We say that a pair $(\{P^{s,t}\}, \omega_0)$, where $\omega_0 \in S$ is an initial state, forms a *quantum quadratic stochastic process (q.q.s.p)*, if every operator $P^{s,t}$ is ultra-weakly continuous and the following conditions hold:

(i) each operator $P^{s,t}$ is a unital (i.e. preserves the identity operators) completely positive mapping with $UP^{s,t} = P^{s,t}$;

(ii) An analogue of Kolmogorov-Chapman equation is satisfied: for initial state $\omega_0 \in S$ and arbitrary numbers $s, \tau, t \in \mathbb{R}_+$ with $\tau - s \geq 1$, $t - \tau \geq 1$ one has either

(ii)$_A$ $P^{s,t}x = P^{s,\tau}(E_{\omega_\tau}(P^{\tau,t}x)), \quad x \in \mathcal{M}$

or

(ii)$_B$ $P^{s,t}x = E_{\omega_s}P^{s,\tau} \otimes E_{\omega_s}P^{s,\tau}(P^{\tau,t}x), \quad x \in \mathcal{M},$

where $\omega_\tau(x) = \omega_0 \otimes \omega_0(P^{0,\tau}x)$, $x \in \mathcal{M}$.

If q.q.s.p. satisfies one of the fundamental equations either (ii)$_A$ or (ii)$_B$ then we say that q.q.s.p. has *type (A)* or *type (B)*, respectively.

Remark 2.1. By using the q.q.s.p., we can specify a law of interaction of states. For $\varphi, \psi \in S$, we set

$$V^{s,t}(\varphi, \psi)(x) = \varphi \otimes \psi(P^{s,t}x), \qquad x \in \mathcal{M}.$$

This equality gives a rule according to which the state $V^{s,t}(\varphi, \psi)$ appears at time t as a result of the interaction of states φ and ψ at time s. From the physical point of view, the interaction of states can be explained as follows: Consider two physical systems separated by a barrier and assume that one of these systems is in the state φ. and the other one is in the state ψ. Upon the removal of the barrier, the new physical system is in the state $\varphi \otimes \psi$ and, as a result of the action of the operator $P^{s,t}$, a new state is formed. This state is exactly the result of the interaction of the states φ and ψ.

Remark 2.2. If $\mathcal{M}$ is an ℓ^∞, i.e., $\mathcal{M} = \ell^\infty$, then a q.q.s.p. $(\{P^{s,t}\}, \omega_0)$ defined on ℓ^∞ coincides with a quadratic stochastic process. Indeed, we set

$$p_{ij,k}^{[s,t]} = P^{s,t}(\chi_{\{k\}})(i,j), \quad i,j,k \in \mathbb{N},$$

where χ_A is the indicator of a set A. Then, by Definition 2.1, the family of functions $p_{ij,k}^{[s,t]}$ forms a quadratic stochastic process.

Conversely, if we have a quadratic stochastic process $(\{p_{ij,k}^{[s,t]}\}, \mu^{(0)})$ then we can define a quantum quadratic stochastic process on ℓ^∞ as follows:

$$(P^{s,t}\mathbf{f})(i,j) = \sum_{k=1}^{\infty} f_k p_{ij,k}^{[s,t]}, \quad \mathbf{f} = \{f_k\} \in \ell^\infty.$$

As the initial state, we take the following state

$$\omega_0(\mathbf{f}) = \sum_{k=1}^{\infty} f_k \mu_k^{(0)}.$$

One can easily check the conditions of Definition 2.1 are satisfied. Thus, a notion of quantum quadratic stochastic process generalizes the notion of quadratic stochastic process.

Remark 2.3. Certain examples of q.q.s.p were given in Ref. 3.

Let $(\{P^{s,t}\}, \omega_0)$ be a q.q.s.p. Then by $P_*^{s,t}$ we denote the linear operator, mapping from $(\mathcal{M} \otimes \mathcal{M})_*$ into $\mathcal{M}_*$, given by

$$P_*^{s,t}(\varphi)(x) = \varphi(P^{s,t}x), \quad \varphi \in (\mathcal{M} \otimes \mathcal{M})_*, \quad x \in \mathcal{M}.$$

Definition 2.2. A q.q.s.p. $(\{P^{s,t}\}, \omega_0)$ is said to satisfy *the ergodic principle*, if for every $\varphi, \psi \in S^2$ and $s \in \mathbb{R}_+$

$$\lim_{t\to\infty} \|P_*^{s,t}\varphi - P_*^{s,t}\psi\|_1 = 0,$$

where $\|\cdot\|_1$ is the norm on $\mathcal{M}^*$.

Let us note that Kolmogorov was the first who introduced the concept of an ergodic principle for Markov processes (see, for example, Ref. 7). For quadratic stochastic processes such a concept was introduced and studied in Refs. 4, 12.

3. Marginal Markov Processes and Ergodic Principle

In this section we are going to consider relation between q.q.s.p. and Markov processes.

First recall that a family $\{Q^{s,t} : \mathcal{M} \to \mathcal{M}, s, t \in \mathbb{R}_+,\ t-s \geq 1\}$ of unital completely positive operators is called *Markov process* if

$$Q^{s,t} = Q^{s,\tau}Q^{\tau,t}$$

holds for any $s, \tau, t \in \mathbb{R}_+$ with $t - \tau \geq 1$, $\tau - s \geq 1$.

A Markov process $\{Q^{s,t}\}$ is said to satisfy the *ergodic principle* if for every $\varphi, \psi \in S$ and $s \in \mathbb{R}_+$ one has

$$\lim_{t\to\infty} \|Q_*^{s,t}\varphi - Q_*^{s,t}\psi\|_1 = 0.$$

Here $Q_*^{s,t}$ is the conjugate operator to $Q^{s,t}$ defined by $Q_*^{s,t}(\varphi)(x) = \varphi(Q^{s,t}x)$, for any $\varphi \in \mathcal{M}_*$, $x \in \mathcal{M}$.

Let $(\{P^{s,t}\}, \omega_0)$ be a q.q.s.p. Then define a new process $Q_P^{s,t} : \mathcal{M} \to \mathcal{M}$ by

$$Q_P^{s,t} = E_{\omega_s} P^{s,t}. \tag{2}$$

Then according to Proposition 4.3[10] $\{Q_P^{s,t}\}$ is a Markov process associated with q.q.s.p. It is evident that the defined process satisfies the ergodic principle if the q.q.s.p. satisfies one. An interesting question is about the converse. Corollary 4.4[10] states the following important result:

Theorem 3.1. *Let $(\{P^{s,t}\}, \omega_0)$ be a q.q.s.p. on a von Neumann algebra $\mathcal{M}$ and let $\{Q_P^{s,t}\}$ be the corresponding Markov process. Then the following conditions are equivalent*

(i) $(\{P^{s,t}\}, \omega_0)$ satisfies the ergodic principle;

(ii) $\{Q_P^{s,t}\}$ satisfies the ergodic principle;

(iii) There is a number $\lambda \in [0,1)$ such that, given any states $\varphi, \psi \in S^2$ and a number $s \in \mathbb{R}_+$ one has

$$\|Q_{P,*}^{s,t}\varphi - Q_{P,*}^{s,t}\psi\|_1 \leq \lambda \|\varphi - \psi\|_1$$

for at least one $t \in \mathbb{R}_+$.

3.1. *Case type A*

In this subsection we assume that q.q.s.p. $(\{P^{s,t}\}, \omega_0)$ has type (A).

Now define another process $\{H_P^{s,t} : \mathcal{M}\otimes\mathcal{M} \to \mathcal{M}\otimes\mathcal{M},\ s,t \in \mathbb{R}_+,\ t-s \geq 1\}$ by

$$H_P^{s,t}\mathbf{x} = P^{s,t}E_{\omega_t}\mathbf{x}, \qquad \mathbf{x} \in \mathcal{M} \otimes \mathcal{M}. \tag{3}$$

It is clear that every $H_P^{s,t}$ is a unital completely positive operator. It turns out that $\{H^{s,t}\}$ is a Markov process. Indeed, using $(\text{ii})_A$ of Def. 2.1 one has

$$H_P^{s,t}\mathbf{x} = P^{s,t}E_{\omega_t}\mathbf{x} = P^{s,\tau}E_{\omega_\tau}(P^{\tau,t}E_{\omega_t}\mathbf{x}) = H_P^{s,\tau}H_P^{\tau,t}\mathbf{x},$$

which is the assertion.

The defined two Markov processes $Q_P^{s,t}$ and $H_P^{s,t}$ are related with each other by the following equality

$$E_{\omega_s}(H_P^{s,t}\mathbf{x}) = E_{\omega_s}(P^{s,t}(E_{\omega_t}(\mathbf{x})) = Q_P^{s,t}(E_{\omega_t}(\mathbf{x}))$$

for every $\mathbf{x} \in \mathcal{M} \otimes \mathcal{M}$. Moreover, $H_P^{s,t}$ has the following properties

$$H_P^{s,t}\mathbf{x} = P^{s,t}(E_{\omega_t}(\mathbf{x})) = P^{s,t}E_{\omega_t}E_{\omega_t}(\mathbf{x}) = H_P^{s,t}(E_{\omega_t}(\mathbf{x})\otimes \mathbb{1}) \tag{4}$$

$$UH_P^{s,t} = H_P^{s,t}, \quad H_P^{s,t}(x\otimes \mathbb{1}) = P^{s,t}x, \quad x \in \mathcal{M}.$$

from (4) one gets $H_P^{s,t}(\mathbb{1} \otimes x) = \omega_t(x)\, \mathbb{1} \otimes \mathbb{1}$. Here we can represent

$$\omega_t(x) = \omega_0 \otimes \omega_0(P^{0,t}x) = \omega_0(Q_P^{0,t}x),$$
$$\omega_t(x) = \omega_0 \otimes \omega_0(H_P^{0,t}(x\otimes \mathbb{1})).$$

Now we are interested in the following question: can such kind of two Markov processes (i.e. with above properties) determine uniquely a q.q.s.p.? To answer to this question we need to introduce some notations.

Let $\{Q^{s,t} : \mathcal{M} \to \mathcal{M}\}$ and $\{H^{s,t} : \mathcal{M} \otimes \mathcal{M} \to \mathcal{M} \otimes \mathcal{M}\}$ be two Markov processes with an initial state $\omega_0 \in S$. Denote

$$\varphi_t(x) = \omega_0(Q^{0,t}x), \quad \psi_t(x) = \omega_0 \otimes \omega_0(H^{0,t}(x\otimes \mathbb{1})).$$

Assume that the given processes satisfy the following conditions:

(i) $UH^{s,t} = H^{s,t}$;

(ii) $E_{\psi_s}H^{s,t} = Q^{s,t}E_{\varphi_t}$;

(iii) $H^{s,t}\mathbf{x} = H^{s,t}(E_{\psi_t}(\mathbf{x})\otimes \mathbb{1})$ for all $\mathbf{x} \in \mathcal{M} \otimes \mathcal{M}$.

First note that if we take $\mathbf{x} = \mathbb{1} \otimes x$ in (iii) then we get

$$\begin{aligned} H^{s,t}(\mathbb{1} \otimes x) &= H^{s,t}(E_{\psi_t}(\mathbb{1} \otimes x)\otimes \mathbb{1}) \\ &= H^{s,t}(\psi_t(x)\, \mathbb{1} \otimes \mathbb{1}) \\ &= \psi_t(x)\, \mathbb{1} \otimes \mathbb{1} \end{aligned} \tag{5}$$

Now from (ii) and (5) we have

$$\begin{aligned} E_{\psi_s}H^{s,t}(\mathbb{1} \otimes x) &= E_{\psi_s}(\psi_t(x)\, \mathbb{1} \otimes \mathbb{1}) \\ &= \psi_t(x)\, \mathbb{1} \\ &= Q^{s,t}E_{\varphi_t}(\mathbb{1} \otimes x) \\ &= \varphi_t(x)\, \mathbb{1}\,. \end{aligned} \tag{6}$$

This means that $\varphi_t = \psi_t$, therefore in the sequel we denote $\omega_t := \varphi_t = \psi_t$.

Now we are ready to formulate the result.

Theorem 3.2. *Let $\{Q^{s,t}\}$ and $\{H^{s,t}\}$ be two Markov Processes with (i)-(iii). Then by the equality $P^{s,t}x = H^{s,t}(x\otimes \mathbb{1})$ one defines a q.q.s.p. of type (A). Moreover, one has*

(a) $P^{s,t} = H^{s,\tau}P^{\tau,t}$ for any $s, \tau, t \in \mathbb{R}_+$ with $\tau - s \geq 1$, $t - \tau \geq 1$,

(b) $Q^{s,t} = E_{\omega_s}P^{s,t}$.

Proof. We have to check only the condition $(\mathrm{ii})_A$ of Def. 2.1. Take any $s, \tau, t \in \mathbb{R}_+$ with $\tau - s \geq 1$, $t - \tau \geq 1$. Then using the assumption (iii) we derive

$$\begin{aligned} P^{s,\tau}E_{\omega_\tau}(P^{\tau,t}x) &= H^{s,\tau}(E_{\omega_\tau}H^{\tau,t}(x\otimes \mathbb{1})\otimes \mathbb{1}) \\ &= H^{s,\tau}H^{\tau,t}(x\otimes \mathbb{1}) \\ &= H^{s,t}(x\otimes \mathbb{1}) \\ &= P^{s,t}x, \quad x \in \mathcal{M}. \end{aligned}$$

From the markovianity of $H^{s,t}$ we immediately get (a).

If we put $\mathbf{x} = x\otimes \mathbb{1}$ to (iii) then from (1) one finds

$$E_{\omega_s}P^{s,t}x = E_{\omega_s}H^{s,t}(x\otimes \mathbb{1}) = Q^{s,t}E_{\omega_t}(x\otimes \mathbb{1}) = Q^{s,t}x.$$

This completes the proof. □

These two $\{Q^{s,t}\}$ and $\{H^{s,t}\}$ Markov processes are called *marginal Markov processes* associated with q.q.s.p. $\{P^{s,t}\}$. So, according to Theorem 3.2 the marginal Markov processes uniquely define q.q.s.p.

Now define an other process $\{Z^{s,t} : \mathcal{M}\otimes\mathcal{M} \to \mathcal{M}\otimes\mathcal{M}\}$ by

$$Z^{s,t}\mathbf{x} = E_{\omega_s}H^{s,t}(\mathbf{x})\otimes \mathbb{1}, \quad \mathbf{x} \in \mathcal{M}\otimes\mathcal{M}. \tag{7}$$

From (ii) one gets $Z^{s,t}\mathbf{x} = Q^{s,t}E_{\omega_t}\mathbf{x}\otimes \mathbb{1}$. In particular,

$$\begin{aligned} Z^{s,t}(x\otimes \mathbb{1}) &= Q^{s,t}x\otimes \mathbb{1}, \\ Z^{s,t}(\mathbb{1}\otimes x) &= \omega_t(x)\, \mathbb{1}\otimes\mathbb{1}\,. \end{aligned}$$

Proposition 3.1. *The process $\{Z^{s,t}\}$ is a Markov one.*

Proof. Take any $s, \tau, t \in \mathbb{R}_+$ with $\tau - s \geq 1$, $t - \tau \geq 1$. Then the assumption (iii) and the markovianity of $H^{s,t}$ imply that

$$\begin{aligned} Z^{s,\tau} Z^{\tau,t}\mathbf{x} &= E_{\omega_s} H^{s,\tau}(E_{\omega_\tau} H^{\tau,t}(\mathbf{x})\otimes \mathbb{1})\otimes \mathbb{1} \\ &= E_{\omega_s} H^{s,\tau} H^{\tau,t}(\mathbf{x})\otimes \mathbb{1} \\ &= E_{\omega_s} H^{s,t}(\mathbf{x})\otimes \mathbb{1} \\ &= Z^{s,t}\mathbf{x}, \end{aligned}$$

for every $\mathbf{x} \in \mathcal{M} \otimes \mathcal{M}$, which is the assertion. □

Remark 3.1. Consider a q.q.s.p. $(\{P^{s,t}\}, \omega_0)$ of type (A). Let $H^{s,t}$, $Z^{s,t}$ be the associated Markov processes. Take any $\varphi \in S^2$ then from (3) with taking into account (1), one concludes that

$$\varphi(H^{s,t}\mathbf{x}) = P_*^{s,t}\varphi(E_{\omega_t}(\mathbf{x})) = P_*^{s,t}\varphi \otimes \omega_t(\mathbf{x}), \tag{8}$$

for any $\mathbf{x} \in \mathcal{M} \otimes \mathcal{M}$.

Similarly, using (7), for $Z^{s,t}$ we have

$$Z_*^{s,t}(\sigma \otimes \psi) = \psi(\mathbb{1}) P_*^{s,t}(\sigma \otimes \omega_s) \otimes \omega_t, \tag{9}$$

for every $\sigma, \psi \in \mathcal{M}_*$.

From Theorem 3.1 and using (8),(9) one can prove the following

Corollary 3.1. *Let $(\{P^{s,t}\}, \omega_0)$ be a q.q.s.p. of type (A) on $\mathcal{M}$ and let $\{Q^{s,t}\}$, $\{H^{s,t}\}$ be its marginal processes. Then the following conditions are equivalent*

(i) $(\{P^{s,t}\}, \omega_0)$ satisfies the ergodic principle;

(ii) $\{Q^{s,t}\}$ satisfies the ergodic principle;

(iii) $\{H^{s,t}\}$ satisfies the ergodic principle;

(iv) $\{Z^{s,t}\}$ satisfies the ergodic principle;

3.2. *Case type B*

In this subsection we suppose that a q.q.s.p. $(\{P^{s,t}\},\omega_0)$ has type (B).

Like (3) let us define a process $\mathfrak{h}_P^{s,t}:\mathcal{M}\otimes\mathcal{M}\to\mathcal{M}\otimes\mathcal{M}$ by

$$\mathfrak{h}_P^{s,t}\mathbf{x}=P^{s,t}E_{\omega_t}\mathbf{x},\qquad \mathbf{x}\in\mathcal{M}\otimes\mathcal{M}. \tag{10}$$

The defined process $\{\mathfrak{h}_P^{s,t}\}$ is not Markov, but satisfies another equation. Namely, using $(\mathrm{ii})_B$ of Def. 2.1 and (2) we get

$$\mathfrak{h}_P^{s,t}\mathbf{x}=E_{\omega_s}P^{s,\tau}\otimes E_{\omega_s}P^{s,\tau}(P^{\tau,t}E_{\omega_t}\mathbf{x})=Q_P^{s,\tau}\otimes Q_P^{s,t}(\mathfrak{h}_P^{\tau,t}\mathbf{x}),$$

where $\mathbf{x}\in\mathcal{M}\otimes\mathcal{M}$.

Note that the process $\{\mathfrak{h}_P^{s,t}\}$ has the same properties like $\{H_P^{s,t}\}$.

Similarly to Theorem 3.2 we can formulate the following

Theorem 3.3. *Let $\{Q^{s,t}\}$ be a Markov process and $\{\mathfrak{h}^{s,t}\}$ be another process, which satisfy (i)-(iii) and*

$$\mathfrak{h}^{s,t}=Q^{s,\tau}\otimes Q^{s,\tau}\circ\mathfrak{h}^{\tau,t} \tag{11}$$

for any $s,\tau,t\in\mathbb{R}_+$ with $\tau-s\geq 1$, $t-\tau\geq 1$. Then by the equality $P^{s,t}x=\mathfrak{h}^{s,t}(x\otimes\mathbb{1})$ one defines a q.q.s.p. of type (B). Moreover, one has $Q^{s,t}=E_{\omega_s}P^{s,t}$.

Proof. We have to check only the condition $(\mathrm{ii})_B$. Note that the assumption (iii) implies that

$$E_{\omega_s}\mathfrak{h}^{s,t}(x\otimes\mathbb{1})=Q^{s,t}E_{\omega_t}(\cdot\otimes\mathbb{1})=Q^{s,t}x,\quad x\in\mathcal{M}.$$

Using this equality with (11) for any $s,\tau,t\in\mathbb{R}_+$ with $\tau-s\geq 1$, $t-\tau\geq 1$ one finds

$$\begin{aligned}E_{\omega_s}P^{s,\tau}\otimes E_{\omega_s}P^{s,\tau}(P^{\tau,t}x)&=E_{\omega_s}\mathfrak{h}^{s,\tau}(\cdot\otimes\mathbb{1})\otimes E_{\omega_s}\mathfrak{h}^{s,\tau}(\cdot\otimes\mathbb{1})(\mathfrak{h}^{\tau,t}(x\otimes\mathbb{1}))\\&=Q^{s,\tau}\otimes Q^{s,\tau}(\mathfrak{h}^{\tau,t}(x\otimes\mathbb{1}))\\&=\mathfrak{h}^{s,t}(x\otimes\mathbb{1})\\&=P^{s,t}x\end{aligned}$$

for any $x\in\mathcal{M}$.

This completes the proof. □

These two processes $\{Q^{s,t}\}$ and $\{\mathfrak{h}^{s,t}\}$ we call *marginal processes* associated with q.q.s.p. $\{P^{s,t}\}$.

Similarly like (7), define a process $\{z^{s,t} : \mathcal{M}\otimes\mathcal{M} \to \mathcal{M}\otimes\mathcal{M}\}$ by

$$z^{s,t}\mathbf{x} = E_{\omega_s}\mathfrak{h}^{s,t}(\mathbf{x})\otimes \mathbb{1}, \quad \mathbf{x}\in\mathcal{M}\otimes\mathcal{M}. \tag{12}$$

Note that for this process the equality (9) holds as well.

Proposition 3.2. *The process $z^{s,t}$ is a Markov one.*

Proof. First from Theorem 3.3 and Propsition 4.3[10] we conclude that

$$E_{\omega_s}Q^{s,t} = E_{\omega_t}. \tag{13}$$

Let us take any $s,\tau,t\in\mathbb{R}_+$ with $\tau - s \geq 1$, $t-\tau\geq 1$. Then from (12) with (11),(13) one gets

$$\begin{aligned} z^{s,t}\mathbf{x} &= E_{\omega_s}(Q^{s,\tau}\otimes Q^{s,\tau}(h^{\tau,t}(\mathbf{x}))\otimes \mathbb{1} \\ &= Q^{s,\tau}E_{\omega_s}Q^{s,\tau}(h^{\tau,t}(\mathbf{x}))\otimes \mathbb{1} \\ &= Q^{s,\tau}E_{\omega_\tau}(\mathfrak{h}^{\tau,t}(\mathbf{x}))\otimes \mathbb{1}\,. \end{aligned} \tag{14}$$

On the other hand, using conditions (ii),(iii) we obtain

$$\begin{aligned} z^{s,\tau}z^{\tau,t}\mathbf{x} &= E_{\omega_s}h^{s,\tau}(E_{\omega_\tau}\mathfrak{h}^{\tau,t}(\mathbf{x})\otimes \mathbb{1})\otimes \mathbb{1} \\ &= E_{\omega_s}h^{s,\tau}\mathfrak{h}^{\tau,t}(\mathbf{x})\otimes \mathbb{1} \\ &= Q^{s,\tau}E_{\omega_\tau}\mathfrak{h}^{\tau,t}(\mathbf{x})\otimes \mathbb{1} \end{aligned}$$

for every $\mathbf{x}\in\mathcal{M}\otimes\mathcal{M}$. This relation with (14) proves the assertion. □

Corollary 3.2. *Let $(\{P^{s,t}\},\omega_0)$ be a q.q.s.p. of type (B) on $\mathcal{M}$ and let $\{Q^{s,t}\}$, $\{\mathfrak{h}^{s,t}\}$ be its marginal processes. Then the following conditions are equivalent*

(i) $(\{P^{s,t}\},\omega_0)$ satisfies the ergodic principle;

(ii) $\{Q^{s,t}\}$ satisfies the ergodic principle;

(iii) $\{\mathfrak{h}^{s,t}\}$ satisfies the ergodic principle;

(iv) $\{z^{s,t}\}$ satisfies the ergodic principle;

Acknowledgement

The author thanks the MOHE grant FRGS0308-91 and the MOSTI grant 01-01-08-SF0079.

References

1. S. N. Bernshtein, *Uchen. Zapiski NI Kaf. Ukr. Otd. Mat.* **1** 83115 (1924) (Russian).
2. N. N. Ganikhodjaev, *J. Theor. Probab.* **4** 639-653 (1991).
3. N.N. Ganikhodjaev, F.M. Mukhamedov, *Izvestiya Math.* **64** (2000), 873-890.
4. N.N. Ganikhodjaev, H. Akin, F.M. Mukhamedov, *Lin. Algebra Appl.* **416** (2006), 730–741.
5. R.D. Jenks, *J. Diff. Equations* **5** (1969), 497-514.
6. H. Kesten, *Adv. Appl. Probab.* **2** 1-82 (1970), 179-228.
7. A. N. Kolmogorov, in *Selected works of A. N. Kolmogorov*, vol. II, article 9 (Kluwer, Dortrecht 1992).
8. Yu.I. Lyubich, *Mathematical structures in population genetics* (Springer-Verlag, Berlin, 1992).
9. V. M. Maksimov, *Theory Probab. Appl.* **41** (1996), 5569.
10. F.M. Mukhamedov, *Izvestiya Math.* **68** (2004), 10091024.
11. T. A. Sarymsakov, N. N. Ganikhodjaev, *J. Theor. Probab.* **3** (1990), 51-70.
12. T. A. Sarymsakov, N. N. Ganikhodjaev, *Soviet Math. Dokl.* **43** (1991), 279-283.
13. M.Takesaki, *Theory of operator algebars I*, (Springer-Verlag, Berlin, 1979).
14. S. M. Ulam, *A collection of mathematical problems* (Interscience, New York-London 1960).
15. S. S. Vallander, *Soviet Math. Dokl.* **13** (1973), 123126

On the Applicability of Multiplicative Renormalization Method for Certain Power Functions

Izumi Kubo

Department of Mathematics

Graduate School of Science

Hiroshima University

Higashi-Hiroshima, 739-8526, Japan

E-mail: izumi_kubo@ccv.ne.jp

Hui-Hsiung Kuo

Department of Mathematics

Louisiana State University

Baton Rouge, LA 70803, USA

E-mail: kuo@math.lsu.edu

Suat Namli

Texas Gulf Foundation

9441 W. Sam Houston Parkway, Suite 100

Houston, TX 77099, USA

E-mail: namli@hotmail.com

We characterize the set of probability measures with density functions which we can apply the multiplicative renormalization method for the function $h(x) = (1-x)^{-\kappa}$ with $\kappa = \frac{1}{2}$ and $\kappa = 2$. For the case $\kappa = \frac{1}{2}$, we only have uniform distributions on intervals. For the case $\kappa = 2$, we have three types of density functions.

1. Multiplicative renormalization method

In this short article we will briefly explain new ideas introduced in the papers by Kubo-Kuo-Namli[7,8] for characterizing probability measures by the *multiplicative renormalization method* (MRM). Then we will show how to modify the ideas with rather complicated computation to cover the other cases in Kubo-Kuo-Namli.[9–11] This leads to the concept of MRM-factor and its characterization, which will be carried out in the more recent papers by Kubo-Kuo.[5,6]

Consider a probability measure μ on $\mathbb{R}$ with infinite support and finite moments of all orders. Upon applying the Gram-Schmidt process to the sequence $\{x^n\}_{n=0}^{\infty}$ we get a μ-orthogonal sequence $\{P_n\}_{n=0}^{\infty}$ of polynomials with $P_0 = 1$ and P_n being of degree n with leading coefficient 1. Moreover, there are sequences $\alpha_n, \omega_n, n \geq 0$ of real numbers such that

$$(x - \alpha_n)P_n(x) = P_{n+1}(x) + \omega_{n-1}P_{n-1}(x), \quad n \geq 0,$$

where $\omega_{-1} = 1$ and $P_{-1} = 0$ by convention.

A method, called the multiplicative renormalization method, has been introduced by Asai-Kubo-Kuo[2,3] for the derivation of the three sequences $\{P_n, \alpha_n, \omega_n\}_{n=0}^{\infty}$. The crucial idea is the derivation of a function $\psi(t, x)$, called an *orthogonal polynomial* (*or OP-*)*generating function*, which has the following series expansion

$$\psi(t, x) = \sum_{n=0}^{\infty} c_n P_n(x) t^n,$$

where P_n's are the above polynomials and $c_n \neq 0$ for all n's. Once we have such an OP-generating function $\psi(t, x)$, then we can find the sequences $\{P_n, \alpha_n, \omega_n\}_{n=0}^{\infty}$, (see Asai-Kubo-Kuo,[2,3] Kubo-Kuo-Namli,[7–9] or Kuo[12] for the derivation procedure.)

Here is a brief review of this method. Start with a suitable function $h(x)$ and define two functions θ and $\widetilde{\theta}$ by

$$\theta(t) = \int_{\mathbb{R}} h(tx)\, d\mu(x),$$

$$\widetilde{\theta}(t, s) = \int_{\mathbb{R}} h(tx)h(sx)\, d\mu(x).$$

Then we have the following theorem due to Asai–Kubo–Kuo.[2,3]

Theorem 1.1. *Let $\rho(t)$ be an analytic function at 0 with $\rho(0) = 0$ and $\rho'(0) \neq 0$. Then the multiplicative renormalization*

$$\psi(t,x) := \frac{h(\rho(t)x)}{\theta(\rho(t))} = \frac{h(\rho(t)x)}{E_\mu h(\rho(t)\,\cdot\,)} \tag{1}$$

is an OP-generating function for μ if and only if

$$\Theta_\rho(t,s) := \frac{\widetilde{\theta}(\rho(t),\rho(s))}{\theta(\rho(t))\theta(\rho(s))} \tag{2}$$

is a function of the product ts in some neighborhood of $(0,0)$.

The essential part of the theorem is to find a function $\rho(t)$ so that the function $\Theta_\rho(t,s)$ in Equation (2) is a function of ts. If such a function $\rho(t)$ exists, then we say that μ is *multiplicative renormalization method* (MRM)-*applicable* for the function $h(x)$.

The multiplicative renormalization method can be used not only for the derivation of an OP-generating function $\psi(t,x)$, but also to discover new probability measures as well. It has been shown in Asai-Kubo-Kuo[2,3] that the classical probability measures are MRM-applicable for functions of the form either $h(x) = e^x$ or $h(x) = (1-x)^{-\kappa}$. On the other hand, we have a characterization problem, i.e., given a function $h(x)$, characterize all probability measures μ which are MRM-applicable for $h(x)$.

For $h(x) = e^x$, Kubo[4] has proved that the class of MRM-applicable probability measures coincides with the Meixner class.[1,13] For the function $h(x) = (1-x)^{-1}$, Kubo-Kuo-Namli[7,8] have obtained the corresponding class of MRM-applicable probability measures, which contains many interesting distributions beside the classical arcsine and semi-circle distributions. On the other hand, for the function $h(x) = (1-x)^{-1/2}$, it has recently been proved by Kubo-Kuo-Namli[9] that the class of MRM-applicable probability measures consists of only uniform distributions on intervals.

2. Power function of order $\kappa = 1/2$

In this section we consider the power function $h(x) = (1-x)^{-\kappa}$ of order $\kappa = 1/2$, namely, $h(x) = (1-x)^{-1/2}$, and describe the essential ideas from the paper by Kubo-Kuo-Namli.[9]

By direct differentiation, we find that the function

$$\widetilde{\theta}(t,s) = \int_{\mathbb{R}} \frac{1}{\sqrt{1-tx}}\frac{1}{\sqrt{1-sx}}\,d\mu(x)$$

satisfies the following partial differential equation

$$\frac{\partial\widetilde{\theta}}{\partial t} - \frac{\partial\widetilde{\theta}}{\partial s} = 2(t-s)\frac{\partial^2\widetilde{\theta}}{\partial t\partial s}. \tag{3}$$

A key step to derive all probability measures that are MRM-applicable for $h(x)$ is to find a function $\rho(t)$ for Theorem 1.1. In order to do so, define an associated function $f(t)$ by

$$f(t) := \theta\big(\rho(t)\big) = \int_{\mathbb{R}} \frac{1}{\sqrt{1-\rho(t)x}}\, d\mu(x). \tag{4}$$

We can use the information on $f(t)$ to derive $\rho(t)$. Once $\rho(t)$ is found, we can compute $f(t)$ and then $\theta(t)$, which will produce μ. The procedure is very simple, but the calculations are extremely complicated.

Assume that the function $J(ts) := \Theta_\rho(t,s)$ is a function of ts as in Theorem 1.1. Then we can use the PDE in Equation (3) to show that the function $f(t)$ defined by Equation (4) must satisfy the following three Fundamental Equations:

$$\frac{f'(t)}{f(t)} = \frac{2b\rho(t) + (a+bt)\rho'(t)}{1-2(a+bt)\rho(t)}, \tag{5}$$

$$\frac{f'(t)}{f(t)} = \frac{-3b + 4(ab+c_3t)\rho(t) + (c_2 + 2abt + c_3t^2)\rho'(t)}{3a + c_1 + 3bt - 2(c_2 + 2abt + c_3t^2)\rho(t)}, \tag{6}$$

$$\frac{f'(t)}{f(t)} = \frac{A\big(\rho(t), \rho'(t), t\big)}{B\big(\rho(t), t\big)}, \tag{7}$$

where the constants a, b, c_i's, and the functions A, B are given by

$$a = \theta'(0),\ b = \frac{J'(0)}{\rho'(0)},\ c_1 = \frac{\rho''(0)}{\rho'(0)^2},\ c_2 = \frac{f''(0)}{\rho'(0)^2},$$

$$c_3 = \frac{J''(0)}{\rho'(0)^2},\ c_4 = \frac{\rho'''(0)}{\rho'(0)^3},\ c_5 = \frac{f'''(0)}{\rho'(0)^3},\ c_6 = \frac{J'''(0)}{\rho'(0)^3},$$

$$\begin{aligned} A(X,Y,t) &= -4bc_1 - 10ab - 10c_3t + 6\Big\{bc_2 + 2ac_3t + c_6t^2\Big\}X \\ &\quad + \Big\{c_5 + 3bc_2t + 3ac_3t^2 + c_6t^3\Big\}Y, \\ B(X,t) &= c_4 + 4ac_1 + 5c_2 + (4bc_1 + 10ab)t + 5c_3t^2 \\ &\quad - 2\Big\{c_5 + 3bc_2t + 3ac_3t^2 + c_6t^3\Big\}X. \end{aligned}$$

Recall that μ is assumed to have infinite support. By using the three Fundamental Equations in Equations (5)–(7), we can derive the following relationships regarding to the above constants and the function $\rho(t)$:

$$b \neq 0, \quad 4a + c_1 = 0, \quad a^2 + c_2 > 0, \quad c_3 = \frac{18}{5}b^2,$$

$$c_4 = 6(3a^2 - c_2), \quad c_5 = -3ac_2, \quad c_6 = \frac{162}{7}b^2,$$

$$\rho(t) = \frac{3bt}{a^2 + c_2 + 6abt + 9b^2t^2}.$$

The derivation of these relationships is rather complicated, see the proof of Theorem 4.1 in Kubo-Kuo-Namli.[9] Thus we have found the function $\rho(t)$. Note that it is specified by three parameters a, b, and c_2.

Make a change of the parameters by

$$\alpha = \frac{2(a^2 + c_2)}{3b}, \quad \beta = 2a, \quad \gamma = 6b.$$

It is easy to see that the condition for the new parameters is $\alpha\gamma > 0$. Then the function $\rho(t)$ can be rewritten as

$$\rho(t) = \frac{2t}{\alpha + 2\beta t + \gamma t^2}, \tag{8}$$

Next put this function $\rho(t)$ into Equation (5) to get the initial-valued differential equation

$$\frac{f'(t)}{f(t)} = \frac{\beta + \gamma t}{\alpha + 2\beta t + \gamma t^2}, \quad f(0) = 1,$$

which has a unique solution given by

$$f(t) = \sqrt{1 + \frac{2\beta}{\alpha}t + \frac{\gamma}{\alpha}t^2}.$$

Recall that $f(t) = \theta\big(\rho(t)\big)$. Therefore, we have

$$\theta\big(\rho(t)\big) = \sqrt{1 + \frac{2\beta}{\alpha}t + \frac{\gamma}{\alpha}t^2}. \tag{9}$$

On the other hand, we can show that the inverse function of $\rho(t)$ in Equation (8) is given by

$$\rho^{-1}(s) = \frac{\alpha s}{1 - \beta s + \sqrt{(1 - \beta s)^2 - \alpha\gamma s^2}}. \tag{10}$$

It follows from Equations (9) and (10) that

$$\theta(t) = \sqrt{\frac{2}{1 - \beta t + \sqrt{(1 - \beta t)^2 - \alpha\gamma t^2}}}.$$

Then we use the following algebraic identity

$$\sqrt{\frac{2}{a+\sqrt{a^2-b^2}}} = \frac{2}{\sqrt{a+b}+\sqrt{a-b}}, \qquad a>0,\ |b|\le a,$$

to rewrite the expression of the function $\theta(t)$ as

$$\theta(t) = \frac{2}{\sqrt{1-(\beta+\sqrt{\alpha\gamma})t}+\sqrt{1-(\beta-\sqrt{\alpha\gamma})t}}. \tag{11}$$

Finally we will derive μ from this function $\theta(t)$. Before doing so, let us give an example.

Example 2.1. Let ν be the uniform probability measure on the $[-1,1]$ and $h(x)=(1-x)^{-1/2}$. Then the associated function $\theta(t)$ is given by

$$\theta(t) = \frac{2}{\sqrt{1-t}+\sqrt{1+t}}. \tag{12}$$

Note that this function is given by $\theta(t)$ in Equation (11) with $\alpha=\gamma=1$ and $\beta=0$. Thus we have the corresponding function from Equation (8),

$$\rho(t) = \frac{2t}{1+t^2}.$$

With this function $\rho(t)$, we can show that the function in Equation (2) is given by

$$\Theta_\rho(t,s) = \frac{1}{2\sqrt{ts}}\log\frac{1+\sqrt{ts}}{1-\sqrt{ts}} = \sum_{n=0}^{\infty}\frac{1}{2n+1}(ts)^n.$$

Therefore, by Theorem 1.1, the uniform probability measure ν on $[-1,1]$ is MRM-applicable for the function $h(x)=(1-x)^{-1/2}$. Moreover, we have the OP-generating function from Equation (1) for ν,

$$\psi(t,x) = \frac{1}{\sqrt{1-2tx+t^2}},$$

which yields the following corresponding sequences for $n \geq 0$,

$$P_n(x) = \frac{n!}{(2n-1)!!} L_n(x),$$

$$\alpha_n = 0,$$

$$\omega_n = \frac{(n+1)^2}{(2n+3)(2n+1)}, \tag{13}$$

where $(-1)!! = 1$ by convention and $L_n(x)$ is the Legendre polynomial of degree n. Note that obviously $\alpha_n = 0$ for all n since ν is symmetric.

Now we proceed to use the information from Example 2.1 to derive μ which has the $\theta(t)$ function given by Equation (11). Consider the affine transformation

$$T_{v,q}(x) = vx + q, \quad v \neq 0,$$

and the transformation of the uniform probability measure ν defined by

$$\nu_{v,q} = \nu \circ T_{v,q}, \quad v \neq 0.$$

Obviously, $\nu_{v,q}$ is a uniform probability measure on an interval. By using the information in Example 2.1, we can check that $\nu_{v,q}$ is MRM-applicable for $h(x) = (1-x)^{-1/2}$. In particular, the $\theta(t)$-function, α_n-sequence, and ω_n-sequence of $\nu_{v,q}$ are given by

$$\theta_{v,q}(t) = \frac{2}{\sqrt{1 + \frac{q-1}{v}t} + \sqrt{1 + \frac{q+1}{v}t}}, \tag{14}$$

$$(\alpha_{v,q})_n = -\frac{q}{v}, \quad n \geq 0,$$

$$(\omega v, q)_n = \frac{1}{v^2} \frac{(n+1)^2}{(2n+3)(2n+1)}, \quad n \geq 0. \tag{15}$$

Compare Equations (11) and (14) and set the equations

$$\frac{q-1}{v} = -\beta - \sqrt{\alpha\gamma}, \quad \frac{q+1}{v} = -\beta + \sqrt{\alpha\gamma},$$

which have the solution

$$v = \frac{1}{\sqrt{\alpha\gamma}}, \quad q = -\frac{\beta}{\sqrt{\alpha\gamma}}.$$

Thus μ and $\nu_{\frac{1}{\sqrt{\alpha\gamma}}, -\frac{\beta}{\sqrt{\alpha\gamma}}}$ have the same $\theta(t)$-function. Does the function $\theta(t)$ uniquely determine a probability measure? We can use the $(\omega_{v,q})_n$-sequence in Equation (15) and Theorem 1.11 in the book by Shohat-Tamarkin[14] to verify that the function $\theta_{v,q}(t)$ in Equation (14) does uniquely determine a probability measure. Hence we can conclude that $\mu = \nu_{\frac{1}{\sqrt{\alpha\gamma}}, -\frac{\beta}{\sqrt{\alpha\gamma}}}$.

The above discussion proves the next theorem from Kubo-Kuo-Namli.[9]

Theorem 2.1. *A probability measure μ on $\mathbb{R}$ with infinite support and finite moments of all orders is MRM-applicable for $h(x) = (1-x)^{-1/2}$ if and only if it is a uniform probability measure on an interval.*

In fact, the condition "μ has infinite support" can be weakened to "μ is supported by at least three points" (see Theorem 5.2 in Kubo-Kuo-Namli.[9]) It is quite obvious that a Dirac delta measure is MRM-applicable for the function $h(x) = (1-x)^{-1/2}$. On the other hand, it does take some effort to prove that a probability measure being supported by two points is also MRM-applicable for $h(x) = (1-x)^{-1/2}$.

Consider the power function $h(x) = (1-x)^{-\kappa}$ of order κ. When $\kappa = 1/2$, the results are somewhat surprising because uniform probability measures on intervals and probability measures being supported by one or two points are the only probability measures which are MRM-applicable. On the other hand, for the case $\kappa = 1$, the class of MRM-applicable probability measures consists of many new interesting probability measures in addition to the classical arcsine and semi-circle distributions.

3. Power functions of other order

Consider the power function $h(x) = (1-x)^{-\kappa}$ of order κ. As pointed out in Section 2, the values $\kappa = 1$ and $\kappa = 1/2$ are the extreme cases. For other nonzero value of κ, there are several types of probability measures.

Here we explain the special case $\kappa = 2$ from Kubo-Kuo-Namli.[10] In that case we have

$$\widetilde{\theta}(t,s) = \int_{\mathbb{R}} \frac{1}{(1-tx)^2} \frac{1}{(1-sx)^2}\, d\mu(x),$$

which can be shown to satisfy the partial differential equation

$$\frac{\partial \widetilde{\theta}}{\partial t} - \frac{\partial \widetilde{\theta}}{\partial s} = \frac{1}{2}(t-s)\frac{\partial^2 \widetilde{\theta}}{\partial t \partial s}. \tag{16}$$

Suppose μ is MRM-applicable for the function $h(x) = (1-x)^{-2}$. Then we can use Equation (16) to show that the function

$$f(t) := \theta\big(\rho(t)\big) = \int_{\mathbb{R}} \frac{1}{\big(1-\rho(t)x\big)^2}\, d\mu(x)$$

satisfies the following three Fundamental Equations:

$$\frac{f'(t)}{f(t)} = \frac{b\rho(t) + 2(a+bt)\rho'(t)}{2-(a+bt)\rho(t)},$$

$$\frac{f'(t)}{f(t)} = \frac{-3b + 2(ab+c_3t)\rho(t) + 2(c_2+2abt+c_3t^2)\rho'(t)}{3a+2c_1+3bt-(c_2+2abt+c_3t^2)\rho(t)},$$

$$\frac{f'(t)}{f(t)} = \frac{A\big(\rho(t),\rho'(t),t\big)}{B\big(\rho(t),t\big)},$$

where a, b, c_i's are constants and the functions A, B are given by

$$\begin{aligned} A(X,Y,t) &= -5bc_1 - 8ab - 8c_3t + 3\Big\{bc_2 + 2ac_3t + c_6t^2\Big\}X \\ &\quad + 2\Big\{c_5 + 3bc_2t + 3ac_3t^2 + c_6t^3\Big\}Y, \\ B(X,t) &= 2c_4 + 5ac_1 + 4c_2 + (5bc_1+8ab)t + 4c_3t^2 \\ &\quad - \Big\{c_5 + 3bc_2t + 3ac_3t^2 + c_6t^3\Big\}X. \end{aligned}$$

Now assume in addition that μ has a density function. Then we can use the above three Fundamental Equations to derive the conclusion that there are only three possible cases below. For the proof, see Kubo-Kuo-Namli.[10]

Case 1. $b \neq 0$, $c_1 = -a$, $c_3 = \frac{45}{32}b^2$.

In this case, we have the following values and functions,

$$\begin{aligned} c_2 > \frac{1}{2}a^2, \quad c_4 &= \frac{3}{4}(a^2-2c_2), \quad c_5 = \frac{3}{4}a(2c_2-a^2), \quad c_6 = \frac{81}{32}b^3, \\ \rho(t) &= \frac{24bt}{8(2c_2-a^2)+12abt+9b^2t^2}, \\ f(t) &= \Big(\frac{8(2c_2-a^2)+12abt+9b^2t^2}{8(2c_2-a^2)}\Big)^2, \\ \theta(t) &= \frac{16}{\big[2-at+\sqrt{4-4at-(4c_2-3a^2)t^2}\,\big]^2}. \end{aligned} \tag{17}$$

Case 2. $b \neq 0$, $c_1 = -a$, $c_3 = \frac{9}{8}b^2$.

In this case, we have the following values and functions,

$$
\begin{aligned}
c_2 &> \frac{1}{2}a^2, \quad c_4 = 2a^2 - c_2, \quad c_5 = a(2c_2 - a^2), \quad c_6 = \frac{3}{2}b^3,\\
\rho(t) &= \frac{12bt}{4(2c_2 - a^2) + 6abt + 3b^2t^2},\\
f(t) &= \frac{4(2c_2 - a^2)}{4(2c_2 - a^2) - 3b^2t^2}\left(\frac{4(2c_2 - a^2) + 6abt + 3b^2t^2}{4(2c_2 - a^2)}\right)^2,\\
\theta(t) &= \frac{24}{p(t) + \sqrt{3}(2 - at)\sqrt{p(t)}},
\end{aligned}
\tag{18}
$$

where $p(t) = 12 - 12at - (8c_2 - 7a^2)t^2$.

Case 3. $b \neq 0$, $c_1 \neq -a$, $c_3 = \frac{4}{3}b^2$.

In this case, we have the following values and functions,

$$
\begin{aligned}
c_2 &= \frac{1}{2}(12a^2 + 20ac_1 + 9c_1^2),\\
c_4 &= -\frac{3}{2}(2a + c_1)(2a + 3c_1),\\
c_5 &= \frac{1}{2}(36a^3 + 36a^2c_1 - 3ac_1^2 - c_1^3 + 48a^2c_1 + 66ac_1^2 + 16c_1^3),\\
c_6 &= \frac{20}{9}b^3,
\end{aligned}
$$

$$
\begin{aligned}
\rho(t) &= \frac{6bt}{18(a + c_1)^2 - 3bc_1t + 2b^2t^2},\\
f(t) &= \frac{\{18(c_1 + a)^2 - 3bc_1t + 2b^2t^2\}^2}{108(c_1 + a)^3\big[3(c_1 + a) - bt\big]},\\
\theta(t) &= \frac{8}{\big[2 + (2a + 3c_1)t\big]\big[2 - (4a + 3c_1)t\big] + \big[2 - (2a + c_1)t\big]\sqrt{q(t)}},
\end{aligned}
\tag{19}
$$

where $q(t) = \big[2 + (4a + 5c_1)t\big]\big[2 - (4a + 3c_1)t\big]$.

Finally we can apply the Hilbert transform

$$(\mathcal{H}g)(t) = \text{p.v.} \int_{\mathbb{R}} \frac{g(x)}{1 - tx}\, dx$$

and use Equations (17), (18), and (19) to derive the probability measures given in the next theorem.

Theorem 3.1. *Let a probability measure $d\mu = g(x)\,dx$ be MRM-applicable for $h(x) = (1 - x)^{-2}$. Then the density function $g(x)$ must be one of the* *following three types:*

$$g(x) = \frac{4}{3\pi(2c_2 - a^2)^2}\left\{2(2c_2 - a^2) - (2x - a)^2\right\}^{3/2},$$

$$g(x) = \frac{\sqrt{3}}{\pi(2c_2 - a^2)}\sqrt{4(2c_2 - a^2) - 3(2x - a)^2},$$

$$g(x) = \frac{\sqrt{(4a + 5c_1 + 2x)^3(4a + 3c_1 - 2x)}}{16\pi|a + c_1|^3},$$

where $g(x)$ is understood to be zero outside the interval where the expression in the right hand side makes sense.

Observe that the first two cases have parameters a and c_2, while the third case has parameters a and c_1. Moreover, note that the constant b appears only as a scaling of t in the functions $\rho(t), f(t), \theta(t)$ and does not effect the density function $g(x)$.

In the chart below we give three examples corresponding to the three cases with particular values of the parameters.

	$a = 0,\ b = \frac{4}{3},\ c_2 = 1$	$a = 0,\ b = 2,\ c_2 = \frac{3}{2}$	$a = \frac{1}{2},\ b = \frac{3}{2},\ c_1 = 0$
$\rho(t)$	$\frac{2t}{1+t^2}$	$\frac{2t}{1+t^2}$	$\frac{2t}{1+t^2}$
$f(t)$	$(1+t^2)^2$	$\frac{(1+t^2)^2}{1-t^2}$	$\frac{(1+t^2)^2}{1-t}$
$\psi(t,x)$	$\frac{1}{(1-2xt+t^2)^2}$	$\frac{1-t^2}{(1-2xt+t^2)^2}$	$\frac{1+t}{(1-2xt+t^2)^2}$
μ	$\beta\left(\frac{5}{2}, \frac{5}{2}\right)$	$\beta\left(\frac{3}{2}, \frac{3}{2}\right)$	$\beta\left(\frac{5}{2}, \frac{3}{2}\right)$
$g(x)$	$\frac{8}{3\pi}(1+x)^{\frac{3}{2}}(1-x)^{\frac{3}{2}}$	$\frac{2}{\pi}\sqrt{1-x^2}$	$\frac{2}{\pi}(1+x)^{\frac{3}{2}}(1-x)^{\frac{1}{2}}$
$P_n(x)$	$\frac{G_n^{(2)}(x)}{2^n(n+1)}$	$\frac{G_n^{(1)}(x)}{2^n}$	$\frac{G_n^{(2)}(x) - G_{n-1}^{(1)}(x)}{2^n(n+1)}$
α_n	0	0	$\frac{1}{2(n+1)(n+2)}$
ω_n	$\frac{(n+1)(n+4)}{4(n+2)(n+3)}$	$\frac{1}{4}$	$\frac{(n+1)(n+3)}{4(n+2)^2}$

Here $\beta(p,q)$ denotes the *beta distribution* on the interval $[-1,1]$, namely, its density function is given by

$$\frac{\Gamma(p+q)}{2^{p+q-1}\Gamma(p)\Gamma q)}(1+x)^{p-1}(1-x)^{q-1}, \quad -1<x<1,$$

and $G_n^{(\kappa)}(x)$ are the Gegenbauer polynomials with parameter κ defined by the following series expansion

$$\frac{1}{(1-2xt+t^2)^\kappa}=\sum_{n=0}^{\infty}G_n^{(\kappa)}(x)t^n.$$

It can be easily seen that

$$G_n^{(\kappa)}(x)=\frac{2^n}{n!}\frac{\Gamma(\kappa+n)}{\Gamma(\kappa)}x^n+\cdots\cdots$$

so that the leading coefficient of $G_n^{(2)}(x)$ is $2^n(n+1)$. Moreover, observe that $G_n^{(1)}(x)$ is the Chebyshev polynomial $U_n(x)$ of the second kind,

$$G_n^{(1)}(x)=U_n(x)=\frac{\sin\left[(n+1)\cos^{-1}x\right]}{\sin(\cos^{-1}x)}=2^nx^n+\cdots\cdots.$$

Note that for the second example, the $\beta\left(\frac{3}{2},\frac{3}{2}\right)$ distribution is the same as the semi-circle distribution. It is also worthwhile to point out that for $a=-\frac{1}{2}, b=\frac{3}{2}, c_1=0$, we have the $\beta\left(\frac{3}{2},\frac{5}{2}\right)$ distribution, which is just the reflection of the $\beta\left(\frac{5}{2},\frac{3}{2}\right)$ distribution in the third example, and thus does not provide a new typical example.

Finally we mention that for the power function $h(x)=(1-x)^{-\kappa}$ of general order κ, the computations are much more complicated although the ideas are somewhat similar to the case $\kappa=2$. We refer the detail to the paper by Kubo-Kuo-Namli.[11]

References

1. L. Accardi, *Meixner classes and the square of white noise*, in: Finite and Infinite Dimensional Analysis in Honor of Leonard Gross, Contemporary Mathematics **317** (2003), 1–13, Amer. Math. Soc.
2. N. Asai, I. Kubo, and H.-H. Kuo, *Multiplicative renormalization and generating functions I*, Taiwanese Journal of Mathematics **7** (2003), 89–101.
3. N. Asai, I. Kubo, and H.-H. Kuo, *Multiplicative renormalization and generating functions II*, Taiwanese Journal of Mathematics **8** (2004), 593–628.

4. I. Kubo, *Generating functions of exponential type for orthogonal polynomials*, Inifinite Dimensional Analysis, Quantum Probability and Related Topics **7** (2004), 155–159.
5. I. Kubo and H.-H. Kuo, *MRM-applicable orthogonal polynomials for certain hypergeometric functions*, Communications on Stochastic Analysis (2009) (to appear)
6. I. Kubo and H.-H. Kuo, *MRM-factors for special probability measures*, (submitted for publication)
7. I. Kubo, H.-H. Kuo, and S. Namli, *Interpolation of Chebyshev polynomials and interacting Fock spaces*, Infinite Dimensional Analysis, Quantum Probability and Related Topics **9** (2006), 361–371.
8. I. Kubo, H.-H. Kuo, and S. Namli, *The characterization of a class of probability measures by multiplicative renormalization*, Communications on Stochastic Analysis **1** (2007), 455–472.
9. I. Kubo, H.-H. Kuo, and S. Namli, *Applicability of multiplicative renormalization method for a certain function*; Communications on Stochastic Analysis **2** (2008), 405–422.
10. I. Kubo, H.-H. Kuo, and S. Namli, *MRM-applicable measures for the power function of order* 2; *Preprint* 2008.
11. I. Kubo, H.-H. Kuo, and S. Namli, *MRM-applicable measures for power functions of general order*; *Preprint* 2008.
12. H.-H. Kuo, *Multiplicative renormalization method for orthogonal polynomials*; in: Quantum Probability and Related Topics, QP-PQ: Quantum Prob. White Noise Analysis **23** (2008), 165–175, R. Quezada et al. (eds.), World Scientific
13. J. Meixner, *Orthogonale polynomsysteme mit einen besonderen gestalt der erzeugenden funktion*, J. Lond. Math. Soc. **9** (1934), 6–13.
14. J. Shohat and J. Tamarkin, *The Problem of Moments*, Math. Surveys **1**,

Amer. Math. Soc., 1943.

Convolution Equation: Solution and Probabilistic Representation

José L. Da Silva

Mathematics and Engineering Department, University of Madeira, Funchal, 9000-390, Madeira, Portugal

E-mail: luis@uma.pt

http://dme.uma.pt/luis

Mohamed Erraoui

Département de Mathématiques, Université Cadi Ayyad, Marrakech, BP 2390, Maroc

E-mail: erraoui@ucam.ac.ma

Habib Ouerdiane

Department of Mathematics, University of Tunis El Manar, 1060 Tunis, Tunisia

E-mail: habib.ouerdiane@fst.rnu.tn

In this paper we study a Cauchy problem associated to Δ_K^*, the adjoint of Δ_K which is related to the Gross Laplacian for certain choice of the operator K. We show that the solution is a well defined generalized function in an appropriate space. Finally, using infinite dimensional stochastic calculus we give a probabilistic representation of the solution in terms of K-Wiener process W.

Keywords: Generalized functions, convolution, adjoint operator, Cauchy problem, stochastic integrals in Hilbert space

1. Introduction

The Gross Laplacian Δ_G was introduced by L. Gross in Ref. 7 in order to study the heat equation in infinite dimensional spaces. It has been shown that the solution is represented as an integral with respect to Gaussian measure, see Refs. 7 and 13. There exists a huge literature dedicated to the Gross Laplacian with different points of view. We would like to mention the white noise analysis approach, see Refs. 2, 8, 9, 11 and references therein.

In this paper we study the study the Cauchy problem

$$\frac{\partial}{\partial u}U(u) = \frac{1}{2}\Delta_K^* U(t), \qquad U(0) = \Phi, \tag{1}$$

where Φ is a generalized functions and Δ_K^* is the adjoint operator of Δ_K which is related to the Gross Laplacian for certain choice of the operator K, see (7) for more details. As the main tool we use the Laplace transform and the fact that Δ_K^* is a convolution operator. It is straightforward for show that the solution of (1) is an well defined element in an appropriate space generalized functions, see Section 2 for details. Thus, the main result of the paper is to give a probabilistic representation for that solution. This entails, between others things, a stochastic calculus in infinite dimensions such as the Itô formula, see Theorem 3.1 below.

The paper is organized as follows. In Section 2 we provide the mathematical background needed to solve the Cauchy problem (1); namely we construct the appropriate test functions space $\mathcal{F}_\theta(N')$ and the associated generalized functions $\mathcal{F}'_\theta(N')$. The elements in $\mathcal{F}_\theta(N')$ are entire functions on the co-nuclear space N' with exponential growth of order θ (a Young function) and of minimal type and $\mathcal{F}'_\theta(N')$ is the topological dual. The main tools we use are the Laplace transform which characterizes the space $\mathcal{F}'_\theta(N')$ in terms of holomorphic functions with certain growth conditions. We introduce the convolution $\Phi * \varphi$ between a generalized function $\Phi \in \mathcal{F}'_\theta(N')$ and a test function $\varphi \in \mathcal{F}_\theta(N')$ which generalizes the convolution of a measure and a function. We then introduce the convolution of two generalized functions as an extension of the convolution of two measures. It turns out that, indeed, the operator Δ_K^* is given as a convolution. Finally in Section 3 we prove the existence of the the Cauchy problem (1) and give a probabilistic representation of it. We use the stochastic integration in Hilbert spaces, as developed in[3] and[12] with respect to K-Wiener process W and the Itô formula for $\mathcal{F}'_\theta(N')$-valued processes $t_{W(u)}\Phi$, where $t_x\Phi$ is the translation of Φ by x.

2. Preliminaries

In this section we will introduce the framework which is necessary later on. Let X be a real nuclear Fréchet space with topology given by an increasing family $\{|\cdot|_k;\, k \in \mathbb{N}_0\}$ of Hilbertian norms, $\mathbb{N}_0 := \{0, 1, 2, \ldots\}$. Then X is represented as

$$X = \bigcap_{k\in\mathbb{N}_0} X_k,$$

where X_k is the completion of X with respect to the norm $|\cdot|_k$. We use X_{-k} to denote the dual space of X_k. Then the dual space X' of X can be represented as

$$X' = \bigcup_{k\in\mathbb{N}_0} X_{-k}$$

which is equipped with the inductive limit topology.

Let $N = X + iX$ and $N_k = X_k + iX_k$, $k \in \mathbb{Z}$, be the complexifications of X and X_k, respectively. For $n \in \mathbb{N}_0$, we denote by $N^{\hat{\otimes} n}$ the n-fold symmetric tensor product of N equipped with the π-topology and by $N_k^{\hat{\otimes} n}$ the n-fold symmetric Hilbertian tensor product of N_k. We will preserve the notation $|\cdot|_k$ and $|\cdot|_{-k}$ for the norms on $N_k^{\hat{\otimes} n}$ and $N_{-k}^{\hat{\otimes} n}$, respectively.

Functional spaces

Let $\theta : \mathbb{R}_+ \longrightarrow \mathbb{R}_+$ be a continuous, convex, increasing function satisfying

$$\lim_{t\to\infty} \frac{\theta(t)}{t} = \infty \qquad \text{and} \quad \theta(0) = 0.$$

Such a function is called a Young function. For a Young function θ we define

$$\theta^*(x) := \sup_{t\geq 0}\{tx - \theta(t)\}, \qquad x \geq 0.$$

This is called the polar function associated to θ. It is known that θ^* is again a Young function and $(\theta^*)^* = \theta$, see Ref. 10 for more details and general results.

For a Young function θ, we denote by $\mathcal{F}_\theta(N')$ the space of holomorphic functions on N' with exponential growth of order θ and of minimal type. Similarly, let $\mathcal{G}_\theta(N)$ denote the space of holomorphic functions on N with exponential growth of order θ and of arbitrary type. Moreover, for each $k \in \mathbb{Z}$ and $m > 0$, define $\mathcal{F}_{\theta,m}(N_k)$ to be the Banach space of entire functions f on N_k satisfying the condition

$$|f|_{\theta,k,m} := \sup_{x\in N_k} |f(x)|e^{-\theta(m|x|_k)} < \infty.$$

Then the spaces $\mathcal{F}_\theta(N')$ and $\mathcal{G}_\theta(N)$ may be represented as

$$\mathcal{F}_\theta(N') = \bigcap_{k\in\mathbb{N}_0, m>0} \mathcal{F}_{\theta,m}(N_{-k}),$$
$$\mathcal{G}_\theta(N) = \bigcup_{k\in\mathbb{N}_0, m>0} \mathcal{F}_{\theta,m}(N_k)$$

which are equipped with the projective limit topology and the inductive limit topology, respectively. The space $\mathcal{F}_\theta(N')$ is called the space of test functions on N'. Its dual space $\mathcal{F}'_\theta(N')$, equipped with the strong topology, is called the space of generalized functions. The dual pairing between $\mathcal{F}'_\theta(N')$ and $\mathcal{F}_\theta(N')$ is denoted by $\langle\langle \cdot, \cdot \rangle\rangle$.

For $k \in \mathbb{N}_0$ and $m > 0$, we define the Hilbert spaces

$$F_{\theta,m}(N_k) = \left\{ \vec{\varphi} = (\varphi_n)_{n=0}^{\infty};\ \varphi_n \in N_k^{\hat{\otimes} n},\ \sum_{n=0}^{\infty} \theta_n^{-2} m^{-n} |\varphi_n|_k^2 < \infty \right\},$$
$$G_{\theta,m}(N_{-k}) = \left\{ \vec{\Phi} = (\Phi_n)_{n=0}^{\infty}\ ;\ \Phi_n \in N_k^{\hat{\otimes} n},\ \sum_{n=0}^{\infty} (n!\theta_n)^2 m^n |\Phi_n|_{-k}^2 < \infty \right\},$$

where

$$\theta_n = \inf_{x>0} \frac{e^{\theta(x)}}{x^n}, \qquad n \in \mathbb{N}_0.$$

We define

$$F_\theta(N) := \bigcap_{k\in\mathbb{N}_0, m>0} F_{\theta,m}(N_k)$$
$$G_\theta(N') := \bigcup_{k\in\mathbb{N}_0, m>0} G_{\theta,m}(N_{-k}).$$

The space $F_\theta(N)$ equipped with the projective limit topology is a nuclear Fréchet space, see [6, Proposition 2]. The space $G_\theta(N')$ carries the dual topology of $F_\theta(N)$ with respect to the bilinear pairing given by

$$\langle\langle \vec{\Phi}, \vec{\varphi} \rangle\rangle = \sum_{n=0}^{\infty} n! \langle \Phi_n, \varphi_n \rangle, \tag{2}$$

where $\vec{\Phi} = (\Phi_n)_{n=0}^{\infty} \in G_\theta(N')$ and $\vec{\varphi} = (\varphi_n)_{n=0}^{\infty} \in F_\theta(N)$.

The Taylor map defined by

$$\mathfrak{T} : \mathcal{F}_\theta(N') \longrightarrow F_\theta(N),\ \varphi \mapsto \left(\frac{1}{n!} \varphi^{(n)}(0) \right)_{n=0}^{\infty}$$

is a topological isomorphism. The same is true between $\mathcal{G}_{\theta^*}(N)$ and $G_\theta(N')$. The action of a distribution $\Phi \in \mathcal{F}'_\theta(N')$ on a test function $\varphi \in \mathcal{F}_\theta(N')$ can be expressed in terms of the Taylor map as follows:

$$\langle\langle \Phi, \varphi \rangle\rangle = \langle\langle \vec{\Phi}, \vec{\varphi} \rangle\rangle, \tag{3}$$

where $\vec{\Phi} = (\mathfrak{T}^*)^{-1}\Phi$ and $\vec{\varphi} = \mathfrak{T}\varphi$.

Laplace transform

It is easy to see that for each $\xi \in N$, the exponential function

$$e_\xi(z) = e^{\langle z, \xi \rangle}, \quad z \in N',$$

is a test function in the space $\mathcal{F}_\theta(N')$ for any Young function θ, cf. [6, Lemme 2]. Thus the Laplace transform of a generalized function $\Phi \in \mathcal{F}'_\theta(N')$

$$\mathcal{L}\Phi(\xi) := \langle\langle \Phi, e_\xi \rangle\rangle, \quad \xi \in N, \tag{4}$$

is well defined. The Laplace transform is a topological isomorphism between $\mathcal{F}'_\theta(N')$ and $\mathcal{G}_{\theta^*}(N)$ (cf. [6, Théorème 1]).

Convolution

For $\varphi \in \mathcal{F}_\theta(N')$, the translation $t_x\varphi$ of φ by $x \in N'$ is defined by

$$(t_x\varphi)(y) = \varphi(y - x), \quad y \in N'.$$

It is easy to check that, for any $x \in N'$, t_x is a continuous linear operator from $\mathcal{F}_\theta(N')$ into itself , cf. [5, Proposition 2.1]. We may define the translation t_x on $\mathcal{F}'_\theta(N')$ as follows: Let $\Phi \in \mathcal{F}'_\theta(N')$ be given, then $t_x\Phi \in \mathcal{F}'{}_\theta(N')$ is defined by

$$\langle\langle t_x\Phi, \varphi \rangle\rangle := \langle\langle \Phi, t_{-x}\varphi \rangle\rangle, \quad \forall \varphi \in \mathcal{F}_\theta(N').$$

We define the convolution $\Phi * \varphi$ of a generalized function $\Phi \in \mathcal{F}'_\theta(N')$ and a test function $\varphi \in \mathcal{F}_\theta(N')$ to be the test function

$$(\Phi * \varphi)(x) := \langle\langle \Phi, t_{-x}\varphi \rangle\rangle, \quad x \in N'.$$

For the proof, see [5, Lemme 2.1].

Remark 2.1. The definition of $\Phi * \varphi$ does not generalize the notion of convolution of two test functions even if we have the injection: $\mathcal{F}_\theta(N') \hookrightarrow$

$\mathcal{F}'_\theta(N')$. On the other hand, it does generalize the notion of convolution between a measure and a function.

For any $\Phi, \Psi \in \mathcal{F}'_\theta(N')$ we define the convolution $\Phi * \Psi \in \mathcal{F}'_\theta(N')$ by

$$\langle\langle \Phi * \Psi, \varphi \rangle\rangle := \langle\langle \Phi, \Psi * \varphi \rangle\rangle, \quad \varphi \in \mathcal{F}_\theta(N').$$

Remark 2.2. We notice that the above definition does generalize the notion of convolution of measures. It is not surprising that the commutative and associative laws holds because it does for measures.

As a consequence we have the following

Lemma 2.1. *Let $\Phi, \Psi \in \mathcal{F}'_\theta(N')$ be given, then we have*

$$\mathcal{L}(\Phi * \Psi) = \mathcal{L}\Phi\mathcal{L}\Psi. \tag{5}$$

Operator Δ_K and Δ_K^*

Let $K \in L(N, N')$ be given, where $L(N, N')$ is the set of continuous linear operators from N to N'. We denote by $\tau(K)$ the kernel associated to K in $(N \otimes N)'$ (which is isomorphic to $L(N, N')$, see Ref. 1) which is defined by

$$\langle \tau(K), \xi \otimes \eta \rangle = \langle K\xi, \eta \rangle.$$

For $\varphi \in \mathcal{F}_\theta(N')$ of the form

$$\varphi(x) = \sum_{n=0}^{\infty} \langle x^{\otimes n}, \varphi^{(n)} \rangle, \tag{6}$$

we define the operator Δ_K of φ at $x \in N'$ by

$$(\Delta_K \varphi)(x) := \sum_{n=0}^{\infty} (n+2)(n+1) \langle x^{\otimes n}, \langle \tau(K), \varphi^{(n+2)} \rangle \rangle, \tag{7}$$

where the contraction $\langle \tau(K), \varphi^{(n+2)} \rangle$ is defined by

$$\langle x^{\otimes n}, \langle \tau(K), \varphi^{(n+2)} \rangle \rangle := \langle x^{\otimes n} \hat{\otimes} \tau(K), \varphi^{(n+2)} \rangle.$$

In particular, for $K = I$ (embedding of N in N'), $\tau(I)$ is the trace operator and Δ_K is the Gross Laplacian.

We state the following useful

Lemma 2.2. *Let $\varphi \in \mathcal{F}_\theta(N')$ be given and let K be an operator as described above. Then Δ_K is a convolution operator, namely*

$$\Delta_K \varphi = \mathcal{T}(\tau(K)) * \varphi, \quad \varphi \in \mathcal{F}_\theta(N'). \tag{8}$$

where $\mathcal{T}(\tau(K)) \in \mathcal{F}'_\theta(N')$ is associated with $\overrightarrow{\mathcal{T}(\tau(K))} = (0, 0, \tau(K), 0, \ldots) \in G_\theta(N')$.

Proof. We take φ of the form: $\varphi(x) = e^{\langle x,\xi\rangle} = \sum_{n=0}^{\infty} \frac{1}{n!}\langle x^{\otimes n}, \varphi^{(n)}\rangle$, $\varphi^{(n)} = \xi^{\otimes n}$. Then we have

$$\begin{aligned}(\Delta_K\varphi)(x) &= \sum_{n=0}^{\infty}(n+2)(n+1)\langle x^{\otimes n}, \langle \tau(K), \frac{1}{(n+2)!}\xi^{\otimes(n+2)}\rangle\rangle \\ &= \sum_{n=0}^{\infty}\frac{1}{n!}\langle x^{\otimes n}, \langle \tau(K), \xi^{\otimes(n+2)}\rangle\rangle \\ &= \langle K\xi, \xi\rangle\varphi(x).\end{aligned}$$

Noting that $\varphi^{(2)}(\cdot + x) = \frac{1}{2}\xi^{\otimes 2}e^{\langle x,\xi\rangle}$ then we have

$(\mathcal{T}(\tau(K))*\varphi)(x) = \langle\langle \mathcal{T}(\tau(K)), t_{-x}\varphi\rangle\rangle = 2\langle \tau(K), \varphi^{(2)}(\cdot + x)\rangle = \langle K\xi, \xi\rangle\varphi(x)$.

The result follows by density of the exponential functions on $\mathcal{F}_\theta(N')$. □

Let A be the operator defined for any $\Phi \in \mathcal{F}'_\theta(N')$ by

$$A\Phi := \mathcal{T}(\tau(K)) * \Phi.$$

It follows that for all $\varphi \in \mathcal{F}_\theta(N')$ we have

$$\begin{aligned}\langle\langle A\Phi, \varphi\rangle\rangle &= \langle\langle \mathcal{T}(\tau(K)) * \Phi, \varphi\rangle\rangle \\ &= \langle\langle \Phi, \mathcal{T}(\tau(K)) * \varphi\rangle\rangle \\ &= \langle\langle \Phi, \Delta_K\varphi\rangle\rangle \\ &=: \langle\langle \Delta_K^*\Phi, \varphi\rangle\rangle\end{aligned}$$

which proves that A is the adjoint operator Δ_K^*.

It is clear, using (5), that

$$(\mathcal{L}(\Delta_K^*\Phi))(\xi) = \langle K\xi, \xi\rangle(\mathcal{L}\Phi)(\xi). \tag{9}$$

3. Convolution equation: existence and probabilistic representation

Now we are able to investigate the following Cauchy problem

$$\frac{\partial}{\partial u}U(u) = \frac{1}{2}\Delta_K^* U(u), \quad u \in [0,T], \quad U(0) = \Phi \in \mathcal{F}'_\theta(N'). \tag{10}$$

Applying the Laplace transform to (10) we obtain

$$\frac{\partial}{\partial u}\mathcal{L}U(u) = \frac{1}{2}\mathcal{L}\mathcal{T}(\tau(K))\mathcal{L}U(u), \quad u \in [0,T], \quad \mathcal{L}U(0) = \mathcal{L}\Phi \in \mathcal{G}_{\theta^*}(N). \tag{11}$$

Therefore the unique solution of (11) is given by

$$\mathcal{L}U(u) = (\mathcal{L}\Phi)\exp\left(\frac{u}{2}\mathcal{L}\mathcal{T}(\tau(K))\right). \tag{12}$$

Finally, the solution of (10) is obtained using the characterization theorem as

$$U(u) = \Phi * e^{*u\mathcal{T}(\tau(K))/2}. \tag{13}$$

We proceed in order to give a probabilistic representation of the solution (13). First we keep the notation K for its extension to X_{-p} ($p \in \mathbb{N}_0$ fixed) into itself. Moreover we assume that K is a symmetric, nonnegative linear operator with finite trace. We follow closely the ideas from Refs. 4 and 14. Let $(\Omega, \mathcal{F}, (\mathcal{F}_u)_{u\in[0,T]}, P)$ be a filtered probability space with a filtration $(\mathcal{F}_u)_{u\in[0,T]}$ satisfying the usual conditions. By a K-Wiener process $W = (W(u))_{u\in[0,T]}$ we mean a X_{-p}-valued process on $(\Omega, \mathcal{F}, P)$ such that

- $W(0) = 0$,
- W has P-a.s. continuous trajectories,
- the increments of W are independent,
- the increments $W(u) - W(v)$, $0 < v \leq u$ have the following Gaussian law:

  $$P \circ (W(u) - W(v))^{-1} = N(0, (u-v)K),$$

  where $N(0,(u-v)K)$ denotes the Gaussian distribution with zero mean and covariance operator $(u-v)K$.

A K-Wiener process with respect to the filtration $(\mathcal{F}_u)_{u\in[0,T]}$ is a K-Wiener process such that

- $W(u)$ is $\mathcal{F}_u$-adapted,
- $W(u) - W(v)$ is independent of $\mathcal{F}_u$ for all $v \in [0,u]$.

Later on we need to define stochastic integrals of $\mathcal{F}'_\theta(N')$-valued process. We use the theory of stochastic integration in Hilbert spaces developed in Refs. 3 and 12. Before we introduce some notations. Given two separable Hilbert spaces $\mathcal{H}_1, \mathcal{H}_2$ we denote by $L(\mathcal{H}_1, \mathcal{H}_2)$ (resp. $L_2(\mathcal{H}_1, \mathcal{H}_2)$) the space of all bounded linear operators from $\mathcal{H}_1$ to $\mathcal{H}_2$ (resp. Hilbert-Schmidt operators). The Hilbert-Schmidt norm of an element $S \in L_2(\mathcal{H}_1, \mathcal{H}_2)$ is denoted by $\|S\|_{HS}$.

Definition 3.1. Let $(\Phi(u))_{0\le u\le T}$ be a given $L(X_{-p}, \mathcal{F}'_\theta(N'))$-valued, $\mathcal{F}_u$-adapted continuous stochastic process. Assume that there exist $m > 0$ and $p \in \mathbb{N}_0$ such that $\mathfrak{T} \circ \mathcal{L}\Phi(u) \in L(X_{-p}, G_{\theta,m}(N_{-p}))$ and

$$P\left(\int_0^T \left\|(\mathfrak{T} \circ \mathcal{L}\Phi(u)) \circ K^{1/2}\right\|_{HS}^2 du < \infty\right) = 1. \tag{14}$$

Then, for $u \in [0, T]$, we define the generalized stochastic integral

$$\int_0^u \Phi(v)dW(v) \in \mathcal{F}'_\theta(N')$$

by

$$\mathfrak{T}\left(\mathcal{L}\left(\int_0^u \Phi(v)dW(v)\right)(\xi)\right) := \int_0^u \mathfrak{T}((\mathcal{L}\Phi(v))(\xi))\, dW(v). \tag{15}$$

Notice that the right hand side of (15) is a well defined stochastic integral in a Hilbert space by the condition (14), see Ref. 3.
We are going to derive the Itô formula for $t_{W(u)}\Phi$, $\Phi \in \mathcal{F}'_\theta(N')$. Before we give a technical lemma. By a $\mathcal{F}'_\theta(N')$-valued continuous $\mathcal{F}_u$-semimartingale $(Z(u))_{u\in[0,T]}$, we mean processes of the form

$$Z(u) = Z(0) + \int_0^u \Phi(v)dW(v) + \int_0^u \Psi(v)dv,$$

when all terms in the right hand side are well defined. We state to following lemma which the simple proof is left to the interested reader.

Lemma 3.1. *Let* $\Phi \in \mathcal{F}'_\theta(N')$, $\xi \in N$ *and* $g : X_{-p} \longrightarrow \mathbb{C}$, $g(x) := \langle\langle t_x\Phi, e_\xi\rangle\rangle = (\mathcal{L}(t_x\Phi))(\xi)$ *be given. Then* g *is twice continuously differen-*

tiable and

$$(Dg)(x)(y) = \langle\langle \Phi, t_{-x}e_\xi \rangle\rangle \langle y, \xi \rangle, \qquad x, y \in X_{-p}. \tag{16}$$

$$(D^2g)(x)(y_1, y_2) = \langle\langle \Phi, t_{-x}e_\xi \rangle\rangle \langle y_1, \xi \rangle \langle y_2, \xi \rangle, \qquad y_1, y_2 \in X_{-p}. \tag{17}$$

Moreover g, Dg *and* D^2g *are uniformly continuous on bounded sets of* X_{-p}.

Remark 3.1.

(1) In the conditions of Lemma 3.1 we may rewrite (16) and (17) as

$$(Dg)(x) = -\langle\langle D(t_x\Phi), e_\xi \rangle\rangle$$
$$(D^2g)(x) = \langle\langle D^2(t_x\Phi), e_\xi \rangle\rangle.$$

(2) Since the family of exponential functions $\{e_\xi,\ \xi \in N\}$ is dense in $\mathcal{F}_\theta(N')$ the same result holds for $g(x) = \langle\langle t_x\Phi, \varphi \rangle\rangle$ with $\varphi \in \mathcal{F}_\theta(N')$.

Now we are able to prove the announced Itô formula.

Theorem 3.1. *Let* $(W(u))_{u\in[0,T]}$ *be a* K*-Wiener process with respect to the filtration* $(\mathcal{F}_u)_{u\in[0,T]}$ *and let* $\Phi \in \mathcal{F}'_\theta(N')$ *be given. Then* $t_{W(u)}\Phi$ *is a* $\mathcal{F}'_\theta(N')$*-valued continuous* $\mathcal{F}_u$*-semimartingale which has the following decomposition*

$$t_{W(u)}\Phi = t_{W(0)}\Phi - \int_0^u D(t_{W(v)}\Phi)dW(v) + \frac{1}{2}\int_0^u \Delta_K^*(t_{W(v)}\Phi)dv.$$

Proof. By Lemma 3.1 the function $g : X_{-p} \longrightarrow \mathbb{C}$, $g(x) := \langle\langle t_x\Phi, e_\xi \rangle\rangle$ for $\xi \in N$ is twice continuously differentiable. Then applying Itô's formula we get

$$g(W(u)) = g(W(0)) + \int_0^u Dg(W(v))dW(v) + \frac{1}{2}\int_0^u \mathrm{tr}[D^2g(W(v))K]dv.$$

The explicit representation for g gives

$$\langle\langle t_{W(u)}\Phi, e_\xi\rangle\rangle = \langle\langle t_{W(0)}\Phi, e_\xi\rangle\rangle - \int_0^u \langle\langle D(t_{W(v)}\Phi), e_\xi\rangle\rangle dW(v) + \frac{1}{2}\int_0^u \mathrm{tr}[\langle\langle D^2(t_{W(v)}\Phi), e_\xi\rangle\rangle K]dv.$$

The trace in the last integral above may be written as

$$\begin{aligned}\mathrm{tr}[\langle\langle D^2(t_{W(v)}\Phi), e_\xi\rangle\rangle K] &= \sum_{i\geq 1}\langle\langle D^2(t_{W(v)}\Phi), e_\xi\rangle\rangle\langle Kf_i,\xi\rangle\langle f_i,\xi\rangle \\ &= \langle\langle D^2(t_{W(v)}\Phi), e_\xi\rangle\rangle\langle K\xi,\xi\rangle,\end{aligned}$$

where we have used the symmetry of K and the fact $\sum_{i\geq 1}\langle f_i, K\xi\rangle\langle f_i,\xi\rangle = \langle K\xi,\xi\rangle$ for one (hence every) orthonormal basis $\{f_i;\, i\geq 1\}$ in X_{-p}. The result follows by (9). □

In order to show that the solution of (10) is given in terms of an expectation, first we prove the following

Lemma 3.2. *Let $\Phi \in \mathcal{F}'_\theta(N')$, $\xi\in N$ be given and $g: X_{-p}\longrightarrow \mathbb{C}$, $g(x) = \langle\langle t_x\Phi, e_\xi\rangle\rangle$. Then we have*

$$\mathbb{E}\left(\int_0^T \left\|Dg(W(u))\circ K^{1/2}\right\|^2_{HS} du\right) < \infty.$$

Thus, $\{\int_0^u Dg(W(v))dW(v),\, u\in[0,T]\}$ is a $\mathcal{F}'_\theta(N')$-valued continuous, square integrable martingale.

Proof. For every orthonormal basis $\{f_i,\, i\geq 1\}$ in X_{-p} we have

$$\left\|(Dg)(W(u))\circ K^{1/2}\right\|^2_{HS} \leq |\langle\langle t_{W(u)}\Phi, e_\xi\rangle\rangle|^2|\xi|_p^2\sum_{i\geq 1}|K^{1/2}f_i|^2_{-p}.$$

Notice that $\langle\langle t_{W(u)}\Phi, e_\xi\rangle\rangle = \langle\langle\Phi, e_\xi\rangle\rangle e^{\langle W(u),\xi\rangle}$ and $\sum_{i\geq 1}|K^{1/2}f_i|^2_{-p} = ||K^{1/2}||^2_{HS}$. Therefore we obtain

$$\left\|(Dg)(W(u))\circ K^{1/2}\right\|^2_{HS} \leq C(\xi,\Phi,K,\alpha)e^{\alpha|W(u)|^2_{-p}},$$

where the constant $C(\xi,\Phi,K,\alpha)$ is given by

$$C(\xi,\Phi,K,\alpha) := |\xi|_p^2|e^{|\xi|_p^2/\alpha}|\langle\langle\Phi, e_\xi\rangle\rangle|^2||K^{1/2}||^2_{HS}$$

and we have used, for any $\alpha > 0$,

$$\begin{cases} 2|\langle W(u), \xi\rangle| \leq |W(u)|_{-p}|\xi|_p \leq \dfrac{|\xi|_p^2}{\alpha} + \alpha|W(u)|_{-p}^2, \\ |e^{\langle W(u),\xi\rangle}| \leq e^{|\langle W(u),\xi\rangle|} \leq e^{|\xi|_p^2/\alpha} e^{\alpha|W(u)|_{-p}^2}. \end{cases}$$

Since $W(u)$ has Gaussian law $N(0, uK)$ it follows that

$$\mathbb{E}\left(\int_0^T \|(Dg)\,(W(u)) \circ K^{1/2}\|_{HS}^2\, du\right)$$
$$\leq C\,(\xi, \Phi, K, \alpha) \int_0^T \int_{X_{-p}} e^{\alpha u|x|_{-p}^2} dN(0, K) du.$$

For $\alpha \in [0, \frac{1}{2T\mathrm{tr}[K]}[$ the stochastic integral above admits a Fernique estimation of Gaussian measure (cf. [3, Proposition 2.16, p. 56]) so that

$$\int_{X_{-p}} e^{\alpha u|x|_{-p}^2} dN(0, K) \leq \frac{1}{\sqrt{1 - 2u\alpha\mathrm{tr}[K]}}.$$

So, we obtain

$$\mathbb{E}\left(\int_0^T \|(Dg)\,(W(u)) \circ K^{1/2}\|_{HS}^2\, du\right)$$
$$\leq C\,(\xi, \Phi, K, \alpha) \int_0^T \frac{du}{\sqrt{1 - 2u\alpha\mathrm{tr}[K]}} < +\infty. \qquad \square$$

As consequence of the above lemma for each $\xi \in N$ the process

$$\left\{\int_0^u \langle\langle D(t_{W(v)}\Phi), e_\xi\rangle\rangle dW(v),\ u \in [0, T]\right\},$$

is a $L^2(P)$-bounded martingale. Therefore we have

Corollary 3.1. *The following stochastic integral*

$$\left\{\int_0^v \mathfrak{T} \circ \mathcal{L}(D(t_{W(u)}\Phi))dW(u),\ v \in [0, T]\right\},$$

is a $L^2(P)$-bounded martingale.

Now we give a probabilistic representation formula of the solution of the Cauchy problem (10).

Theorem 3.2. *The solution of the Cauchy problem (10) is given by*

$$U(u) = \mathbb{E}_{P^x}(t_{W(u)}\Phi). \tag{18}$$

where P^x is the probability law of W starting at $x \in X_{-p}$.

Proof. To check that $U(u) = \mathbb{E}_{P^x}(t_{W(u)}\Phi)$ is the solution of the Cauchy problem (10), it suffices to show that its Laplace transform $\mathcal{L}U(u)$ satisfies the Cauchy problem (11). It follows from Itô's formula, with $\xi \in N$, that

$$\langle\langle t_{W(u)}\Phi, e_\xi\rangle\rangle = \langle\langle t_{W(0)}\Phi, e_\xi\rangle\rangle - \int_0^u \langle\langle D(t_{W(v)}\Phi), e_\xi\rangle\rangle dW(v) + \frac{1}{2}\int_0^u \mathrm{tr}[\langle\langle D^2(t_{W(v)}\Phi), e_\xi\rangle\rangle K]dv.$$

Taking expectation and using the fact that $\left(\int_0^u \langle\langle D(t_{W(v)}\Phi), e_\xi\rangle\rangle dW(v)\right)$ is a $L^2(P)$-bounded martingale yields

$$\begin{aligned}\mathbb{E}_{P^x}\langle\langle t_{W(u)}\Phi, e_\xi\rangle\rangle &= \mathbb{E}_{P^x}\langle\langle t_{W(0)}\Phi, e_\xi\rangle\rangle + \frac{1}{2}\int_0^u \mathbb{E}_{P^x}\mathrm{tr}[\langle\langle D^2(t_{W(v)}\Phi), e_\xi\rangle\rangle K]dv.\\ &= \mathbb{E}_{P^x}\langle\langle t_{W(0)}\Phi, e_\xi\rangle\rangle + \frac{1}{2}\int_0^u \mathbb{E}_{P^x}\Delta_K^*(t_{W(v)}\Phi)(\xi)dv.\end{aligned}$$

Using the definition of Laplace transform (7), the last equation can be written as

$$\mathcal{L}U(u)(\xi) = \mathcal{L}U(0)(\xi) + \frac{1}{2}\int_0^u \mathcal{L}\Delta_K^* U(v)(\xi)dv$$

or making use of the explicit form (9) as

$$\mathcal{L}U(u)(\xi) = \mathcal{L}U(0)(\xi) + \frac{1}{2}\int_0^u \mathcal{L}U(v)(\xi)\langle K\xi, \xi\rangle dv$$

which implies that $\mathcal{L}U(u)(\xi)$ solves the Cauchy problem (11). □

Acknowledgment. We would like to thanks the referee for the constructive comments. In particular Prof. H.-H. Kuo for his helpful suggestions concerning the presentation of this paper and mathematical comments. Financial support from by FCT, POCTI-219, FEDER through GRICES, Proc. 4.1.5/Marroc, PTDC/MAT/67965/2006, and are gratefully acknowledged.

References

1. Yu. M. Berezansky and Yu. G. Kondratiev, *Spectral methods in infinite-dimensional analysis*, vol. 1, Kluwer Academic Publishers, Dordrecht, 1995.

2. D. M. Chung and U. C. Ji, *Some Cauchy problems in white noise analysis and associated semigroups of operators*, Stochastic Anal. Appl. **17** (1999), no. 1, 1–22.
3. G. Da Prato and J. Zabczyk, *Stochastic equations in infinite dimensions*, Cambridge University Press, London, 1992.
4. M. Erraoui, Y. Ouknine, H. Ouerdiane, and J. L. Silva, *Probabilistic representation of heat equation of convolution type*, Random Oper. and Stoch. Equ. **13** (2005), no. 4, 325–340.
5. R. Gannoun, R. Hachaichi, P. Kree, and H. Ouerdiane, *Division de fonctions holomorphes a croissance θ-exponentielle*, Tech. Report E 00-01-04, BiBoS University of Bielefeld, 2000.
6. R. Gannoun, R. Hachaichi, H. Ouerdiane, and A. Rezgui, *Un théoréme de dualité entre espaces de fonctions holomorphes à croissance exponentielle*, J. Funct. Anal. **171** (2000), no. 1, 1–14.
7. L. Gross, *Potential theory on Hilbert space*, J. Funct. Anal. **1** (1967), 123–181.
8. T. Hida, H. H. Kuo, J. Potthoff, and L. Streit, *White noise. an infinite dimensional calculus*, Kluwer Academic Publishers, Dordrecht, 1993.
9. T. Hida, N. Obata, and K. Saitô, *Infinite-dimensional rotations and Laplacians in terms of white noise calculus*, Nagoya Math. J. **128** (1992), 65–93.
10. M-A. Krasnosel'ski and Ya-B. Ritickili, *Convex functions and orliez spaces*, P. Nordhoff. Itd, Groningen, The Nethelands, 1961.
11. H.H. Kuo, *On Laplacian operators of generalized Brownian functionals*, Stochastic processes and their applications (Nagoya, 1985), Lecture Notes in Math., vol. 1203, Springer, Berlin, 1986, pp. 119–128.
12. M. Métivier, *Seminartingales: A case on stochastic processes*, Walter De Gruyter Inc, 1982.
13. M. A. Piech, *A fundamental solution of the parabolic equation on Hilbert*

space. II. The semigroup property, Trans. Amer. Math. Soc. **150** (1970), 257–286.

14. B. Rajeev and S. Thangavelu, *Probabilistic representations of solutions to the heat equation*, Proc. Indian Acad. Sci. Math. Sci. **113** (2003), no. 3, 321–332.

FROM CLASSICAL TO QUANTUM ENTROPY PRODUCTION

F. FAGNOLA

Politecnico di Milano

Dipartimento di Matematica "F. Brioschi"

Piazza Leonardo da Vinci 32, I-20133 Milano, Italy

E-mail: franco.fagnola@polimi.it

R. REBOLLEDO

Centro de Análisis Estocástico y Aplicaciones

Facultad de Matemáticas

Pontificia Universidad Católica de Chile

Casilla 306, Santiago 22, Chile

E-mail: rrebolle@uc.cl

A definition of entropy production for quantum Markov semigroups is proposed, based on the classical probabilistic concept, and illustrated by an example.

Keywords: quantum Markov semigroup; entropy production; detailed balance; local reversibility.

1. Introduction

This paper starts an investigation on the Markovian approach to non equilibrium quantum dynamics, a terrain which includes the analysis of a number of different concepts like entropy production, irreversibility, KMS states and detailed balance conditions among others.

The analysis of entropy production appears in the celebrated 1931 paper of Onsager[23] on reciprocal relations in irreversible processes, who quotes himself Boltzmann and Thomson, among the founders of Thermodynamics and the Kinetic Theory of gases, to defend the principle of local reversibility which should be the cornerstone of the so called reciprocal relations: while the convergence towards the equilibrium is irreversible, at a microscopic level each particle could reveal a reversible dynamics.

Since then, the subject has been explored in several different frameworks by a number of authors. Jaksic and Pillet[12] propose a definition of quantum entropy production in the framework of algebraic quantum statistical mechanics, and prove the Onsager reciprocity relations for heat and charge fluxes for locally interacting open fermionic systems in Ref. 13. They continue their analysis of the linear response theory in Ref. 11. In Ref. 7, Esposito investigates on the fluctuation of entropy in driven open systems, Breuer[10] computes the entropy production for quantum jumps models; similarly, in a series of papers C. Maes and collaborators address a list of examples on the computation of the entropy production (see Refs. 6,15,16,18–21). In Ref. 26 Prez Madrid writes on the quantum theory of irreversibility based on a generalization of the Gibbs-von Neumann entropy inspired in the Bogolyubov-Born-Green-Kirkwood-Yvon hierarchy. Finally Zia (cf. Ref. 29) stresses the importance of the concept of *probability current* as the principal characteristics in the statistical mechanics of non-equilibrium steady states. The above list of references does not pretend to be exhaustive but illustrates the main theoretical features involved in the concept of entropy production.

A common idea in a number of the above quoted papers, based on axiomatic or phenomenological approaches, is that non-equilibrium states are characterized by non-zero entropy production. Moreover, when the evolution is given by a Markov process, equilibrium states are characterized by the detailed balance condition (see Ref. 2).

In our approach these are the two main inspirations. That is, we start discussing the entropy production for a classical Markov process (see for instance[14]). In addition we extract the main concepts of detailed balance condition from Ref. 9 and references therein. After that, we go through the extension of entropy production to the quantum framework, where the dynamics is given by a Quantum Markov Semigroup (QMS). So that we propose a new genuinely non-commutative notion of entropy production.

2. Detailed balance and entropy production

2.1. *The commutative detailed balance condition*

Let $(E, \mathcal{E}, \mu)$ be a measure space, where μ is a σ-finite measure and $\mathcal{A} = L^\infty(E, \mathcal{E}, \mu)$. Consider an E-valued Markov process $(X_t)_{t\in\mathbb{R}^+}$ with $(T_t)_{t\in\mathbb{R}^+}$ as associated semigroup defined on $\mathcal{A}$. We suppose that there is a transition density function $p_t(x, y)$ such that $T_t f(x) = \int_E p_t(x, y) f(y)\mu(dy)$, for any $f \in \mathcal{A}$, $x \in E$. Moreover, we suppose that there exists an invariant probability measure with a density $\pi(x) > 0$ for all $x \in E$.

The process (or the semigroup) is called *reversible* if the *classical detailed balance condition* holds

$$\pi(x)p_t(x, y) = \pi(y)p_t(y, x), \tag{1}$$

for all t, $x, y \in E$.

Under this condition it follows that for any two $f, g \in \mathcal{A}$ one obtains:

$$\begin{aligned}\int_E \pi g T_t f d\mu &= \int_{E\times E} \mu(dx)\mu(dy)g(x)\pi(x)p_t(x, y)f(y)\\ &= \int_{E\times E} \mu(dx)\mu(dy)g(x)\pi(y)p_t(y, x)f(y)\\ &= \int_E \pi T_t g f d\mu.\end{aligned}$$

That is, the semigroup is self-adjoint in the L^2 space of the invariant measure $\pi d\mu$. Conversely, if the semigroup is self-adjoint in the above L^2 space, equation (1) follows.

We recall the construction of the time reverted process as follows. Fix $t > 0$ and consider any sequence of ordered times $0 = t_0 < t_1 < \ldots < t_n = t$. The Markov property entails

$$\mathbb{E}_x[f_0(X_{t_0}) \cdots f_n(X_{t_n})] = T_{t_0}\left(f_0 T_{t_1-t_0}\left(f_1 \ldots\right)\right)(x). \tag{2}$$

The t-time-reversed process $\overleftarrow{X}^t$ satisfies:

$$\begin{aligned}\mathbb{E}_x\left[f_0\left(\overleftarrow{X}^t_{t_0}\right)\ldots f_n\left(\overleftarrow{X}^t_{t_n}\right)\right] &= \mathbb{E}_x[f_0(X_{t-t_n})\ldots f_n(X_{t-t_0})] \\ &= T_{t-t_n}\left(f_0 T_{t_n-t_{n-1}}\left(f_1 T_{t_{n-1}-t_{n-2}}\cdots\right)\right)(x).\end{aligned} \tag{3}$$

Given any invariant density probability measure π the processes X and $\overleftarrow{X}^t$ become stationary and time homogeneous. For any finite partition $0 \leq t_1 < t_2 < \ldots < t_n = t$ they determine probability densities on the algebra

$\mathcal{A}^{n+1}$ as follows:

$$\overrightarrow{\mathbb{P}}_{\pi,t}(f_0 \otimes \ldots \otimes f_n) =$$
$$\int_E \pi(x_0)p_{t_1}(x_0,x_1)\ldots p_{t-t_{n-1}}(x_{n-1},x_n)f_0(x_0)\ldots f_n(x_n)\mu(dx_0)\ldots\mu(dx_n).$$

$$\overleftarrow{\mathbb{P}}_{\pi,t}(f_0 \otimes \ldots \otimes f_n) =$$
$$\int_E \pi(x_n)p_{t-t_{n-1}}(x_n,x_{n-1})\ldots p_{t_1}(x_1,x_0)f_0(x_n)\ldots f_n(x_0)\mu(dx_0)\ldots\mu(dx_n).$$

These probabilities can be extended to the whole cylindrical σ–algebra through Kolmogorov's Theorem and we denote them as $\overrightarrow{\mathbb{P}}_{\pi}(\cdot)$, $\overleftarrow{\mathbb{P}}_{\pi}(\cdot)$ respectively.

2.2. *Entropy production for classical Markov processes*

Lemma 2.1. *Suppose that π is a stationary density and let $r, t > 0$. Then*

$$\overrightarrow{\mathbb{P}}_{\pi}\left(\log\frac{\pi(X_0)p_r(X_0,X_r)p_t(X_r,X_{t+r})}{\pi(X_{t+r})p_t(X_{t+r},X_r)p_r(X_r,X_0)}\right) =$$
$$\overrightarrow{\mathbb{P}}_{\pi}\left(\log\frac{\pi(X_0)p_r(X_0,X_r)}{\pi(X_r)p_r(X_r,X_0)}\right) + \overrightarrow{\mathbb{P}}_{\pi}\left(\log\frac{\pi(X_r)p_t(X_r,X_{t+r})}{\pi(X_{t+r})p_t(X_{t+r},X_r)}\right)$$

Proof. Use the fact that under the stationary density π, the couples (X_0, X_r) and (X_r, X_{t+r}) have the same distribution, thus the decomposition

$$\overrightarrow{\mathbb{P}}_{\pi}\left(\log\frac{\pi(X_0)p_r(X_0,X_r)p_t(X_r,X_{t+r})}{\pi(X_{t+r})p_t(X_{t+r},X_r)p_r(X_r,X_0)}\right) =$$
$$\overrightarrow{\mathbb{P}}_{\pi}\left(\log\frac{\pi(X_0)p_r(X_0,X_r)}{\pi(X_r)p_r(X_r,X_0)} + \log\frac{\pi(X_r)p_t(X_r,X_{t+r})}{\pi(X_{t+r})p_t(X_{t+r},X_r)}\right),$$

yields the result. □

In particular, the previous lemma shows that the relative entropy between the laws $\overrightarrow{\mathbb{P}}_{\pi,t}(\cdot)$ and $\overleftarrow{\mathbb{P}}_{\pi}(\cdot)$, depends only on t. We stress this fact in the following notation:

$$\mathbf{S}(\overrightarrow{\mathbb{P}}_{\pi,t}, \overleftarrow{\mathbb{P}}_{\pi,t}) = \overrightarrow{\mathbb{P}}_{\pi,t}\left(\log\left(\frac{d\overrightarrow{\mathbb{P}}_{\pi,t}}{d\overleftarrow{\mathbb{P}}_{\pi,t}}\right)\right). \tag{4}$$

And moreover, given $r, t > 0$, we have

$$\mathbf{S}\left(\overrightarrow{\mathbb{P}}_{\pi,t+r}, \overleftarrow{\mathbb{P}}_{\pi,t+r}\right) = \mathbf{S}\left(\overrightarrow{\mathbb{P}}_{\pi,t}, \overleftarrow{\mathbb{P}}_{\pi,t}\right) + \mathbf{S}\left(\overrightarrow{\mathbb{P}}_{\pi,r}, \overleftarrow{\mathbb{P}}_{\pi,r}\right) \tag{5}$$

In the sequel, we suppose that for all $t > 0$, the above relative entropy is finite. So that for each $t > 0$ and $n \in \mathbb{N}$, $\mathbf{S}\left(\overrightarrow{\mathbb{P}}_{\pi,t}, \overleftarrow{\mathbb{P}}_{\pi,t}\right) = n\mathbf{S}\left(\overrightarrow{\mathbb{P}}_{\pi,t/n}, \overleftarrow{\mathbb{P}}_{\pi,t/n}\right)$ is bounded and letting $n \to \infty$ this shows that $\mathbf{S}\left(\overrightarrow{\mathbb{P}}_{\pi,t/n}, \overleftarrow{\mathbb{P}}_{\pi,t/n}\right) \to 0$, so that $t \mapsto \mathbf{S}\left(\overrightarrow{\mathbb{P}}_{\pi,t}, \overleftarrow{\mathbb{P}}_{\pi,t}\right)$ is continuous at 0. As a result, (5) implies that

$$\mathbf{S}\left(\overrightarrow{\mathbb{P}}_{\pi,t}, \overleftarrow{\mathbb{P}}_{\pi,t}\right) = \text{Const}.t.$$

Definition 2.1. The *entropy production* of the initial density π is given by

the constant

$$\mathbf{ep}(\pi) = \frac{d}{dt}\mathbf{S}\left(\overrightarrow{\mathbb{P}}_{\pi,t}, \overleftarrow{\mathbb{P}}_{\pi,t}\right)\Big|_{t=0}.$$

As a matter of fact, one notices that the entropy production does not depend on the whole family of finite-dimensional distributions but only on joint distributions of pairs (X_0, X_t) and $(\overleftarrow{X}_0^t, \overrightarrow{X}_t^t)$.

2.3. *The quantum detailed balance condition*

Various concepts of Quantum Detailed Balance condition are found in the literature. Let $\mathcal{L}$ denote the generator of a norm-continuous quantum Markov semigroup $\mathcal{T}$ and $\widetilde{\mathcal{L}}$ the adjoint of $\mathcal{L}$ defined by $tr\left(\rho\widetilde{\mathcal{L}}(x)y\right) = tr\left(\rho x\mathcal{L}(y)\right)$. Probably the best known quantum detailed balance condition is due to Alicki (see for instance Refs. 5, 4) and Kossakowski, Frigerio and Verri.[2] Namely, the detailed balance condition in the sense of AFGKV is satisfied if there exists a bounded self-adjoint operator $K \in \mathcal{B}(\mathbf{h})$ such that

$$\mathcal{L}(a) - \widetilde{\mathcal{L}}(a) = 2i[K, a], \tag{6}$$

for all $a \in \mathcal{B}(\mathbf{h})$.

Other properties under the same name of "detailed balance" can also be found in the literature, for instance that of Agarwal[1], Majewski,[22] Talkner[28] involving a notion of time reversal.

Here we adopt a definition detailed balance which is more natural as it is explained in Ref. 9.

Definition 2.2. We say that $\mathcal{L}$ satisfies the Standard Quantum Detailed Balance (SQDB-θ) condition with respect to ρ and an antiunitary operator θ on $\mathcal{B}(\mathbf{h})$ if

$$\mathrm{tr}\left(\rho^{1/2}\theta a^*\theta\rho^{1/2}\mathcal{T}_t(b)\right) = \mathrm{tr}\left(\rho^{1/2}\theta\mathcal{T}_t(a^*)\theta\rho^{1/2}b\right), \tag{7}$$

for all $a, b \in \mathcal{B}(\mathbf{h})$.

This condition is equivalent to

$$\mathrm{tr}\left(\rho^{1/2}\theta a^*\theta\rho^{1/2}\mathcal{L}(b)\right) = \mathrm{tr}\left(\rho^{1/2}\theta\mathcal{L}(a^*)\theta\rho^{1/2}b\right), \tag{8}$$

for all $a, b \in \mathcal{B}(\mathbf{h})$.

Throughout this paper θ will always denote an antiunitary operator. The typical example of such a θ is the conjugation with respect to a given orthonormal basis of the Hilbert space.

We recall that for a norm continuous semigroup $\mathcal{T}$ the above generator $\mathcal{L}$ can be represented in the Lindblad form:

$$\mathcal{L}(a) = i[H, a] + \frac{1}{2}\sum_\ell (L_\ell^* L_\ell a - 2L_\ell^* a L_\ell + a L_\ell^* L_\ell). \tag{9}$$

It can be shown (see for instance Ref. 25, Thm.30.16) that the operators L_ℓ can be chosen such that

(1) $\mathrm{tr}(\rho L_\ell) = 0$ for all ℓ.
(2) $\sum_\ell L_\ell^* L_\ell$ strongly converges.
(3) If $\sum_\ell |c_\ell|^2 < \infty$ and $c_0 + \sum_{\ell\geq 1} c_\ell L_\ell = 0$, then $c_\ell = 0$ for all ℓ.

In Ref. 9 Thm. 21, it is proved that a QMS satisfies the SQDB-θ if and only if a representation of the generator can be chosen like the above and the following conditions are satisfied

(a) $\rho^{1/2}\theta G^*\theta = G\rho^{1/2}$,
(b) $\rho^{1/2}\theta L_\ell^*\theta = \sum_j u_{\ell j} L_j \rho^{1/2}$, where $(u_{\ell j})$ is a unitary self-adjoint operator.

The close connection between the Quantum Detailed Balance (QDB) for quantum Markov semigroups obtained via a weak coupling limit and the KMS condition for the associated reservoirs was clarified in Ref. 2.

3. Forward and backward two point states

To start the quantum setup for the entropy production, we consider the von Neumann algebra $\mathcal{B}(\mathbf{h})$ of all linear bounded operators on a given complex separable Hilbert space $\mathbf{h}$.

To rephrase the analysis of the classical case we need to consider the product algebra $\mathcal{B}(\mathbf{h}) \otimes \mathcal{B}(\mathbf{h})$ (initial and final steps in two-point evolution) and two normalised linear functionals $\overrightarrow{\Omega}_t(a \otimes b)$, $\overleftarrow{\Omega}_t(a \otimes b)$. Suppose that we have a semi-finite trace $\mathrm{tr}(\cdot)$ defined on $\mathcal{B}(\mathbf{h})$, so that any normal state is determined by a density ρ. Suppose that ρ is $\mathcal{T}$–invariant.

At time 0, the classical state on the product algebra given by

$$\overrightarrow{\mathbb{P}}_\pi(f \otimes g) = \int \mu(dx)\pi(x)f(x)g(x),$$

is obviously positive while naive quantizations of the above are not, as it is shown in the following example.

Example. Let ω be a state on $M_d(\mathbb{C})$ (identified with its density) and let Ω be the linear functional on $M_d(\mathbb{C}) \otimes M_d(\mathbb{C})$:

$$\Omega(a \otimes b) = \mathrm{tr}\left(\omega^{1/2} a\, \omega^{1/2} b\right)$$

where $\mathrm{tr}(\cdot)$ denotes the usual trace on $M_d(\mathbb{C})$

$$\mathrm{tr}(x) = \sum_{j=1}^{d} \langle e_j, x e_j\rangle$$

($(e_j)_{1\le j\le d}$ o.n. basis of $\mathbb{C}^d$).

Let F be the unitary flip on $\mathbb{C}^d \otimes \mathbb{C}^d$ defined by

$$F u \otimes v = v \otimes u$$

and let $\mathrm{Tr}(\cdot)$ be the usual trace on $M_d(\mathbb{C}) \otimes M_d(\mathbb{C})$

$$\mathrm{Tr}(a \otimes b) = \sum_{j,k=1}^{d} \langle e_j \otimes e_k, (a \otimes b)\, e_j \otimes e_k\rangle = \mathrm{tr}(a)\mathrm{tr}(b).$$

Notice that $\mathrm{Tr}(F(a \otimes b)) = \mathrm{Tr}((a \otimes b)F) = \mathrm{tr}(ab)$. Indeed

$$\begin{aligned}\mathrm{Tr}(F(a \otimes b)) &= \sum_{j,k=1}^{d} \langle e_k \otimes e_j, (a \otimes b)\, e_j \otimes e_k\rangle \\ &= \sum_{j,k=1}^{d} a^k_j b^j_k = \mathrm{tr}(ab)\end{aligned}$$

Proposition 3.1. *The density of the functional* Ω *is*

$$F(\omega^{1/2} \otimes \omega^{1/2}) = (\omega^{1/2} \otimes \omega^{1/2})F.$$

Proof. Indeed, by the previous formula

$$\begin{aligned}\Omega(a \otimes b) &= \mathrm{tr}\left((\omega^{1/2}a)(\omega^{1/2}b)\right) \\ &= \mathrm{Tr}\left(F(\omega^{1/2}a \otimes \omega^{1/2}b)\right) \\ &= \mathrm{Tr}\left(F(\omega^{1/2} \otimes \omega^{1/2})(a \otimes b)\right).\end{aligned}$$

Clearly F commutes with $(\omega^{1/2} \otimes \omega^{1/2})$. □

Unfortunately, since F is not positive ($F(e_1 \otimes e_2 - e_2 \otimes e_1) = -(e_1 \otimes e_2 - e_2 \otimes e_1)$), one obtains that the density has a non trivial negative part. So that Ω is not a state. Similarly, it is easy to see that other choices like $\Omega(a \otimes b) = \mathrm{tr}(\rho^s ab\rho^{1-s})$, where $0 \leq s \leq 1$, do not work either.

Let us return to our previous general framework.

Definition 3.1. We associate with ρ a state $\overrightarrow{\Omega}$ on the tensor product $\mathcal{B}(\mathbf{h}) \otimes \mathcal{B}(\mathbf{h})$ as follows. First, denote $(e_n)_{n\in\mathbb{N}}$ an orthonormal basis of $\mathbf{h}$ which gives a diagonal representation of ρ and call $|r\rangle = \sum_k \rho_k^{1/2}\, \theta e_k \otimes e_k \in \mathbf{h} \otimes \mathbf{h}$; $D = |r\rangle\langle r|$. Now define

$$\overrightarrow{\Omega}_t(a \otimes b) = \mathrm{tr}\left(\rho^{1/2}\theta a \theta \rho^{1/2} \mathcal{T}_t b\right), \tag{10}$$

for any two observables $a, b \in \mathcal{B}(\mathbf{h})$. This will be called the *forward two point state.*

And the *backward two point state* is given by

$$\overleftarrow{\Omega}_t(a \otimes b) = \mathrm{tr}\left(\rho^{1/2}\theta \mathcal{T}_t a \theta \rho^{1/2} b\right) \tag{11}$$

It is worth noticing that the density of $\Omega_0 = \overrightarrow{\Omega}_0 = \overleftarrow{\Omega}_0$ is the operator D defined before.

Proposition 3.2. *The densities of* $\overrightarrow{\Omega}_t$ *and* $\overleftarrow{\Omega}_t$ *are respectively*

$$\overrightarrow{D}_t = (\mathbf{1} \otimes \mathcal{T}_{*t})\,(D)$$

$$\overleftarrow{D}_t = (\mathcal{T}_{*t} \otimes \mathbf{1})\,(D).$$

Proof. Let $a, b \in \mathcal{B}(\mathbf{h})$, then

$$\begin{aligned}\overrightarrow{\Omega}_t(a \otimes b) &= \Omega_0(a \otimes \mathcal{T}_t(b))\\ &= \operatorname{Tr}(D(\mathbf{1} \otimes \mathcal{T}_t)(a \otimes b))\\ &= \operatorname{Tr}((\mathbf{1} \otimes \mathcal{T}_{*t})(D)(a \otimes b))\end{aligned}$$

Similarly,

$$\begin{aligned}\overleftarrow{\Omega}_t(a \otimes b) &= \Omega_0(\mathcal{T}_t(a) \otimes b)\\ &= \operatorname{Tr}(D(\mathcal{T}_t \otimes \mathbf{1})(a \otimes b))\\ &= \operatorname{tr}((\mathcal{T}_{*t} \otimes \mathbf{1})(D)(a \otimes b))\end{aligned}$$

□

4. Entropy production for a QMS

Definition 4.1. The *relative entropy* of $\overrightarrow{\Omega}_t$ with respect to $\overrightarrow{\Omega}_t$ is given by

$$S(\overrightarrow{\Omega}_t, \overleftarrow{\Omega}_t) = \overrightarrow{\Omega}_t(\log \overrightarrow{D}_t - \log \overleftarrow{D}_t) = \operatorname{Tr}\left(\overrightarrow{D}_t(\log \overrightarrow{D}_t - \log \overleftarrow{D}_t)\right),$$

if the support of $\overrightarrow{D}_t$ is included in that of $\overleftarrow{D}_t$ and it is ∞ otherwise.

The *entropy production* of the density matrix ρ is defined as

$$\mathbf{ep}(\rho) = \limsup_{t \to 0+} \frac{S(\overrightarrow{\Omega}_t, \overleftarrow{\Omega}_t)}{t} \tag{12}$$

Remark 4.1. Notice that $\mathbf{ep}(\rho)$ is non-negative since $S(\overrightarrow{\Omega}_t, \overleftarrow{\Omega}_t) \geq 0$ for all $t \geq 0$, and $S(\Omega_0, \Omega_0) = 0$. Moreover, if the limit exists it coincides with

$$\mathbf{ep}(\rho) = \frac{d}{dt} S(\overrightarrow{\Omega}_t, \overleftarrow{\Omega}_t)|_{t=0}. \tag{13}$$

Theorem 4.1 (Petz,Uhlman). *Assume that each normal state Ω_k has a density matrix $D_k \in \mathfrak{J}_1(\mathbf{h} \otimes \mathbf{h})$, $(k = 1, 2)$. If $\Phi : \mathcal{B}(\mathbf{h}) \otimes \mathcal{B}(\mathbf{h}) \to \mathcal{B}(\mathbf{h})$ is a completely positive map which preserves the unit, then*

$$S(\Omega_1 \circ \Phi, \Omega \circ \Phi) \leq S(\Omega_1, \Omega_2). \tag{14}$$

From now on we restrict ourselves to norm-continuous QMS to avoid technicalities.

Lemma 4.1. *With the previous notations and assumptions, the following propositions are equivalent:*

(a) $\overrightarrow{\Omega}_t = \overleftarrow{\Omega}_t$, *for all* $t \geq 0$.

(b) $\frac{d}{dt}\overrightarrow{D}_t|_{t=0} = (\mathbf{1} \otimes \mathcal{L}_*)(D) = \frac{d}{dt}\overleftarrow{D}_t|_{t=0} = (\mathcal{L}_* \otimes \mathbf{1})(D)$.

Proof. Suppose that (a) holds to obtain (b) it suffices to take derivatives in both sides of the equality evaluated at time $t = 0$.

Conversely, if (b) holds, this means that $(\mathbf{1} \otimes \mathcal{L}_*)(D) = (\mathcal{L}_* \otimes \mathbf{1})(D)$. Since $\mathbf{1} \otimes \mathcal{L}_*$ and $\mathcal{L}_* \otimes \mathbf{1}$ commute, we obtain:

$$\begin{aligned}(\mathbf{1} \otimes \mathcal{L}_*)^2(D) &= (\mathbf{1} \otimes \mathcal{L}_*)(\mathcal{L}_* \otimes \mathbf{1})(D) \\ &= (\mathcal{L}_* \otimes \mathbf{1})(\mathbf{1} \otimes \mathcal{L}_*)(D) \\ &= (\mathcal{L}_* \otimes \mathbf{1})^2(D).\end{aligned}$$

Thus, by induction it follows that $(\mathbf{1} \otimes \mathcal{L}_*)^n(D) = (\mathcal{L}_* \otimes \mathbf{1})^n(D)$, for all $n \in \mathbb{N}$. So that

$$\overrightarrow{D}_t = \sum_{n\geq 0} \frac{t^n}{n!}(\mathbf{1} \otimes \mathcal{L}_*)^n(D) = \sum_{n\geq 0} \frac{t^n}{n!}(\mathcal{L}_* \otimes \mathbf{1})^n(D) = \overleftarrow{D}_t,$$

for all $t \geq 0$. □

Theorem 4.2. *A quantum Markov semigroup has zero entropy production if and only if* $S(\overrightarrow{\Omega}_t, \overleftarrow{\Omega}_t) = 0$ *for all* $t \geq 0$. *In particular it has zero entropy production if and only if it satisfies the detailed balance condition.*

Proof. Let φ be any faithful normal state and Π any one-dimensional projection on $\mathbf{h}\otimes\mathbf{h}$ such that $1 > \mathrm{Tr}(\Pi D) > 0$. We associate a conditional expectation $\mathbb{E}^{\varphi}x$ given by:

$$\mathbb{E}^{\varphi}(x) = \Pi x\Pi + \frac{\varphi\left(\Pi^{\perp}x\Pi^{\perp}\right)}{\varphi(\Pi^{\perp})}\Pi^{\perp}, \tag{15}$$

for any $x \in \mathcal{B}(\mathbf{h})\otimes\mathcal{B}(\mathbf{h})$. $\mathbb{E}^{\varphi}$ is by construction an identity preserving completely positive map. So that, due to the monotonicity property of the relative von Neumann entropy (Thm. 4.1) we obtain

$$0 \le S(\overrightarrow{\Omega}_t\circ\mathbb{E}^{\varphi}, \overleftarrow{\Omega}_t\circ\mathbb{E}^{\varphi}) \le S(\overrightarrow{\Omega}_t, \overleftarrow{\Omega}_t).$$

So that the hypothesis implies that $\lim_{t\to 0}\frac{1}{t}S(\overrightarrow{\Omega}_t\circ\mathbb{E}^{\varphi}, \overleftarrow{\Omega}_t\circ\mathbb{E}^{\varphi}) = 0$. The density of the state $\overrightarrow{\Omega}_t\circ\mathbb{E}^{\varphi}$ (respectively $\overleftarrow{\Omega}_t\circ\mathbb{E}^{\varphi}$) is represented by $\overrightarrow{\lambda}_t\Pi + (1-\overrightarrow{\lambda}_t)\Pi^{\perp}$ (respectively $\overleftarrow{\lambda}_t\Pi + (1-\overleftarrow{\lambda}_t)\Pi^{\perp}$) and $t\mapsto\overrightarrow{\lambda}_t = \mathrm{Tr}\left(\Pi\overrightarrow{D}_t\right)$, $t\mapsto\overleftarrow{\lambda}_t = \mathrm{Tr}\left(\Pi\overleftarrow{D}_t\right)$ are differentiable. Since the above densities commute, the relative entropy is easily obtained as

$$\overrightarrow{\lambda}_t\log\frac{\overrightarrow{\lambda}_t}{\overleftarrow{\lambda}_t} + (1-\overrightarrow{\lambda}_t)\log\frac{1-\overrightarrow{\lambda}_t}{1-\overleftarrow{\lambda}_t} \tag{16}$$

To compute the derivative in $t=0$, choose Π such that $0<\lambda_0<1$ and notice that an elementary Taylor expansion yields

$$\log\frac{\overrightarrow{\lambda}_t}{\overleftarrow{\lambda}_t} \approx t\frac{\overrightarrow{\lambda}'_0 - \overleftarrow{\lambda}'_0}{\lambda_0} + o(t)$$
$$\log\frac{1-\overrightarrow{\lambda}_t}{1-\overleftarrow{\lambda}_t} \approx -t\frac{\overrightarrow{\lambda}'_0 - \overleftarrow{\lambda}'_0}{1-\lambda_0} + o(t).$$

So that the expansion of (16) is

$$\frac{1}{t}\left((\lambda_0+o(1))t\frac{\overrightarrow{\lambda}'_0 - \overleftarrow{\lambda}'_0}{\lambda_0} - (1-\lambda_0+o(1))t\frac{\overrightarrow{\lambda}'_0 - \overleftarrow{\lambda}'_0}{1-\lambda_0}\right)$$
$$= (2\lambda_0-1)(\overrightarrow{\lambda}'_0 - \overleftarrow{\lambda}'_0) + o(1).$$

If $\lambda_0 = \mathrm{Tr}(\Pi D) \neq 1/2$, and the entropy production is zero, it follows that $\overrightarrow{\lambda}'_0 - \overleftarrow{\lambda}'_0$ or equivalently,

$$\mathrm{Tr}(\Pi(\mathcal{L}_*\otimes\mathbf{1})(D)) = \mathrm{Tr}(\Pi(\mathbf{1}\otimes\mathcal{L}_*)(D)).$$

By a trivial density argument on one-dimensional projections Π the above equality holds for any one-dimensional projection and therefore $(\mathcal{L}_*\otimes\mathbf{1})(D) = (\mathbf{1}\otimes\mathcal{L}_*)(D)$, for all $t\ge 0$. □

5. An example of a quantum 3-level system

Consider the QMS on $\mathcal{B}(\mathbb{C}^3)$ generated by

$$\mathcal{L}(a) = \alpha S^* a S + (1-\alpha) S a S^* - a$$

where S is the unitary right shift defined on the orthonormal basis $(e_j)_{0\le j\le 2}$ of $\mathbb{C}^3$ by $Se_j = e_{j+1}$ (the sum must be understood mod 3) and $\alpha \in]0,1[$.

This QMS may arise in the stochastic (weak coupling) limit of a three-level system dipole-type interacting with two reservoirs under the generalised rotating wave approximation.

The structure of this QMS is clear:

(1) $\rho = \mathbf{1}/3$ is a faithful invariant state,
(2) the quantum detailed balance condition is satisfied if and only if $\alpha = 1/2$ (in this case it is trace-symmetric).

A complete study of the qualitative behavior of this evolution can be done by applying our methods as in Ref. 8.

In view of these properties it is a good candidate for exhibiting a non-zero entropy production.

We start by computing explicitly the density $\overrightarrow{D}_t$ (resp. $\overleftarrow{D}_t$) of $\overrightarrow{\Omega}_t$, $\overleftarrow{\Omega}_t$. Recalling that

$$\overrightarrow{D}_t = \sum_{jk} E_j^k \otimes \mathcal{T}_{*t}(E_j^k)$$
$$\overleftarrow{D}_t = \sum_{jk} \mathcal{T}_{*t}(E_j^k) \otimes E_j^k$$

it is clear that we must compute explicitly the action of $\mathcal{T}_{*t}$ on the $E_j^k = |e_j\rangle\langle e_k|$. Differentiating (sums on j, k are mod 3)

$$\begin{aligned}\frac{d}{dt}\mathcal{T}_{*t}(E_j^k) &= \mathcal{T}_{*t}\left(\mathcal{L}_*\left(E_j^k\right)\right)\\ &= \alpha\mathcal{T}_{*t}\left(E_{j+1}^{k+1}\right) - \mathcal{T}_{*t}\left(E_j^k\right) + (1-\alpha)\mathcal{T}_{*t}\left(E_{j-1}^{k-1}\right)\end{aligned}$$

This system can be solved exponentiating the 3×3 matrix

$$A = \begin{pmatrix} -1 & \alpha & 1-\alpha \\ 1-\alpha & -1 & \alpha \\ \alpha & 1-\alpha & -1 \end{pmatrix}.$$

The explicit computation yields

$$e^{tA} = \begin{bmatrix} \varphi_0(t) & \varphi_1(t) & \varphi_2(t) \\ \varphi_2(t) & \varphi_0(t) & \varphi_1(t) \\ \varphi_1(t) & \varphi_2(t) & \varphi_0(t) \end{bmatrix}$$

where,

$$\varphi_k(t) = \frac{1}{3}\left[1 + 2e^{-3t/2}\cos\left(\gamma t - \frac{2k\pi}{3}\right)\right],$$

with $\gamma := \sqrt{3}(2\alpha - 1)/2$, $k = 0, 1, 2$. Therefore

$$\mathcal{T}_{*t}\left(E_j^k\right) = \varphi_0(t)E_j^k + \varphi_1(t)E_{j+1}^{k+1} + \varphi_2(t)E_{j-1}^{k-1}.$$

and

$$\overrightarrow{D}_t = \sum_{jk} \varphi_0(t)E_j^k \otimes E_j^k + \varphi_1(t)E_j^k \otimes E_{j+1}^{k+1} + \varphi_2(t)E_j^k \otimes E_{j-1}^{k-1}$$

$$\overleftarrow{D}_t = \sum_{jk} \varphi_0(t)E_j^k \otimes E_j^k + \varphi_1(t)E_{j+1}^{k+1} \otimes E_j^k + \varphi_2(t)E_{j-1}^{k-1} \otimes E_j^k$$

The inspection at $\overrightarrow{D}_t$ and $\overleftarrow{D}_t$ reveals their spectral structure. Putting

$$u_0 = \frac{1}{\sqrt{3}}\sum_j e_j \otimes e_j,\; u_1 = \frac{1}{\sqrt{3}}\sum_j e_j \otimes e_{j+1},\; u_2 = \frac{1}{\sqrt{3}}\sum_j e_j \otimes e_{j-1},$$

we find the spectral representations:

$$\overrightarrow{D}_t = \varphi_0(t)|u_0\rangle\langle u_0| + \varphi_1(t)|u_1\rangle\langle u_1| + \varphi_2(t)|u_2\rangle\langle u_2|,$$
$$\overleftarrow{D}_t = \varphi_0(t)|u_0\rangle\langle u_0| + \varphi_2(t)|u_1\rangle\langle u_1| + \varphi_1(t)|u_2\rangle\langle u_2|.$$

The above eigenvalues are obviously non negative for $t \geq 0$ and the relative entropy is

$$\mathrm{Tr}\left(\overrightarrow{D}_t\left(\log\left(\overrightarrow{D}_t\right) - \log\left(\overleftarrow{D}_t\right)\right)\right) = \varphi_0(t)\log\left(\frac{\varphi_0(t)}{\varphi_0(t)}\right) + (\varphi_1(t) - \varphi_2(t))\log\left(\frac{\varphi_1(t)}{\varphi_2(t)}\right)$$

The first term vanishes and a Taylor expansion gives for $t \to 0^+$,

$$\begin{aligned}
\frac{1 + 2e^{-\frac{3t}{2}}\cos(\gamma t - \frac{2\pi}{3})}{3} &= t\left(-\cos(2\pi/3) - \frac{2\gamma}{3}\sin(2\pi/3)\right) + o(t)\\
&= t\left(\frac{1}{2} - \frac{\gamma\sqrt{3}}{3}\right) + o(t)\\
&= (1-\alpha)t + o(t)\\
\frac{1 + 2e^{-\frac{3t}{2}}\cos(\gamma t - \frac{4\pi}{3})}{3} &= t\left(-\cos(4\pi/3) - \frac{2\gamma}{3}\sin(4\pi/3)\right) + o(t)\\
&= t\left(\frac{1}{2} + \frac{\gamma\sqrt{3}}{3}\right) + o(t)\\
&= \alpha t + o(t)
\end{aligned}$$

It follows that

$$\lim_{t\to 0^+}\frac{\varphi_1(t)}{\varphi_2(t)}=\frac{1-\alpha}{\alpha},\qquad \lim_{t\to 0^+}\frac{\varphi_1(t)-\varphi_2(t)}{t}=2\alpha-1,$$

and

$$\lim_{t\to 0^+}\frac{\mathrm{Tr}\left(\overrightarrow{D}_t\left(\log\left(\overrightarrow{D}_t\right)-\log\left(\overleftarrow{D}_t\right)\right)\right)}{t}=(2\alpha-1)\log\left(\frac{\alpha}{1-\alpha}\right).$$

Therefore, the entropy production is non zero if $\alpha\neq 1/2$ since a "current" remains in the sense of S ("raising") or S^* ("lowering"). Notice that this entropy production coincides with the classical one, when the QMS is restricted to the commutative subalgebra of diagonal matrices.

6. Conclusions and outlook

Within this paper, we just gave the flavor of our approach to entropy production based on the classical stochastic one. We do not look at the open quantum system interacting with any reservoir.

One of the key features has been the understanding of the so called "Two point states" and the role they play in distinguishing between forward and backward dynamics. Another key has been the general approach to the Quantum Detailed Balance Condition studied by Fagnola and Umanità.[9] Moreover, the explicit computation of densities for the above states has lead to the appropriate statement of zero entropy production for a given Quantum Markov Semigroup.

In a forthcoming paper, the explicit expression of the entropy production is given, under suitable hypotheses, as follows:

$$\mathbf{ep}(\rho)=\frac{1}{2}\mathrm{Tr}\left(D^{\perp}(\overrightarrow{D}_0'-\overleftarrow{D}_0')D^{\perp}\left(\log(D^{\perp}\overrightarrow{D}_0'D^{\perp})-\log(D^{\perp}\overleftarrow{D}_0D^{\perp})\right)\right),$$

where the prime denotes a derivative. Indeed this expression involves only the completely positive part of $\mathcal{L}$ because one of the hypotheses is $\rho^{1/2}\theta G^*\theta=G\rho^{1/2}$.

Also, additional examples with physical relevance will be published elsewhere, in particular, a class of quantum interacting particle systems of exclusion type (see Refs. 27, 3, 24), where it will be shown that the conditions of reversibility for the classical exclusion process coincide with those obtained from our approach.

Acknowledgments

The authors kindly acknowledge support of PBCT-ADI 13 grant "Programa Bicentenario de Ciencia y Tecnología", Chile. F. F. thanks also the MIUR-PRIN 2007-2009 project "Quantum Probability and Applications to Information Theory".

References

1. G. S. Agarwal, *Open quantum Markovian systems and the microreversibility.* Z. Physik **258** no. 5 (1973), 409–422.
2. A. Kossakowski, A. Frigerio, V. Gorini and M. Verri, *Quantum detailed balance and KMS condition.* Comm. Math. Phys. **57** no. 2 (1977), 97–110.
3. L. Accardi, F. Fagnola, and S. Hachicha, *Generic q-Markov semigroups and speed of convergence of q-algorithms.* Infin. Dimens. Anal. Quantum Probab. Relat. Top., **9** no. 4 (2006), 567–594.
4. R. Alicki and K. Lendi, *Quantum Dynamical Semigroups and Applications*, Lect. Notes in Physics 286, Springer-Verlag, 1987.
5. R. Alicki, *On the detailed balance condition for non-Hamiltonian systems.* Rep. Math. Phys., **10** no. 2 (1976), 249–258.
6. I. Callens, W. DeRoeck, T. Jacobs, C. Maes and K. Netocny, *Quantum entropy production as a measure of irreversibility.* Phys. D, **187** no. 1-4 (2004), 383–391.
7. M. Esposito, U. Harbola, and S. Mukamel, *Entropy fluctuation theorems in driven open systems: Application to electron counting statistics.* Phys Rev E, 2007.
8. F. Fagnola, R. Rebolledo, *Lectures on the qualitative analysis of quantum Markov semigroups: Quantum interacting particle systems* (Trento, 2000); and QP–PQ: Quantum Probab. White Noise Anal. 14, World Sci. Publ., River Edge, NJ, (2002), 197–239.

9. F. Fagnola and V. Umanità, *Generators of KMS Symmetric Markov Semigroups on $\mathcal{B}(h)$ Symmetry and Quantum Detailed Balance.* http://arxiv.org/abs/0908.0967v1, 2009.
10. H.P. Breuer, *Quantum jumps and entropy production.* Phys. Rev. A (3), **68** (2003), 32–105.
11. V. Jaksic, Y. Ogata, and C. Pillet, *Linear response theory for thermally driven quantum open systems.* Journal of Statistical Physics, 2006.
12. V. Jaksic and C. Pillet, *On entropy production in quantum statistical mechanics.* Communications in Mathematical Physics, 2001.
13. V. Jaksic, Y. Ogata, and C.A. Pillet, *The Green-Kubo formula for locally interacting Fermionic Open Systems.* Ann. Henri Poincare, **8** (2008), 1013–1036.
14. D.Q. Jiang, M. Qian, and F. Zhang, *Entropy production fluctuations of finite Markov chains.* J. Math. Phys., **44** no. 9 (2003), 4176–4188.
15. J.L. Lebowitz and C. Maes, *Entropy: a dialogue.* In *Entropy,* Princeton Ser. Appl. Math., Princeton Univ. Press, Princeton, NJ, (2003), 269–276.
16. C. Maes, F. Redig, and M. Verschuere, *Entropy production for interacting particle systems.* Markov Process. Related Fields, **7** no. 1 (2001), 119–134. Inhomogeneous Random Systems (Cergy-Pontoise, 2000).
17. C. Maes, *Entropy production in driven spatially extended systems.* In *Entropy,* Princeton Ser. Appl. Math., Princeton Univ. Press, Princeton, NJ, (2003), 251–267.
18. C. Maes, *Fluctuation relations and positivity of the entropy production in irreversible dynamical systems.* Nonlinearity, 17 no.4 (2004, 1305–1316.
19. C. Maes and K. Netocny, *Time-reversal and entropy.* J. Statist. Phys., **110** no. 1-2 (2003), 269–310.
20. C. Maes and F. Redig, *Positivity of entropy production.* J. Statist. Phys.,

101 no. 1-2 (2000), 3–15. Dedicated to Grégoire Nicolis on the occasion of his sixtieth birthday (Brussels, 1999).

21. C. Maes, F. Redig, and A. VanMoffaert, *On the definition of entropy production, via examples.* J. Math. Phys., **41** no. 3 (2000), 1528–1554. Probabilistic techniques in equilibrium and nonequilibrium statistical physics.
22. W.A. Majewski, *The detailed balance condition in quantum statistical mechanics.* J Math Phys., **25** (1984), 614–616.
23. Lars Onsager, *Reciprocal relations in irreversible processes.* Phys Rev, **37** (1931), 405–426.
24. L. Pantaleon-Martinez and R. Quezada, *Interacting Particle Quantum Markov Semigroups.* Infin. Dimens. Anal. Quantum Probab. Relat. Top., To appear, 2009.
25. K.R. Parthasarathy, *An Introduction to Quantum Stochastic Calculus*, Monographs in Mathematics 85, Birkhaüser–Verlag, Basel-Boston-Berlin, 1992.
26. A Perez-Madrid, *Quantum theory of irreversibility.* Physica A, **378** (2007), 299–306.
27. R. Rebolledo, *Decoherence of quantum Markov semigroups.* Ann. Inst. H. Poincaré Probab. Statist., **41** no. 3 (2005), 349–373.
28. P. Talkner, *The failure of the quantum regression hypothesis.* Ann. Physics **167** no. 2 (1986), 390–436.
29. R.K.P Zia and B. Schmittmann, *Probability currents as principal characteristics in the statistical mechanics of non-equilibrium steady states.* J Stat Mech-Theory E, 2007.

Extending the Set of Quadratic Exponential Vectors*

Luigi ACCARDI

Centro Vito Volterra, Facoltà di Economia

Università degli Studi di Roma "Tor Vergata"

Via Columbia 2, 00133 Roma, Italy

E-mail: accardi@volterra.uniroma2.it

Homepage: http://www.volterra.uniroma2.it

Ameur DHAHRI

Ceremade, Université Paris Dauphine

Place de Lattre de Tassigny

75775 Paris Cédex 16, France

E-mail: dhahri@ceremade.dauphine.fr

Michael SKEIDE

Dipartimento S.E.G.e S., Università degli Studi del Molise

Via de Sanctis, 86100 Campobasso, Italy

E-mail: skeide@math.tu-cottbus.de

Homepage: http://www.math.tu-cottbus.de/INSTITUT/lswas/_skeide.html

We extend the square of white noise algebra over the step functions on $\mathbb{R}$ to the test function space $L^2(\mathbb{R}^d) \cap L^\infty(\mathbb{R}^d)$, and we show that in the Fock

*LA and MS are supported by Italian MUR (PRIN 2007). MS is supported by research funds of the Dipartimento S.E.G.e S. of University of Molise.

representation the exponential vectors exist for all test functions bounded by $\frac{1}{2}$.

1. Introduction

Modulo minor variations in the choice of the test function space, the square of white noise (SWN) algebra has been introduced by Accardi, Lu and Volovich[3] as follows. Let $\mathcal{L} = L^2(\mathbb{R}^d) \cap L^\infty(\mathbb{R}^d)$ and $c > 0$ a constant. Then the **SWN algebra** $\mathcal{A}$ over $\mathcal{L}$ is the unital *algebra generated by symbols B_f, N_f ($f \in \mathcal{L}$) and the commutation relations

$$[B_f, B_g^*] = 2c\langle f, g\rangle + 4N_{\overline{f}g}, \quad [N_f, B_g^*] = 2B_{fg}^*,$$

($f, g \in \mathcal{L}$) and all other commutators 0. Note that by the first relation, $N_f^* = N_{\overline{f}}$.

A **Fock representation** of $\mathcal{A}$ is a representation ($*$, of course) π of $\mathcal{A}$ on a pre-Hilbert space H with a unit vector $\Phi \in H$, fulfilling $\mathcal{A}\Phi = H$ and $\pi(B_f)\Phi = \pi(N_f)\Phi = 0$ for all $f \in \mathcal{L}$. From the commutation relations it follows that a Fock representation is unique up to unitary equivalence. Existence of a Fock representation has been established by different proofs in[2–4,8] for $d = 1$. They extend easily to general $d \in \mathbb{N}$. Henceforth, we speak about the Fock representation. The Fock representation would be faithful, if we require also that the N_f depend linearly on f. By abuse of notation, we identify $\mathcal{A}$ with its image $\pi(\mathcal{A})$ omitting, henceforth, π.

The **exponential vector** $\psi(f)$ to an element $f \in \mathcal{L}$ is defined as

$$\psi(f) := \sum_{m=0}^{\infty} \frac{B_f^{*m}\Phi}{m!}$$

whenever the series exists. In Accardi and Skeide[5] is has been shown for $d = 1$ that $\psi(\sigma\mathbb{I}_{[0,t]})$ exists for $|\sigma| < \frac{1}{2}$ and that $\langle\psi(\sigma\mathbb{I}_{[0,t]}), \psi(\rho\mathbb{I}_{[0,t]})\rangle = e^{-\frac{ct}{2}\ln(1-4\overline{\sigma}\rho)}$. As noted in Ref. 5, this extends to arbitrary step functions f, g on $\mathbb{R}$ with $\|f\|_\infty < \frac{1}{2}$, with inner product[a]

$$\langle\psi(f), \psi(g)\rangle = e^{-\frac{c}{2}\int \ln(1-4\overline{f(t)}g(t))\,dt}. \tag{1}$$

[a]*The correlation kernel on the right-hand side coincides, modulo scaling, with the correlation kernel in Boukas' representation[6] of Feinsilver's finite difference algebra.[7] In Ref. 5, this observation gave rise to the discovery of an intimate relation between the SWN algebra and the finite difference algebra.*

Our scope is to extend the set of exponential vectors and the formula in (1) for their inner product to test functions $f \in \mathcal{L}$ with $\|f\|_\infty < \frac{1}{2}$.

In the "29th Quantum Probability Conference" in October 2008 in Hammamet, Tunisia, Dhahri explained that the extension can be done for exponential vectors to all elements f in $\mathcal{L}$ with $\|f\|_\infty < \frac{1}{2}$. This a part of the work Accardi and Dhahri[1] (in preparation) on the *second quantization functor* for the square of white noise. Here we give a simple proof of this partial result.

2. The result

Theorem 2.1. *The exponential vector $\psi(f)$ exists for every $f \in \mathcal{L}$ with $\|f\|_\infty < \frac{1}{2}$ and the inner product of two such exponential vectors is given by (1).*

Proof. (i) We show that the right-hand side of (1) exists. Indeed, by Taylor expansion we have $|\ln(1+x)| \leq M_\delta |x|$ for $|x| \leq 1-\delta$ for every $\delta \in (0,1)$, where M_δ may depend on δ but not on x. Choose $\delta = 1 - 4\|f\|_\infty \|g\|_\infty \in (0,1)$. Then

$$|\ln(1 - 4\overline{f(t)}g(t))| \leq M_\delta |4\overline{f(t)}g(t)|.$$

Since $|\overline{f(t)}g(t)|$ is integrable, so is $\ln(1 - 4\overline{f(t)}g(t))$.

(ii) The function $x \mapsto \ln x$ is increasing on the whole half line $(0,\infty)$. It follows that also the function $x \mapsto -\ln(1-x)$ is increasing on $(-1,1)$. We conclude that $\frac{1}{2} > |f| \geq |g|$ implies $-\ln(1-4|f(t)|^2) \geq -\ln(1-4|g(t)|^2)$. Choose for f an L^2–approximating sequence of step functions $(f_n)_{n\in\mathbb{N}}$ in such a way that $|f| \geq |f_n|$ for all $n \in \mathbb{N}$. By the dominated convergence theorem, $\lim_{n\to\infty} e^{-\frac{c}{2}\int \ln(1-4|f_n(t)|^2)\,dt} = e^{-\frac{c}{2}\int \ln(1-4|f(t)|^2)\,dt}$.

(iii) In precisely the same way as in Ref. 5, one shows that (1) is true for all step functions strictly bounded by $\frac{1}{2}$. It follows that $\lim_{n\to\infty} \|\psi(f_n)\|^2 = e^{-\frac{c}{2}\int \ln(1-4|f(t)|^2)\,dt}$.

(iv) Since $\langle B_f^{*m}\Phi, B_f^{*m}\Phi\rangle$ is a polynomial (of degree m) in $\langle f, f\rangle$, it depends continuously in L^2-norm on f. So, for every $M \in \mathbb{N}$ there is an

$n \in \mathbb{N}$ such that

$$\Big\langle \sum_{m=0}^{M} \frac{B_f^{*m}\Phi}{m!}, \sum_{m=0}^{M} \frac{B_f^{*m}\Phi}{m!} \Big\rangle \leq \Big\langle \sum_{m=0}^{M} \frac{B_{f_n}^{*\ m}\Phi}{m!}, \sum_{m=0}^{M} \frac{B_{f_n}^{*\ m}\Phi}{m!} \Big\rangle + 1$$
$$\leq \Big\langle \sum_{m=0}^{\infty} \frac{B_{f_n}^{*\ m}\Phi}{m!}, \sum_{m=0}^{\infty} \frac{B_{f_n}^{*\ m}\Phi}{m!} \Big\rangle + 1$$
$$= \|\psi(f_n)\|^2 + 1 \leq e^{-\frac{c}{2}\int \ln(1-4|f(t)|^2)\, dt} + 1.$$

By the theorem on exchange of limits under domination, it follows that

$$\lim_{M\to\infty} \Big\langle \sum_{m=0}^{M} \frac{B_f^{*m}\Phi}{m!}, \sum_{m=0}^{M} \frac{B_f^{*m}\Phi}{m!} \Big\rangle = \lim_{M\to\infty}\lim_{n\to\infty} \Big\langle \sum_{m=0}^{M} \frac{B_{f_n}^{*\ m}\Phi}{m!}, \sum_{m=0}^{M} \frac{B_{f_n}^{*\ m}\Phi}{m!} \Big\rangle$$
$$= \lim_{n\to\infty}\lim_{M\to\infty} \Big\langle \sum_{m=0}^{M} \frac{B_{f_n}^{*\ m}\Phi}{m!}, \sum_{m=0}^{M} \frac{B_{f_n}^{*\ m}\Phi}{m!} \Big\rangle$$
$$= \lim_{n\to\infty} \|\psi(f_n)\|^2 = e^{-\frac{c}{2}\int \ln(1-4|f(t)|^2)\, dt}.$$

From this we conclude that $\psi(f)$ exists and that

$$\|\psi(f)\|^2 = e^{-\frac{c}{2}\int \ln(1-4|f(t)|^2)\, dt}.$$

(v) Doing the same sort of computation for the difference $\psi(f) - \psi(f_n)$, it follows that $\lim_{n\to\infty} \psi(f_n) = \psi(f)$. Approximating also g by a sequence of step functions g_n with $|g| \geq |g_n|$, we find $\lim_{n\to\infty} \langle \psi(f_n), \psi(g_n)\rangle = \langle \psi(f), \psi(g)\rangle$ (continuity of the inner product), and

$$\lim_{n\to\infty} e^{-\frac{c}{2}\int \ln(1-4\overline{f_n(t)}g_n(t))\, dt} = e^{-\frac{c}{2}\int \ln(1-4\overline{f(t)}g(t))\, dt}$$

(once more, by dominated convergence for $|\overline{f_n}g_n| \leq |\overline{f}g|$ on the other side). This shows (1) for all f, g as specified. □

References

1. L. Accardi and A. Dhahri, *Second quantization on the square of white noise algebra*, Preprint, Rome (in preparation), 2008.
2. L. Accardi, U. Franz, and M. Skeide, *Renormalized squares of white noise and other non-gaußian noises as Lévy processes on real Lie algebras*, Commun. Math. Phys. **228** (2002), 123–150, (Rome, Volterra-Preprint 2000/0423).

3. L. Accardi, Y.G. Lu, and I.V. Volovich, *White noise approach to classical and quantum stochastic calculi*, Preprint, Rome, 1999, To appear in the lecture notes of the Volterra International School of the same title, held in Trento.
4. L. Accardi and M. Skeide, *Hilbert module realization of the square of white noise and the finite difference algebra*, Math. Notes **86** (2000), 803–818, (Rome, Volterra-Preprint 1999/0384).
5. L. Accardi and M. Skeide, *On the relation of the square of white noise and the finite difference algebra*, Infin. Dimens. Anal. Quantum Probab. Relat. Top. **3** (2000), 185–189, (Rome, Volterra-Preprint 1999/0386).
6. A. Boukas, *An example of quantum exponential process*, Mh. Math. **112** (1991), 209–215.
7. P.J. Feinsilver, *Discrete analogues of the Heisenberg-Weyl algebra*, Mh. Math. **104** (1987), 89–108.
8. P. Sniady, *Quadratic bosonic and free white noise*, Commun. Math. Phys. **211** (2000), 615–628.

ON OPERATOR-PARAMETER TRANSFORMS BASED ON NUCLEAR ALGEBRA OF ENTIRE FUNCTIONS AND APPLICATIONS

Abdessatar BARHOUMI

Department of Mathematics

Higher School of Sci. and Tech. of Hammam-Sousse

University of Sousse, Sousse, Tunisia

E-mail: abdessatar.barhoumi@ipein.rnu

Habib OUERDIANE

Department of Mathematics, Faculty of Sciences of Tunis

University of Tunis El-Manar, Tunis, Tunisia

E-mail: hafedh.rguigui@yahoo.fr

Hafedh RGUIGUI

Department of Mathematics, Faculty of Sciences of Tunis

University of Tunis El-Manar, Tunis, Tunisia

E-mail: hafedh.rguigui@yahoo.fr

Anis RIAHI

Department of Mathematics

Higher Institute of Appl. Sci. and Tech. of Gabes

University of Gabes, Gabes, Tunisia

E-mail: a1riahi@yahoo.fr

Based on nuclear algebra of entire functions, we extend some results about operator-parameter transforms involving the Fourier-Gauss and Fourier-Mehler transforms. We investigate the solution of a initial-value problem associated to infinitesimal generators of these transformations. In particular, by using convolution product, we show to what extent regularity properties can be performed on our setting.

Keywords: Convolution operator, Fourier-Gauss transform, Fourier-Mehler transform, heat equation, K-Gross Laplacian.

1. Introduction

The Fourier transform plays a central role in the theory of distribution on Euclidean spaces. Although Lebesgue measure does not exist in infinite dimensional spaces, the Fourier transform can be introduced in the space of generalized white noise functionals. The Fourier transform has many properties similar to the finite dimensional case; e.g., the Fourier transform carries coordinate differentiation into multiplication and vice versa. It plays an essential role in the theory of differential equations in infinite dimensional spaces. An important example of a partial differential equation in the infinite dimensional space is the heat equation with Gross Laplacian operator studied firstly by Gross.[11]

In a series of papers based on the white noise theory,[17,18] Kuo formulated the Fourier-Mehler transform as continuous linear operator acting on the space of generalized white noise functionals; see also Ref. 19 and references cited therein. Later on, Chung-Ji[5] generalized this transformation as a two parameters transform obtained from the adjoint of the Fourier-Gauss transform studied firstly by Lee.[20,21] Next, by using the symbol transform of operator,[23] Chung-Ji[4] introduced two operator-parameter transforms $\mathcal{G}_{A,B}$ and $\mathcal{F}_{A,B}$ acting on white noise functionals as a generalization of the scaler-parameter transforms $\mathcal{G}_{\alpha,\beta}$ and $\mathcal{F}_{\alpha,\beta}$ studied in Ref. 5. Similarly, in Ref. 22 Luo-Yan introduced Gaussian kernel operators on white noise functional spaces including second quantization, Fourier-Mehler transform, Scaling, renormalization, etc.

In this paper we prove some basic results for the transformations $\mathcal{G}_{A,B}$

and $\mathcal{F}_{A,B}$ in white noise theory within the framework of nuclear algebras of entire functions. The basic tools are the duality theorem for the Taylor series map and characterization theorem for the Laplace transform. By using convolution calculus, we investigate a generalized heat equation associated to infinitesimal generators of our transformations.

The paper is organized as follows. In Section 2 we briefly recall well-known results on nuclear algebras of entire holomorphic functions. In Section 3, we give explicitly the Taylor expansions of the basic transformations, and we show that $\mathcal{G}_{A,B}$ (resp. $\mathcal{F}_{A,B}$) realizes a topological isomorphism from $\mathcal{F}_\theta(N')$ (resp. $\mathcal{F}_\theta^*(N')$) into itself. In Section 4, we investigate the solution of a initial-value problem associated to the K-Gross Laplacian. In particular, regularity properties and integral representation are in our consideration.

2. Preliminaries

In this Section we shall briefly recall some of the concepts, notations and known results on nuclear algebras of entire functions.[6,8,10,19,23] Let H be a real separable (infinite dimensional) Hilbert space with inner product $\langle\cdot,\cdot\rangle$ and norm $|\cdot|_0$. Let $A \geq 1$ be a positive self-adjoint operator in H with Hilbert-Schmidt inverse. Then there exist a sequence of positive numbers $1 < \lambda_1 \leq \lambda_2 \leq \cdots$ and a complete orthonormal basis of H, $\{e_n\}_{n=1}^\infty \subseteq \mathrm{Dom}(A)$ such that

$$Ae_n = \lambda_n e_n\,, \qquad \sum_{n=1}^\infty \lambda_n^{-2} = \|A^{-1}\|_{HS}^2 < \infty.$$

For every $p \in \mathbb{R}$ we define:

$$|\xi|_p^2 := \sum_{n=1}^\infty \langle\xi, e_n\rangle^2 \lambda_n^{2p} = |A^p\xi|_0^2\,, \quad \xi \in H.$$

The fact that, for $\lambda > 1$, the map $p \mapsto \lambda^p$ is increasing implies that:

(i) for $p \geq 0$, the space X_p, of all $\xi \in H$ with $|\xi|_p < \infty$, is a Hilbert space with norm $|\cdot|_p$ and, if $p \leq q$, then $X_q \subseteq X_p$;

(ii) denoting X_{-p} the $|\cdot|_{-p}$-completion of H ($p \geq 0$), if $0 \leq p \leq q$, then $X_{-p} \subseteq X_{-q}$.

This construction gives a decreasing chain of Hilbert spaces $\{X_p\}_{p\in\mathbb{R}}$ with natural continuous inclusions $i_{q,p} : X_q \hookrightarrow X_p$ ($p \leq q$). Defining the countably Hilbert nuclear space (see e.g. Ref. 10):

$$X := \operatorname*{projlim}_{p\to\infty} X_p \cong \bigcap_{p\geq 0} X_p$$

the strong dual space X' of X is:

$$X' := \operatorname*{indlim}_{p\to\infty} X_{-p} \cong \bigcup_{p\geq 0} X_{-p}$$

and the triple

$$X \subset H \equiv H' \subset X' \tag{1}$$

is called a real standard triple.[23] The complexifications of X_p, X and H respectively will be denoted

$$N_p := X_p + iX_p\,; \quad N := X + iX\,; \quad \mathcal{H} := H + iH. \tag{2}$$

Notice that $\{e_n\}_{n=1}^{\infty}$ is also a complete orthonormal basis of $\mathcal{H}$. Thus the complexification of the standard triple (1) is:

$$N \subset \mathcal{H} \subset N'.$$

When dealing with complex Hilbert spaces, we will always assume that the scalar product is linear in the second factor and the duality $\langle N', N\rangle$, also denoted $\langle\cdot,\cdot\rangle$, is defined so to be compatible with the inner product of $\mathcal{H}$.

For $n \in \mathbb{N}$ we denote by $N^{\widehat{\otimes}n}$ the n-fold symmetric tensor product of N equipped with the π-topology and by $N_p^{\widehat{\otimes}n}$ the n-fold symmetric Hilbertian tensor product of N_p. We will preserve the notation $|\cdot|_p$ and $|\cdot|_{-p}$ for the norms on $N_p^{\widehat{\otimes}n}$ and $N_{-p}^{\widehat{\otimes}n}$, respectively.

From Ref. 8 we recall the following background. Let θ be a Young function, i.e., it is a continuous, convex, and increasing function defined on $\mathbb{R}_+$ and satisfies the condition $\lim_{x\to\infty}\theta(x)/x = \infty$. We define the conjugate function θ^* of θ by

$$\theta^*(x) = \sup_{t\geq 0}\big(tx - \theta(t)\big), \quad x \geq 0.$$

For a Young function θ, we denote by $\mathcal{F}_\theta(N')$ the space of holomorphic functions on N' with exponential growth of order θ and of minimal type. Similarly, let $\mathcal{G}_\theta(N)$ denote the space of holomorphic functions on N with exponential growth of order θ and of arbitrary type. Moreover, for each $p \in \mathbb{Z}$ and $m > 0$, define $\mathrm{Exp}(N_p, \theta, m)$ to be the space of entire functions f on N_p satisfying the condition:

$$\|f\|_{\theta,p,m} = \sup_{x\in N_p} |f(x)|e^{-\theta(m|x|_p)} < \infty.$$

Then the spaces $\mathcal{F}_\theta(N')$ and $\mathcal{G}_\theta(N)$ can be represented as

$$\mathcal{F}_\theta(N') = \bigcap_{p\in\mathbb{N},\, m>0} \mathrm{Exp}(N_{-p}, \theta, m),$$

$$\mathcal{G}_\theta(N) = \bigcup_{p\in\mathbb{N},\, m>0} \mathrm{Exp}(N_p, \theta, m),$$

and are equipped with the projective limit topology and the inductive limit topology, respectively. The space $\mathcal{F}_\theta(N')$ is called the space of *test functions* on N'. Its topological dual space $\mathcal{F}^*_\theta(N')$, equipped with the strong topology, is called the space of *distributions* on N'.

For $p \in \mathbb{N}$ and $m > 0$, we define the Hilbert spaces

$$F_{\theta,m}(N_p) = \Big\{\vec{\varphi} = (\varphi_n)_{n=0}^\infty\ ;\ \varphi_n \in N_p^{\widehat{\otimes} n},\ \sum_{n=0}^\infty \theta_n^{-2} m^{-n} |\varphi_n|_p^2 < \infty\Big\},$$

$$G_{\theta,m}(N_{-p}) = \Big\{\vec{\Phi} = (\Phi_n)_{n=0}^\infty\ ;\ \Phi_n \in N_{-p}^{\widehat{\otimes} n},\ \sum_{n=0}^\infty (n!\theta_n)^2 m^n |\Phi_n|_{-p}^2 < \infty\Big\},$$

where

$$\theta_n = \inf_{r>0} e^{\theta(r)}/r^n, \rangle n \in \mathbb{N}. \tag{3}$$

Put

$$F_\theta(N) = \bigcap_{p\in\mathbb{N},m>0} F_{\theta,m}(N_p),$$

$$G_\theta(N') = \bigcup_{p\in\mathbb{N},m>0} G_{\theta,m}(N_{-p}).$$

The space $F_\theta(N)$ equipped with the projective limit topology is a nuclear Frechét space.[8] The space $G_\theta(N')$ carries the dual topology of $F_\theta(N)$ with respect to the $\mathbb{C}$-bilinear pairing given by

$$\langle\langle \vec{\Phi}, \vec{\varphi} \rangle\rangle = \sum_{n=0}^\infty n! \langle \Phi_n, \varphi_n \rangle, \tag{4}$$

where $\vec{\Phi} = (\Phi_n)_{n=0}^\infty \in G_\theta(N')$ and $\vec{\varphi} = (\varphi_n)_{n=0}^\infty \in F_\theta(N)$.

It was proved in Ref. 8 that the *Taylor map* defined by

$$T\colon\ \varphi \longmapsto \Big(\frac{1}{n!}\varphi^{(n)}(0)\Big)_{n=0}^\infty$$

is a topological isomorphism from $\mathcal{F}_\theta(N')$ onto $F_\theta(N)$. The Taylor map T is also a topological isomorphism from $\mathcal{G}_{\theta^*}(N))$ onto $G_\theta(N'))$. The action of a distribution $\Phi \in \mathcal{F}_\theta^*(N')$ on a test function $\varphi \in \mathcal{F}_\theta(N')$ can be expressed in terms of the Taylor map as follows:

$$\langle\langle \Phi, \varphi \rangle\rangle = \langle\langle \vec{\Phi}, \vec{\varphi} \rangle\rangle,$$

where $\vec{\Phi} = (T^*)^{-1}\Phi$ and $\vec{\varphi} = T\varphi$.

For $\vec{\varphi} = (\varphi_n)_{n\geq 0} \in F_\theta(N)$, $\vec{\Phi} = (\Phi_n)_{n\geq 0} \in G_\theta(N')$, we write $\varphi \sim (\varphi_n)_{n\geq 0}$ and $\Phi \sim (\Phi_n)_{n\geq 0}$ for short. The following estimates are useful.

Lemma 2.1. *(See Ref. 8)*

(1) Let $\varphi \sim (\varphi_n)_{n\geq 0}$ in $\mathcal{F}_\theta(N')$. Then, for any $n \geq 0$, $p \geq 0$ and $m > 0$, there exist $q > p$ such that

$$|\varphi_n|_p \leq e^n \theta_n m^n \|i_{q,p}\|_{HS}^n \|\varphi\|_{\theta,q,m}. \tag{5}$$

(2) Let $\Phi \sim (\Phi_n)_{n\geq 0}$ in $\mathcal{F}_\theta^*(N')$. Then, there exist $p \geq 0$ and $m > 0$ such that for any $q > p$ we have

$$|\Phi_n|_{-q} \leq e^n \theta_n^* m^n \|i_{q,p}\|_{HS}^n \|\mathcal{L}\Phi\|_{\theta^*,-p,m}. \tag{6}$$

It is easy to see that for each $\xi \in N$, the exponential function

$$e_\xi(z) = e^{\langle z,\xi\rangle}, \quad z \in N',$$

is a test function in the space $\mathcal{F}_\theta(N')$ for any Young function θ. Thus we can define the *Laplace transform* of a distribution $\Phi \in \mathcal{F}_\theta^*(N')$ by

$$\mathcal{L}\Phi(\xi) = \langle\langle \Phi, e_\xi \rangle\rangle, \quad \xi \in N. \tag{7}$$

From the paper,[8] we have the duality theorem which says that the Laplace transform is a topological isomorphism from $\mathcal{F}_\theta^*(N')$ onto $\mathcal{G}_{\theta^*}(N)$. Moreover, the following useful estimate holds.

Lemma 2.2. *(See Ref. 8) Let $\Phi \sim (\Phi_n)_{n\geq 0}$ in $\mathcal{F}_\theta^*(N')$. Then, there exist $p \geq 0$ and $m > 0$ such that for any $q > p$, $m' < m$ we have*

$$\|\mathcal{L}(\Phi)\|_{\theta^*,p,m} \leq \left\{ \sum_{n=0}^{\infty} \left(\frac{e}{m'\sqrt{m}} \|i_{p,q}\|_{HS}\right)^{2n} \right\}^{1/2} \left\|\vec{\Phi}\right\|_{\theta,-q,m'}. \tag{8}$$

The Borel σ-algebra on X' will be denoted by $\mathcal{B}(X')$. It is well-known[10] that $\mathcal{B}(X')$ coincides with the σ-algebra generated by the cylinder subsets of X'. Let μ be the standard Gaussian measure on $(X', \mathcal{B}(X'))$, i.e., its characteristic function is given by

$$\int_{X'} e^{i\langle y,\xi\rangle}\, d\mu(y) = e^{-|\xi|_0^2/2}, \quad \xi \in X.$$

Suppose that the Young function θ satisfies

$$\lim_{r\to+\infty} \frac{\theta(r)}{r^2} < +\infty \tag{9}$$

then we obtain the nuclear Gel'fand triple[8]

$$\mathcal{F}_\theta(N') \subset L^2(X', \mathcal{B}(X'), \mu) \subset \mathcal{F}_\theta(N')^*.$$

We denote by $\tau(K)$ the corresponding distribution to $K \in \mathcal{L}(N, N')$ under the canonical isomorphism $\mathcal{L}(N, N') \cong (N \otimes N)'$, i.e.

$$\langle \tau(K), \xi \otimes \eta\rangle = \langle K\xi, \eta\rangle, \quad \xi,\ \eta \in N.$$

In particular, $\tau(I)$ is the usual trace τ. It can be easily shown that

$$\tau(K) = \sum_{j=0}^{\infty} (K^* e_j) \otimes e_j,$$

where K^* is the adjoint of K with respect to the dual pairing $\langle N', N\rangle$ and the infinite sum is in the sense of the strong topology on $(N \otimes N)'$.

3. The operator-parameter $\mathcal{G}_{A,B}-$ and $\mathcal{F}_{A,B}-$transforms

For locally convex spaces $\mathfrak{X}$ and $\mathfrak{Y}$ we denote by $\mathcal{L}(\mathfrak{X}, \mathfrak{Y})$ the set of all continuous linear operators from $\mathfrak{X}$ into $\mathfrak{Y}$.

Definition 3.1. Let $B, C \in \mathcal{L}(N, N)$ and put $A = C^*C$. The $\mathcal{G}_{A,B}-$transform is defined by

$$\mathcal{G}_{A,B}\varphi(y) = \int_{X'} \varphi(C^*x + B^*y) d\mu(x), \quad y \in N',\ \varphi \in \mathcal{F}_\theta(N'). \tag{10}$$

Theorem 3.1. *Let B, $C \in \mathcal{L}(N, N)$ and put $A = C^*C$, then $\mathcal{G}_{A,B}$ is a continuous linear operator from $\mathcal{F}_\theta(N')$ into itself. More precisely, for any*

$p \geq 0$, $m > 0$, we have

$$\|\mathcal{G}_{A,B}\varphi\|_{\theta,-p,m'} \leq I_{p,m} \, \|\varphi\|_{\theta,-p,m}$$

where $m' = 2m\|B^*\|$ and

$$I_{p,m} = \int_{X_{-p}} e^{\theta(2m\|C^*\||x|_{-p})} d\mu(x) < +\infty.$$

Proof. We know that, for any $\varphi \in \mathcal{F}_\theta(N')$,

$$\begin{aligned} |\mathcal{G}_{A,B}\varphi(y)| &\leq \int_{X'} |\varphi(C^*x + B^*y)| \, d\mu(x) \\ &\leq \|\varphi\|_{\theta,-p,m} \int_{X'} e^{\theta(m|C^*x+B^*y|_{-p})} d\mu(x). \end{aligned}$$

Since θ is convexe, we have

$$\theta(m|C^*x + B^*y|_{-p}) \leq \frac{1}{2}\theta(2m\|C^*\||x|_{-p}) + \frac{1}{2}\theta(2m\|B^*\||y|_{-p}),$$

and therefore

$$|\mathcal{G}_{A,B}\varphi(y)| \leq \|\varphi\|_{\theta,-p,m} \, e^{\theta(2m\|B^*\||y|_{-p})} \int_{X_{-p}} e^{\theta(2m\|C^*\||x|_{-p})} d\mu(x). \quad (11)$$

Recall that, for $p > 1$, (H, X_{-p}) is an abstract Wiener space. Then, under the condition $\lim_{x\to+\infty} \theta(x)/x^2 < +\infty$, the measure μ satisfies Fernique theorem, i.e., there exist some $\alpha > 0$ such that

$$\int_{X_{-p}} e^{\alpha|x|_{-p}^2} d\mu(x) < +\infty. \quad (12)$$

Hence, in view of (12), we obtain

$$|\mathcal{G}_{A,B}\varphi(y)| \, e^{-\theta(2m\|B^*\||y|_{-p})} \leq I_{p,m} \, \|\varphi\|_{\theta,-p,m}$$

with

$$I_{p,m} = \int_{X_{-p}} e^{\theta(2m\|C^*\||x|_{-p})} d\mu(x) < +\infty.$$

This follows

$$\|\mathcal{G}_{A,B}\varphi\|_{\theta,-p,m'} \leq I_{p,m} \, \|\varphi\|_{\theta,-p,m}$$

with $m' = 2m\|B^*\|$. □

Later on, we need the following Lemma for Taylor expansion.

Lemma 3.1. *For any* $A \in \mathcal{L}(N', N')$, $B \in \mathcal{L}(N', N')$ *and* $n \geq 0$,

$$\int_{X'} (Ax + By)^{\otimes n} d\mu(x) = \sum_{l=0}^{[n/2]} \frac{n!}{(n-2l)!\, 2^l\, l!} (\tau(AA^*))^{\otimes l} \widehat{\otimes} (By)^{\otimes (n-2l)}.$$

Proof. Using the following equality

$$(Ax + By)^{\otimes n} = \sum_{k=0}^{n} \binom{n}{k} (Ax)^{\otimes k} \widehat{\otimes} (By)^{\otimes (n-k)},$$

for $\xi \in N$, we easily obtain

$$\left\langle \int_{X'} (Ax + By)^{\otimes n} d\mu(x), \xi^{\otimes n} \right\rangle$$
$$= \sum_{k=0}^{n} \frac{n!}{(n-k)!\, k!} \left\langle (By)^{\otimes (n-k)}, \xi^{\otimes (n-k)} \right\rangle \int_{X'} \langle x^{\otimes k}, (A^*\xi)^{\otimes k} \rangle\, d\mu(x).$$

We recall the following identity for the Gaussian white noise measure, see Ref. 23,

$$\int_{X'} \langle x^{\otimes k}, (A^*\xi)^{\otimes k} \rangle\, d\mu(x) = \begin{cases} \frac{(2l)!}{2^l\, l!} |A^*\xi|_0^{2l} & \text{if} \quad k = 2l \\ 0 & \text{if} \quad k = 2l+1 \end{cases}$$

from which we deduce

$$\left\langle \int_{X'} (Ax + By)^{\otimes n} d\mu(x), \xi^{\otimes n} \right\rangle$$
$$= \sum_{l=0}^{[n/2]} \frac{n!}{(n-2l)!\, 2^l\, l!} \left\langle (By)^{\otimes (n-2l)}, \xi^{\otimes (n-2l)} \right\rangle \langle A^*\xi, A^*\xi \rangle^l$$
$$= \sum_{l=0}^{[n/2]} \frac{n!}{(n-2l)!\, 2^l\, l!} \left\langle (\tau(AA^*))^{\otimes l} \widehat{\otimes} (By)^{\otimes (n-2l)}, \xi^{\otimes n} \right\rangle$$
$$= \left\langle \sum_{l=0}^{[n/2]} \frac{n!}{(n-2l)!\, 2^l\, l!} (\tau(AA^*))^{\otimes l} \widehat{\otimes} (By)^{\otimes (n-2l)}, \xi^{\otimes n} \right\rangle.$$

The above equalities hold for all $\xi^{\otimes n}$ with $\xi \in N$, thus the statement follows by the polarization identity (see Refs. 19, 23). □

Now we can use Lemma 3.1 to represent $\mathcal{G}_{A,B}$ by Taylor expansion.

Proposition 3.1. *Let $B, C \in \mathcal{L}(N, N)$ and put $A = C^*C$, then for any $\varphi(y) = \sum_{n=0}^{\infty} \langle y^{\otimes n}, \varphi_n \rangle \in \mathcal{F}_\theta(N')$, we have*

$$\mathcal{G}_{A,B}\varphi(y) = \sum_{n=0}^{\infty} \langle y^{\otimes n}, g_n \rangle,$$

where g_n is given by

$$g_n = (B)^{\otimes n} \left(\sum_{l=0}^{\infty} \frac{(n+2l)!}{n!\, 2^l\, l!} (\tau(A))^{\otimes l} \widehat{\otimes}_{2l} \varphi_{n+2l} \right).$$

Proof. Consider $\varphi_\nu = \sum_{n=0}^{\nu} \langle \cdot^{\otimes n}, \varphi_n \rangle$ as an approximating sequence of φ in $\mathcal{F}_\theta(N')$. Then for any $p \in \mathbb{N}$ and $m > 0$ there exist $M \geq 0$ such that $|\varphi_\nu(z)| \leq M e^{\theta(m|z|_{-p})}$. Hence, in view of (12), we can apply the Lebesgue dominated convergence theorem to get

$$\int_{X'} \varphi(C^*x + B^*y) d\mu(x) = \sum_{n=0}^{\infty} \int_{X'} \langle (C^*x + B^*y)^{\otimes n}, \varphi_n \rangle d\mu(x).$$

Then by Lemma 3.1,

$$\mathcal{G}_{A,B}\varphi(y) = \sum_{n=0}^{\infty} \sum_{l=0}^{[n/2]} \frac{n!}{(n-2l)!\, 2^l\, l!} \langle (\tau(A))^{\otimes l} \widehat{\otimes} (B^*y)^{\otimes (n-2l)}, \varphi_n \rangle.$$

By changing the order of summation (which can be justified easily), we get

$$\begin{aligned} \mathcal{G}_{A,B}\varphi(y) &= \sum_{l=0}^{\infty} \sum_{n=2l}^{\infty} \frac{n!}{(n-2l)!\, 2^l\, l!} \langle (\tau(A))^{\otimes l} \widehat{\otimes} (B^*y)^{\otimes (n-2l)}, \varphi_n \rangle \\ &= \sum_{k,l=0}^{\infty} \frac{(k+2l)!}{k!\, 2^l\, l!} \langle (\tau(A))^{\otimes l} \widehat{\otimes} (B^*y)^{\otimes k}, \varphi_{k+2l} \rangle \\ &= \sum_{k=0}^{\infty} \left\langle y^{\otimes k}, B^{\otimes k} \Big(\sum_{l=0}^{\infty} \frac{(k+2l)!}{k!\, 2^l\, l!} (\tau(A))^{\otimes l} \widehat{\otimes}_{2l} \varphi_{k+2l} \Big) \right\rangle. \end{aligned}$$

This proves the desired statement. □

The adjoint of the $\mathcal{G}_{A,B}$−transform with respect to the dual pairing $\langle \mathcal{F}_\theta^*(N'), \mathcal{F}_\theta(N') \rangle$ is denoted by $\mathcal{F}_{A,B}$.

Proposition 3.2. *Let $B, C \in \mathcal{L}(N, N)$ and put $C^*C = A$. Then the $\mathcal{F}_{A,B}-$transform is the unique operator in $\mathcal{L}(\mathcal{F}_\theta^*(N'), \mathcal{F}_\theta^*(N'))$ such that for each $\Phi \in \mathcal{F}_\theta^*(N')$, the Laplace transform of $\mathcal{F}_{A,B}\Phi$ is given by*

$$\mathcal{L}(\mathcal{F}_{A,B}\Phi)(\xi) = \mathcal{L}\Phi(B\xi) \exp\left\{\frac{\langle A\xi, \xi\rangle}{2}\right\}, \quad \xi \in N.$$

Proof. Let $\Phi \in \mathcal{F}_\theta^*(N')$ and $\xi \in N$. By direct computation we easily verify the identity

$$\mathcal{G}_{A,B}e_\xi = \exp\left\{\frac{\langle A\xi, \xi\rangle}{2}\right\} e_{B\xi}.$$

Then one calculate

$$\begin{aligned}
\mathcal{L}(\mathcal{F}_{A,B}\Phi)(\xi) &= \langle\!\langle \mathcal{F}_{A,B}\Phi, e_\xi \rangle\!\rangle \\
&= \langle\!\langle \Phi, \mathcal{G}_{A,B}e_\xi \rangle\!\rangle \\
&= \langle\!\langle \Phi, e_{B\xi} \rangle\!\rangle \exp\left\{\frac{\langle A\xi, \xi\rangle}{2}\right\} \\
&= \mathcal{L}\Phi(B\xi) \exp\left\{\frac{\langle A\xi, \xi\rangle}{2}\right\}
\end{aligned}$$

which completes the proof. □

The following result is refinement of a result by Chung-Ji.[5]

Proposition 3.3. *Let $A \in \mathcal{L}(N, N')$, $B \in \mathcal{L}(N, N)$ and let $\Phi \in \mathcal{F}_\theta^*(N')$ with chaos expansion $(\Phi_n)_{n\geq 0} \in G_\theta(N')$. Then $\mathcal{F}_{A,B}\Phi$ is represented by $(F_n)_{n\geq 0} \in G_\theta(N')$ where*

$$F_n = \sum_{k=0}^{[n/2]} \frac{1}{k!\,2^k}((B^*)^{\otimes(n-2k)}\Phi_{n-2k})\widehat{\otimes}(\tau(A))^{\otimes k}.$$

Proof. Since $\mathcal{L}(\Phi)(\xi) = \sum_{n=0}^{\infty} \langle \Phi_n, \xi^{\otimes n} \rangle$, we have by Proposition 3.2

$$\begin{aligned}
\mathcal{L}(\mathcal{F}_{A,B}\Phi)(\xi) &= \mathcal{L}(\Phi)(B\xi) \exp\left\{ \frac{\langle A\xi, \xi \rangle}{2} \right\} \\
&= \sum_{n=0}^{\infty} \sum_{m=0}^{\infty} \frac{1}{m! 2^m} \langle \Phi_n, (B\xi)^{\otimes n} \rangle \langle A\xi, \xi \rangle^m \\
&= \sum_{n=0}^{\infty} \sum_{m=0}^{\infty} \frac{1}{m!\, 2^m} \langle ((B^*)^{\otimes n} \Phi_n) \widehat{\otimes} (\tau(A))^{\otimes m}, \xi^{\otimes (n+2m)} \rangle \\
&= \sum_{n=0}^{\infty} \sum_{m=0}^{[n/2]} \frac{1}{m!\, 2^m} \langle ((B^*)^{\otimes (n-2m)} \Phi_{n-2m}) \widehat{\otimes} (\tau(A))^{\otimes m}, \xi^{\otimes n} \rangle.
\end{aligned}$$

Hence we obtain the result. □

Theorem 3.2. *$\mathcal{F}_{A,B}$ is a continuous linear operator from $\mathcal{F}_\theta^*(N')$ into itself. More precisely, given $\Phi \in \mathcal{F}_\theta^*(N')$, $p \geq 0$ and $m > 0$ such that $\|\mathcal{L}(\Phi)\|_{\theta^*,p,m} < +\infty$, there exist $q > p$, $\delta < m$ and $m' > 0$ such that*

$$\left\| \overrightarrow{\mathcal{F}_{A,B}\Phi} \right\|_{\theta,-q,m'} \leq C \left\| \vec{\Phi} \right\|_{\theta,-q,\delta}$$

for some constant $C > 0$.

Proof. From (6), there exist $p \geq 0$ and $m > 0$ such that for any $q > p$ we have

$$|\Phi_n|_{-q} \leq e^n \theta_n^* m^n \|i_{q,p}\|_{HS}^n \|\mathcal{L}\Phi\|_{\theta^*,p,m}.$$

Then, by using Proposition 3.3, we get

$$\begin{aligned}
|F_n|_{-q}^2 &\leq ([n/2]+1) \sum_{k=0}^{[n/2]} \left(\frac{1}{k! 2^k} \right)^2 \|B^*\|^{2n-4k} |\Phi_{n-2k}|_{-q}^2 |\tau(A)|_{-q}^{2k} \\
&\leq 2^n \|\mathcal{L}(\Phi)\|_{\theta^*,p,m}^2 \sum_{k=0}^{[n/2]} \frac{(me\|B^*\| \|i_{q,p}\|_{HS})^{2n-4k}}{2^{2k}(k!)^2} (\theta_{n-2k}^*)^2 |\tau(A)|_{-q}^{2k},
\end{aligned}$$

where the obvious inequality $[n/2]+1 \leq 2^n$ is used into account. Therefore, for $m' > 0$, one can estimate

$$\left\|\overrightarrow{\mathcal{F}_{A,B}\Phi}\right\|_{\theta,-q,m'}^2 = \sum_{n=0}^{\infty}(n!\theta_n)^2 m'^n \left|F_n\right|_{-q}^2$$

$$\leq \|\mathcal{L}(\Phi)\|_{\theta^*,p,m}^2 \sum_{n=0}^{\infty}\sum_{k=0}^{[n/2]}(n!\theta_n)^2(2m')^n$$

$$\times \frac{(me\|B^*\|\|i_{q,p}\|_{HS})^{2n-4k}}{2^{2k}(k!)^2}(\theta_{n-2k}^*)^2 \left|\tau(A)\right|_{-q}^{2k}$$

$$= \|\mathcal{L}(\Phi)\|_{\theta^*,p,m}^2 \sum_{k=0}^{\infty}\sum_{n=2k}^{\infty}(n!\theta_n)^2(2m')^n$$

$$\times \frac{(me\|B^*\|\|i_{q,p}\|_{HS})^{2n-4k}}{2^{2k}(k!)^2}(\theta_{n-2k}^*)^2 \left|\tau(A)\right|_{-q}^{2k}$$

$$\leq \|\mathcal{L}(\Phi)\|_{\theta^*,p,m}^2 \sum_{j,k=0}^{\infty}\left(\frac{(j+2k)!\theta_j^*\theta_{j+2k}}{2^k k!}\right)^2 (2m')^{j+2k}$$

$$\times (me\|B^*\|\|i_{q,p}\|_{HS})^{2j} \left|\tau(A)\right|_{-q}^{2k}.$$

On the other hand, there exist constants $\alpha > 0$, $\nu > 0$ such that

$$(j+2k)! \leq 2^{j+4k}(k!)^2 j!\,, \quad \theta_{j+2k} \leq \alpha 2^{j+2k}\left(\frac{\nu e}{k}\right)^k \theta_j.$$

Hence

$$\left\|\overrightarrow{\mathcal{F}_{A,B}\Phi}\right\|_{\theta,-q,m'}^2 \leq \|\mathcal{L}(\Phi)\|_{\theta^*,p,m}^2 \sum_{k=0}^{\infty}\alpha^2(64m'\nu e|\tau(A)|_{-q})^{2k}\left(\frac{k!}{k^k}\right)^2 \quad (13)$$

$$\times \sum_{j=0}^{\infty}(32m'm^2e^2\|B^*\|^2\|i_{q,p}\|_{HS}^2)^j(j!\theta_j\theta_j^*)^2. \quad (14)$$

Then, if we choose m' such that

$$\max\left(64m'\nu e|\tau(A)|_{-q}, 32m'm^2e^4\|B^*\|^2\|i_{q,p}\|_{HS}^2\right) < 1,$$

the series in (13) and (14) converge, respectively, to $C_1 > 0$ and $C_1 > 0$. This follows

$$\left\|\overrightarrow{\mathcal{F}_{A,B}\Phi}\right\|_{\theta,-q,m'} \leq \sqrt{C_1C_2}\,\|\mathcal{L}(\Phi)\|_{\theta^*,p,m}.$$

Finally, from (8), there exist $\delta < m$ such that

$$\|\mathcal{L}(\Phi)\|_{\theta^*,p,m} \leq C_3 \left\|\vec{\Phi}\right\|_{\theta,-q,\delta},$$

where

$$C_3 = \left\{ \sum_{n=0}^{\infty} \left(\frac{e}{\delta\sqrt{m}} \|i_{q,p}\|_{HS} \right)^{2n} \right\}^{1/2}.$$

Therefore, put $C = C_3\sqrt{C_1C_2}$, we get

$$\left\| \overrightarrow{\mathcal{F}_{A,B}\Phi} \right\|_{\theta,-q,m'} \leq C \left\| \vec{\Phi} \right\|_{\theta,-q,\delta}$$

which completes the proof. □

Proposition 3.4. *Let B, C_1, C_2, $D \in \mathcal{L}(N,N)$ and denote $A_1 = C_1^*C_1$, $A_2 = C_2^*C_2$, hen*

$$\mathcal{G}_{A_2,D}\mathcal{G}_{A_1,B} = \mathcal{G}_{A_1+B^*A_2B,DB}.$$

In particular, if B is invertible, then the operator $\mathcal{G}_{A,B}$ is invertible and

$$\mathcal{G}_{A,B}^{-1} = \mathcal{G}_{-(B^*)^{-1}AB^{-1},B^{-1}}.$$

Proof. For any $\xi \in N$, we have

$$\begin{aligned}\mathcal{G}_{A_2,D}(\mathcal{G}_{A_1,B}e_\xi) &= e_{DB\xi}\exp\left\{\frac{\langle A_2B\xi, B\xi\rangle + \langle A_1\xi,\xi\rangle}{2}\right\} \\ &= e_{DB\xi}\exp\left\{\frac{1}{2}\langle (A_1 + B^*A_2B)\xi,\xi\rangle\right\} \\ &= \mathcal{G}_{A_1+B^*A_2B,DB}e_\xi.\end{aligned}$$

In particular, if B is invertible, by using the identity $\mathcal{G}_{0,I} = I$, we find

$$\mathcal{G}_{A,B}^{-1} = \mathcal{G}_{-(B^*)^{-1}AB^{-1},B^{-1}}.$$ □

By duality we have the following results, the proofs of which are immediate.

Proposition 3.5. *Let B, C_1, C_2, $D \in \mathcal{L}(N,N)$ and denote $A_1 = C_1^*C_1$, $A_2 = C_2^*C_2$, then we have*

$$\mathcal{F}_{A_2,D}\mathcal{F}_{A_1,B} = \mathcal{F}_{A_2+D^*A_1D,BD}.$$

In particular, if B is invertible, then the operator $\mathcal{F}_{A,B}$ is also invertible and

$$\mathcal{F}_{A,B}^{-1} = \mathcal{F}_{-(B^{-1})^*AB^{-1},B^{-1}}.$$

Proposition 3.6. *Let $B, C \in \mathcal{L}(N, N)$ such that B is invertible and put $C^*C = A$. Then $\mathcal{G}_{A,B}$ and $\mathcal{F}_{A,B}$ realize two topological isomorphisms from $\mathcal{F}_\theta(N')$ into itself and $\mathcal{F}_\theta^*(N')$ into itself, respectively.*

In the following Section we will study the Cauchy problem associated to the K-Gross Laplacian using convolution calculus. In particular, we shall focus on some regularity properties of the solution.

4. Generalized Gross heat equation

In infinite dimensional complex analysis[6] a convolution operator on the test space $\mathcal{F}_\theta(N')$ is a continuous linear operator from $\mathcal{F}_\theta(N')$ into itself which commutes with translation operator. Let $x \in N'$, we define the translation operator τ_{-x} on $\mathcal{F}_\theta(N')$ by

$$\tau_{-x}\varphi(y) = \varphi(x + y), \quad y \in N', \quad \varphi \in \mathcal{F}_\theta(N').$$

It is easy to see that τ_{-x} is a continuous linear operator from $\mathcal{F}_\theta(N')$ into itself. Now, we define the convolution product of a distribution $\Phi \in \mathcal{F}_\theta^*(N')$ with a test function $\varphi \in \mathcal{F}_\theta(N')$ as follows

$$\Phi * \varphi(x) = \langle\langle \Phi, \tau_{-x}\varphi \rangle\rangle, \quad x \in N'.$$

If Φ is represented by $\vec{\Phi} = (\Phi_n)_{n\geq 0} \in G_\theta(N')$, then

$$\Phi * \varphi(x) = \sum_{k=0}^{\infty} \langle x^{\otimes n}, \psi^{(n)} \rangle, \tag{15}$$

where for every integer $n \in \mathbb{N}$

$$\psi^{(n)} = \sum_{k=0}^{\infty} k! \binom{n+k}{n} \Phi_k \otimes_k \varphi^{(n+k)}.$$

A direct calculation shows that the sequence $(\psi^{(n)})_{n\geq 0}$ is an element of $F_\theta(N)$ and consequently $\Phi * \varphi \in \mathcal{F}_\theta(N')$. It was proved in Ref. 7 that Ξ is a convolution operator on $\mathcal{F}_\theta(N')$ if and only if there exists $\Phi \in \mathcal{F}_\theta^*(N')$ such that

$$\Xi(\varphi) = \Phi * \varphi, \quad \forall \varphi \in \mathcal{F}_\theta(N'). \tag{16}$$

It is well-known that $\mathcal{G}_{\theta^*}(N)$ is closed under the usual multiplication. Then, for Φ, $\Psi \in \mathcal{F}_\theta^*(N')$, we define the convolution $\Phi * \Psi$, via the Laplace transform, as the unique distribution in $\mathcal{F}_\theta^*(N')$, characterized by

$$\mathcal{L}(\Phi * \Psi) = \mathcal{L}(\Phi)\,\mathcal{L}(\Psi).$$

We can easily prove the fact

$$\langle\langle \Phi * \Psi, \varphi \rangle\rangle = \langle\langle \Phi, \Psi * \varphi \rangle\rangle\,, \quad \Phi\,,\, \Psi \in \mathcal{F}_\theta^*(N')\,,\ \varphi \in \mathcal{F}_\theta(N'). \tag{17}$$

Let $C \in \mathcal{L}(N, N)$ and put $K = C^*C \in \mathcal{L}(N, N')$. In the following we investigate the following Cauchy problem:

$$\frac{\partial u(x,t)}{\partial t} = \frac{1}{2}\Delta_G(K)u(x,t), \quad u(x,0) = \varphi(x), \tag{18}$$

where $\varphi \in \mathcal{F}_\theta(N')$ and $\Delta_G(K)$ is the K-Gross Laplacian defined in (6).

Theorem 4.1. *Let θ be a Young function satisfying $\lim_{r\to+\infty} \frac{\theta(r)}{r^2} < +\infty$ and $\varphi \in \mathcal{F}_\theta(N')$. Then the heat equation* (18) *associated with the K-Gross Laplacian has a unique solution in $\mathcal{F}_\theta(N')$ given by*

$$u(x,t) = \mathcal{G}_{tK,I}\varphi(x).$$

Proof. It is well-known[7] that the K-Gross Laplacian $\Delta_G(K)$ is a convolution operator associated to the distribution $\Phi_{\tau(K)} \sim (0, 0, \tau(K), 0, \cdots)$, i.e.,

$$\Delta_G(K)\varphi = \Phi_{\tau(K)} * \varphi\,, \quad \forall \varphi \in \mathcal{F}_\theta(N'). \tag{19}$$

Thus the heat equation (18) is equivalent to

$$\frac{\partial u}{\partial t} = \Phi_{\frac{1}{2}\tau(K)} * u\,, \quad t \geq 0\,, \quad u(0) = \varphi.$$

Since

$$\mathcal{L}(\Phi_{\frac{1}{2}\tau(K)})(\xi) = \frac{1}{2}\langle K\xi, \xi\rangle, \quad \xi \in N,$$

it follows, from Theorem 3 in Ref. 2, that the equation (18) has a unique solution in $\mathcal{F}_\theta(N')$ given by

$$u(x,t) = (e^{*t\Phi_{\frac{1}{2}\tau(K)}} * \varphi)(x), \quad t \geq 0.$$

On the other hand, notice that

$$\mathcal{L}(e^{*t\Phi_{\frac{1}{2}\tau(K)}})(\xi) = \exp\left\{\frac{t}{2}\langle K\xi,\xi\rangle\right\} = \mathcal{L}(\widetilde{\mu}_{tK})(\xi), \quad \xi \in N,\ t \geq 0, \tag{20}$$

where $\widetilde{\mu}_{tK}$ is the distribution in $\mathcal{F}_\theta^*(N')$ associated to the Gaussian measure μ_{tK} with covariance operator tK. Then, the solution $u(x,t)$ can be expressed as

$$u(x,t) = (\widetilde{\mu}_{tK} * \varphi)(x), \quad t \geq 0, \quad x \in N'.$$

Now, for any $\xi \in N$, $x \in N'$, we have

$$\begin{aligned}(\widetilde{\mu}_{tK} * e_\xi)(x) &= \mathcal{L}(\widetilde{\mu}_{tK})(\xi)e_\xi)(x)\\ &= \exp\left\{\frac{t}{2}\langle K\xi,\xi\rangle\right\} e_\xi(x)\\ &= \mathcal{G}_{tK,I}e_\xi(x).\end{aligned}$$

Then, since $\{e_\xi\,;\ \xi \in N\}$ is a total subset of $\mathcal{F}_\theta(N')$, we conclude that

$$u(x,t) = (\widetilde{\mu}_{tK} * \varphi)(x) = (\mathcal{G}_{tK,I}\varphi)(x)$$

as desired. □

Let $\Delta_G^*(K)$ stands for the adjoint of the generalized Gross Laplacian $\Delta_G(K)$ with respect to the dual pairing $\langle \mathcal{F}_\theta^*(N'), \mathcal{F}_\theta(N')\rangle$. Then, for $\Phi \in \mathcal{F}_\theta^*(N')$ and $\varphi \in \mathcal{F}_\theta(N')$, we have

$$\begin{aligned}\langle\!\langle \Delta_G^*(K)\Phi, \varphi\rangle\!\rangle &= \langle\!\langle \Phi, \Delta_G(K)\varphi\rangle\!\rangle\\ &= \left\langle\!\left\langle \Phi, \Phi_{\tau(K)} * \varphi\right\rangle\!\right\rangle\\ &= \left\langle\!\left\langle \Phi * \Phi_{\tau(K)}, \varphi\right\rangle\!\right\rangle,\end{aligned}$$

where identity (17) is used into account. We then deduce the extension of (19) to $\mathcal{F}_\theta^*(N')$:

$$\Delta_G^*(K)\Phi = \Phi_{\tau(K)} * \Phi, \quad \forall\, \Phi \in \mathcal{F}_\theta^*(N').$$

By using duality arguments, the proof of the following Theorem is a slight modification of the previous one.

Theorem 4.2. *The generalized heat equation associated to* $\Delta_G^*(K)$*:*

$$\frac{\partial U_t}{\partial t} = \frac{1}{2}\Delta_G^*(K)U_t, \quad U_0 = \Phi \in \mathcal{F}_\theta^*(N'), \tag{21}$$

has a unique solution in $\mathcal{F}_\theta^*(N')$ *given by*

$$U_t = \mathcal{F}_{tK,I}\Phi. \tag{22}$$

In the remainder of this paper, we give some regularity properties of the solution U_t in (22). By definition, a test function $\varphi \in \mathcal{F}_\theta(N')$ is said to be positive if $\varphi(x + i0) \geq 0$ for all $x \in X'$. We denote by $\mathcal{F}_\theta(N')_+$ the set of all such positive test function. A generalized function $\Phi \in \mathcal{F}_\theta^*(N')$ is called positive if it satisfies the condition:

$$\langle\langle \Phi, \varphi \rangle\rangle \geq 0, \quad \forall \varphi \in \mathcal{F}_\theta(N')_+.$$

The set of all positive generalized functions is denoted by $\mathcal{F}_\theta^*(N')_+$. The next characterization theorem, quoted from the paper,[26] gives the integral representation of positive generalized functions by Radon measures.

Theorem 4.3. *For any* $\Phi \in \mathcal{F}_\theta^*(N')_+$ *there exists a unique Radon measure* μ_Φ *on* X' *such that*

$$\langle\langle \Phi, \varphi \rangle\rangle = \int_{X'} \varphi(y + i0) d\mu_\Phi(y), \quad \forall \varphi \in \mathcal{F}_\theta(N'). \tag{23}$$

Conversely, a finite positive Borel measure ν *on* X' *represents a positive generalized function* $\Phi \in \mathcal{F}_\theta^*(N')_+$ *if and only if there exist* $q > 0$ *and* $\delta > 0$ *such that* $\nu(X_{-q}) = 1$ *and*

$$\int_{X_{-q}} e^{\theta(\delta|y|_{-q})} d\nu(y) < +\infty. \tag{24}$$

We are now ready to give sufficient condition for the solution of the Cauchy problem (21) to be positive.

Theorem 4.4. *Suppose the initial condition* Φ *in Equation (21) is a positive generalized function. Then the solution (22) is a positive generalized* *function. Moreover, there exists a unique positive Radon measure* $\mu_{U_{tK}}$ *on* X' *such that*

$$\langle\langle U_t, \varphi\rangle\rangle = \int_{X'} \varphi(y) d\mu_{U_{tK}}(y) \quad \forall \varphi \in \mathcal{F}_\theta(N'), \tag{25}$$

and $\mu_{U_{tK}}$ *satisfies the following integrability condition*

$$\int_{X_{-q}} e^{\theta(\delta|y|_{-q})} d\mu_{U_{tK}}(y) < +\infty, \tag{26}$$

for some $q > 0$ *and* $\delta > 0$.

Proof. First we prove that the convolution product of two positive generalized functions is also a positive generalized function. Let $\Phi, \Psi \in \mathcal{F}^*_\theta(N')_+$. For any $\varphi \in \mathcal{F}_\theta(N')_+$, we have

$$\langle\langle \Phi * \Psi, \varphi\rangle\rangle = \langle\langle \Phi, \Psi * \varphi\rangle\rangle = [\Phi * (\Psi * \varphi)](0). \tag{27}$$

On the hand, for any $x \in X'$,

$$(t_x\varphi)(y) = \varphi(x+y) \geq 0, \quad \forall y \in X',$$

which implies that $t_x\varphi \in \mathcal{F}_\theta(N')_+$, for all $x \in X'$. Since $(\Psi * \varphi)(x) = \langle\langle \Psi, t_x\varphi\rangle\rangle \geq 0$, we conclude that $\Psi * \varphi \in \mathcal{F}_\theta(N')_+$. Hence, Equation (27) show that the generalized function $\Phi * \Psi$ is positive. It follows that the solution in Eq. (22) is positive. Finally, Theorem 4.3 guarantees the existence and uniqueness of a Radon measure $\mu_{U_{tK}}$ on X' associated with U_t as in Equation (25) and $\mu_{U_{tK}}$ satisfies the integrability condition (26). □

References

1. A. Barhoumi, H.-H. Kuo and H. Ouerdiane, *Generalized heat equation with noises*, Soochow J. Math. 32 (2006), 113-125 .

2. M. Ben Chrouda, M. El Oued and H. Ouerdiane, *Convolution calculus and application to stockastic differential equation*, Soochow J. Math. 28 (2002), 375-388.

3. M. Ben Chrouda and H. Ouerdiane *Algebras of Operators on Holomorphic Functions and Application*, Mathematical Physics, Analysis Geomety 5: 65-76, 2002.

4. D. M. Chung and U.C. Ji, *Transform on white noise functionals with their application to Cauchy problems*, Nagoya Math. J. Vol 147 (1997), 1-23.
5. D. M. Chung and U. C. Ji, *Transformation groups on white noise functionals and their application*, Appl. Math. Optim. 37 No. 2 (1998), 205-223.
6. D. Dineen, *Complex analysis in locally convex space*, Mathematical Studies, North Holland, Amsterdam 75, 1981.
7. R. Gannoun, R. Hachaichi and H. Ouerdiane, *Division de fonctions holomorphes à croissance θ−exponentielle*, Preprint, BiBos No E 00-01-04, (2000).
8. R. Gannoun, R. Hachachi, H. Ouerdiane and A. Rezgui, *Un théorème de dualité entre espaces de fonctions holomorphes á croissance exponentielle*, J. Funct. Anal., Vol. 171 No. 1 (2000), 1-14.
9. I.M. Gel'fand and G.E. Shilov, *Generalized Functions*, Vol.I. Academic Press, Inc., New York, 1968.
10. I.M. Gel'fand and N.Ya. Vilenkin, *Generalized unctions*, Vol. 4, Academic press New York and London 1964.
11. L. Gross, *Abstract Wiener spaces*, Proc. 5-th Berkeley Symp. Math. Stat. Probab. 2 (1967), 31-42.
12. T. Hida, H.-H. Kuo and N. Obata, *Transformations for white noise functionals*, J. Funct. Anal., 111 (1993), 259-277.
13. U. C. Ji, *Integral kernel operators on regular generalized white noise functions.* Bull. Korean Math. Soc. 37 No. 3 (2000), 601-618.
14. U. C. Ji, *Quantum extensions of Fourier-Gauss and Fourier-Mehler transforms*, J. Korean Math. Soc. 45 No.6 (2008), 1785-1801.
15. U. C. Ji, N. Obata and H. Ouerdiane, *Analytic characterization of generalized Fock space operators as two-variable entire functions with growth condition*, World Scientific, Vol. 5 No. 3 (2002), 395-407.
16. P. Krée and R. Raczka, *Kernels and symbols of operators in quantum field*

theory, Ann. Inst. H. Poincaré A 18 (1978), 41-73.

17. H.-H. Kuo, *On Fourier transform of generalized Brownian functionals*, J. Multivariate Anal. 12 (1982), 415-431.
18. H.-H. Kuo, *The Fourier transform in white noise calculus*, J. Multivariate Analysis 31 (1989), 311-327.
19. H.-H. Kuo, *White noise distrubution theory*, CRC Press, Boca Raton 1996.
20. Y. J. Lee, *Analytic version of test functionals, Fourier transform and a characterization of measures in white noise calculus*, J. Funct. Anal. 100 (1991), 359-380.
21. Y. J. Lee, *Integral transform of analytic function on abstract Wiener space.* J. Funct. Anal. 47 No. 2 (1982), 153-164.
22. S. Luo and J. Yan, *Gaussian kernel operators on white noise functional spaces.* Science in china Vol. 43 No. 10 (2000), 1067-1074.
23. N. Obata, *White Noise calculs and Fock Space.* Lecture Notes in Math. 1577, Spring, New York 1994.
24. N. Obata, *Quantum White Noise Calculus based On Nuclear Algebras of Entier Functions* . RIMS Kokyuroku 2002.
25. H. Ouerdiane, *Noyaux et symboles d'opérateurs sur des fonctionelles analytiques gaussiennes.* Japon. J. Math. 21 (1995), 223-234.
26. H. Ouerdiane and A. Rezgui, *Representation intégrales de fonctionnelles analytiques*, Stochastic Processes. Physics and Geometry: New Interplays. A volume in honor of S. Albevrio, canadian Math. Society. Conference Proceeding Series, 1999.
27. J. Potthoff and J. A. Yan, *Some results about test and generalized functionals of white noise*, Proc. Singapore Prob. Conf. L.Y. Chen et al.(eds.) (1989), 121-145.

Dissipative Quantum Annealing

D. de Falco, E. Pertoso, D. Tamascelli

Dipartimento di Scienze dell'Informazione, Università degli Studi di Milano, via Comelico 39/41, 20135 Milano, Italy

* *E-mail: tamascelli@dsi.unimi.it*

In the evolution of a quantum system subject to a nonlinear dissipative term of Kostin type the probability amplitudes associated to excited states of the Hamiltonian are damped; this results in a dynamic convergence toward the ground state. In this paper we dicuss how dissipation can replace adiabatic evolution in the search of the ground state of a target Hamiltonian and discuss applications to Quantum Annealing.

Keywords: Quantum annealing, ground-state process, adiabatic computation, dissipative dynamics.

1. Introduction

Quantum computation studies the power of computing machines whose basic constituents are quantum systems. Obviously, a task which quantum computers will perform better than any classical device is the simulation of quantum systems.[18] However, whether quantum computers are in general more powerful than our classical computers is an unanswered question. What is known, for the time being, is that, at least, a quantum computer would factorize an integer number exponentially faster (the most celebrated Shor factorization[32]) and retrieve a specified element in an unstructured database quadratically faster (Grover search[21]) than any classical computer. When implemented on a classical computer, however,

quantum algorithms perform poorly since, in the general case, the simulation of quantum systems requires an amount of (classical) computational resources which scales exponentially with the dimension of the system.
On the opposite, the analysis of quantum systems by means of stochastic mechanics[26] may suggest *classical* probabilistic algorithms, efficiently implementable on classical computers: hybrid classical-quantum procedures which, by mimicking some useful features of a Nelson process,[10,11] can provide good heuristics for the solution of difficult problems. One enlightening example is provided by *Quantum Annealing (or Quantum Stochastic Optimization)*: it was suggested by the semiclassical ($\hbar \to 0$) behaviour of the trajectories of the stochastic process canonically associated to the ground state of a Hamiltonian. Quantum Annealing has nowadays become a widely known optimization technique used in different fields, from solid state[29,30] to chemical[2,15,17,20] physics.

In this paper, after a brief review on Quantum Annealing, we discuss the role played by dissipation of Kostin type as an alternative to adiabatic evolution toward the ground state of a given Hamiltonian.

The exposition is organized as follows: in section 2 we sketchily present Quantum Annealing. Adiabatic computation is presented in section 3. In section 4 we introduce the Schrödinger-Kostin equation and present some of its features by means of simple toy models in the continuous 1-dimensional and discrete 2-dimensional case. In the fifth and last section we discuss our results and present lines of future research.

2. Quantum annealing

The aim of combinatorial optimization is to find good approximations of the solution(s) of minimization problems, usually formulated as follows: given a cost function $f : \mathcal{S} \to \mathbb{R}$, $\mathcal{S}$ being the set of admissible solutions, we want to find

$$s_{min} : f(s_{min}) = min\{f(s), s \in \mathcal{S}\},$$

that is the element(s) of $\mathcal{S}$ which minimizes the cost function.
Many of the most famous algorithms currently used in this field[27] were inspired by analogies with physical systems. Among them, the most celebrated is *Thermal simulated annealing*[23] proposed in 1983 by Kirkpatrick *et al.*: the space of all admissible solutions is endowed with a potential profile dependent on the cost function associated to the optimization problem. The exploration of this space is represented by a random walk depending

on the potential and on a temperature dependent diffusion. Thermal fluctuations allow the walker to jump, from time to time, from one minimum to another one of lower energy (see Fig. 1(a)). A suitably scheduled temperature lowering (annealing) stabilizes then the walk around a, hopefully global, minimum of the potential profile.

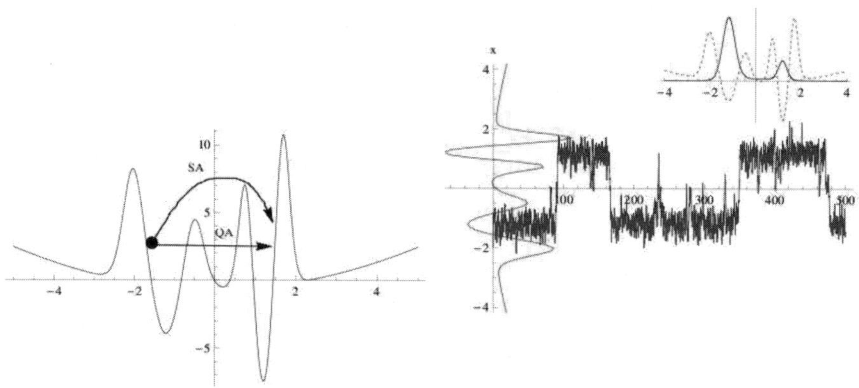

Fig. 1. (Color online) (a) Simulated vs. quantum annealing: the thermal jumps from one minimun to another which allow the exploration of the configuration space in thermal simulated annealing are substituted, in quantum annealing, by tunneling. (b) A sample path $q_\nu(t)$ of the ground state process. The potential $V(x)$ is shown on the left. Inset: the same potential profile (dashed red line) and the invariant measure $\rho_\nu(x) = (\psi_\nu(x))^2$ (solid blue line).

In *Quantum Annealing*,[6] or *Quantum Stochastic Optimization*,[5] the cost function f is encoded in the potential function V_f appearing in the system Hamiltonian:

$$H_\nu = -\frac{\nu^2}{2}\delta + V_f.$$

The physical intuition which stands behind the *Quantum Annealing* approach to combinatorial optimization comes from the deep analysis of the semiclassical limit performed in Ref. 22 and is summarized in figure 1(b):

if one knew the ground state $\psi_\nu(x)$ of Schrödinger Hamiltonian

$$H_\nu = -\frac{1}{2}\nu^2 \frac{d^2}{dx^2} + V(x) \tag{1}$$

then one could construct a stochastic process[1,28] evolving according to

$$dq_\nu(t) = b_\nu(q_\nu(t))dt + \nu^{1/2} dw(t), \tag{2}$$

having $\rho_\nu(x) = (\psi_\nu(x))^2$ as its invariant measure and drift field

$$b_\nu(t) = \frac{1}{2}\nu \frac{d}{dx} \ln(\rho_\nu(x)).$$

The semiclassical behavior of q_ν is characterized by long sojourns around the stable configurations, i.e. minima of $V(x)$ (the dashed line of the inset of Fig. 1(b)), interrupted by rare large fluctuations which carry q_ν from one minimum to another: q_ν in thus allowed to "tunnel" away from local minima to the global minimum of $V(x)$, as Fig. 1(b) shows quite clearly. Indeed, as $\nu \to 0^+$ "*the process will behave much like a Markov chain whose state space is discrete and given by the stable configurations*".[22]
Thermal fluctuations, which in *Simulated Annealing* were responsible for the exploration of the configuration space via a classical quantum walk, are replaced in *Quantum Annealing* by quantum fluctuations, which allow the corresponding *quantum walk* to *tunnel* from one mininum to another (see Fig. 1). Since the ground state relevant support is a contraction of the solution space, which is usually very large, the stochastic process associated to the ground state will visit, sooner or later, the optimal solution.
It is worth noticing here that since many computational problems can be translated in terms of minimization of some cost functional, the possibility of encoding optimization problems into ground state problems has played a key role in quantum complexity (see Ref. 16 and references therein); for simple examples of the encoding of an optimization problem in a potential function on some configuration space, instead, we refer the interested reader to Ref. 5 (Graph Partitioning) and Refs. 13,14 (Satisfiability).

3. Adiabatic quantum computation

A difficulty in the above approach is the fact that the ground state ψ_ν of the Hamiltonian H_ν is seldom exactly known and approximations are required. One of the earliest proposals in this direction was advanced in Ref. 5 and applied in Ref. 6: construct an unnormalized approximation of ψ_ν by acting on a suitably chosen initial condition $\phi_{trial}(x)$ with the Hamiltonian semi-group $\exp(-tH_\nu)$, namely by solving, with the initial condition ϕ_{trial}, the imaginary time Schrödinger equation. Similar ideas appeared in the chemical physics literature[3] and, with more specific reference to the optimization problems considered here, in Refs. 2,15,17.

The unphysical step of imaginary time evolution is, however, not strictly necessary: one could use, as a matter of fact, adiabatic techniques to start with an initial state which is easy to prepare and which is the ground state of an initial time Hamiltonian H_0 and then turn it adiabatically into the ground state of a target Hamiltonian H_T at time T that corresponds, in Jona-Lasinio description, to a stochastic process which visits good minima with high probability. Using the notation of Ref. 13, we consider the time dependent Hamiltonian:

$$H(t) = tH_T + (T - t)H_0, \; 0 \leq t \leq T.$$

Let us indicate with $|t; e_k(t)\rangle$ the *instantaneous eigenvector* of $H(t)$ corresponding to the *instantaneous eigenvalue* $e_k(t)$ with $e_0(t) \leq e_1(t) \leq \ldots \leq e_n(t), \; 0 \leq t \leq T$. The adiabatic theorem states that, if the time T satisfies

$$T \mathfrak{G} \frac{\xi}{g_{min}}, \tag{3}$$

where g_{min} is the minimum gap

$$g_{min} = \min_{0 \leq t \leq T} (e_1(t) - e_0(t))$$

and

$$\xi = \max_{0 \leq t \leq T} \left| \langle t; e_1(t)| \frac{dH}{dt} |t; e_0(t)\rangle \right|,$$

then $|\langle T; e_0|\psi(T)\rangle|$ can be made arbitrarily close to 1. In other words if we start in the state $|0; e_0(0)\rangle$ we will end up in the ground state $|T; e_0(T)\rangle$ of the target Hamiltonian H_T. In practical cases ξ is not too big, thus the size of T is governed by g_{min}^{-2}: the smaller g_{min} the slower must be the change rate of the Hamiltonian if we want to avoid transition from the ground state to excited states.

As an example, let us consider the following toy model.[33] We take the potential function:

$$V(x) = \begin{cases} V_0 \frac{(x^2 - a_+^2)^2}{a_+^4} + \delta x, & \text{for } x \geq 0 \\ V_0 \frac{(x^2 - a_-^2)^2}{a_-^4} + \delta x, & \text{for } x < 0, \end{cases} \tag{4}$$

as the cost function of a given optimization problem. For proper choice of the parameters,[29] the local minimum of the potential $V(x)$ can be made wider than the global one. We insert $V(x)$ in the Hamiltonian:

$$H(t;T) = -\Gamma(t;T)\delta + V(x),$$

where

$$\Gamma(t;T) = \frac{1}{2}(1 - \frac{t}{T}).$$

The kinetic energy coefficient is thus time dependent and decreases linearly

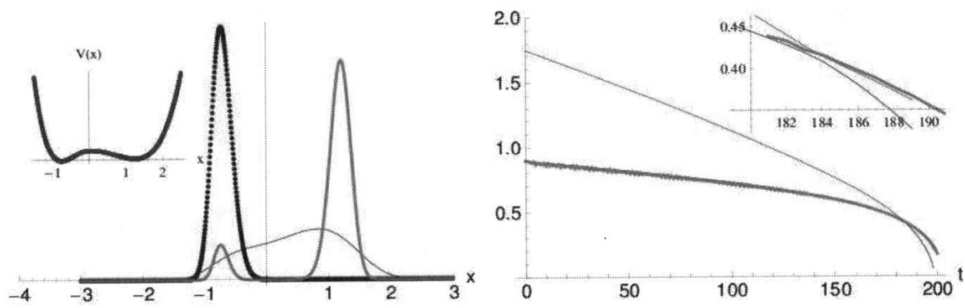

(a) (b)

Fig. 2. (Color online) (a) In the inset the shape of the asymmetric double well potential $V(x)$ of Eq. (4) for the choice $a_+ = 1.25$, $a_- = 0.75$, $\delta = 0.1$, $V_0 = 1$ of the parametrs. The graph shows the probability distribution of the position observable on: the ground state of the Hamiltonian $H(0) = H_0$ (solid thin blue line), the ground state of $H(T)$ (dotted blue line) and the state $\phi(200;T)$ (solid, thick green line). (b) The instantaneous eigenvalues $e_0(t)$ and $e_1(t)$ (respectively blue and red thin lines) of $H(t)$; the expected value of the energy observable $H(t)$ on the state $\phi(200;t)$ is represented as a thick green line. In the inset we show a magnification of the jump happening around time $t = 185$.

in the time interval $(0,T)$. As shown in figure Fig. 2(a) the state $|t;e_0(t)\rangle$ is initially ($t = 0$) distributed on both wells; then, as $\Gamma(t)$ decreases, it gets concentrated in the rightmost well, corresponding to the local minimum. Then, as $\Gamma(t) \to 0$, the state "tunnels" through the potential barrier to the leftmost, global, minimum. It is an easy guess that the ground-state process associated to the state $|T;e_0(T)\rangle$ would visit only the optimal solution. However, if we try to approximate the ground state $|T;e_0(T)\rangle$ by adiabatic evolution of the initial state $|0; e_0(0)\rangle$, T must be chosen according to Eq. (3). From Fig. 2(b) one can see that at $t/T \approx 0.90$ there is an avoided Landau-Zener crossing of the instantaneous eigenvalues $e_0(t)$ and $e_1(t)$ corresponding to $g_{min} \approx 0.05$. The annealing time must then be of order $g_{min}^{-2} \approx 25000$. If the Hamiltonian changes too fast, there is a non-vanishing probability of jumping out of the ground state. In fact, the expected value of the energy operator $H(200;t)$ on the solution $\phi(200;t)$ of the Schrödinger equation

$$i\frac{d}{dt}\psi(t) = -\Gamma(200;t)\Delta + V(x)$$

ceases to follow the value $e_0(t)$ and the projection on $|t; e_0(t)\rangle$ passes from 1 to 0 quite suddenly around the time $t = 185$. In other words, there is a jump to an excited state, which, as Fig. 2(a) shows, is mostly localized on the wrong minimum. What should be stressed here is that since g_{min} is in general not known, the annealing procedure can lead to very poor approximations of the ground state corresponding to very bad answers to the optimization problem.

4. Dissipative Schrödinger equation

An alternative mehod to find the ground state of a given Hamiltonian is provided by dissipative dynamics of Kostin type.[24]
Dissipation arises in quantum mechanics when an observable (sub)system of interest is coupled to an environment (heat-bath, reservoir) and the environmental degrees of freedom are traced out.[19] Effective descriptions of dissipative systems stems from the necessity of reducing the dynamics of many-body problem to an effective one of few- or even one-body problem which, of course, would greatly simplify the simulation of quantum systems. Effective modifications of the Schrödinger equation, by means of time-dependent Hamiltonians as well as the introduction of nonlinear terms, are discussed in the literature (see Ref. 31 and references therein).
In a seminal paper[25] M.D. Kostin proposed a set of conditions that an effective Schrödinger equation should fulfill to determine a norm-preserving dissipative dynamics and propose one such a term himself (notably, this is by no means a unique choice, as we will see in a little while). The Schrödinger-Kostin equation reads:

$$i\frac{\partial\psi(t,x)}{\partial t} = \left(-H_0 + K(\gamma, \psi(t,x))\right)\psi(t,x) \tag{5}$$

with $H_0 = -\frac{\nu^2}{2}\Delta\psi(t,x) + V(x)$ and

$$K(\gamma, \psi(t,x)) = \gamma\ S(t,x), \tag{6}$$

$S(t,x)$ being the phase of the wave function $\psi(t,x) = \sqrt{\rho(t,x)}\exp(S(t,x))$. The energy of a system evolving under this nonlinear Hamiltonian is a monotone non-increasing function of time, that is, given a solution $\phi(t,x)$ of Eq. (5) it satisfies $d/dt\langle\phi|H_0|\phi\rangle \leq 0$, for $\gamma \geq 0$.

In Ref. 12 it has been shown that dissipation, usually seen as detrimental, can be useful in quantum annealing. In fact, the presence of friction allows an exhaustive exploration of the solution space of simple optimization problems by balancing genuinely quantum effects such as Bloch oscillations[7,34] and Anderson localization.[4] Here, we want to show how friction dynamically drives the state of the system toward the ground state of the Hamiltonian H_0, thus implementing the same *contraction* mechanism which stands behind conventional quantum annealing.

4.1. *Continuous case*

For the sake of comparison with the results of the previous section, we still consider the potential $V(x)$ defined in Eq. (4). We take the *time-independent* Hamiltonian H_0 and set $\nu^2/2 = 0.025$: with this choice the ground state is localized around the absolute minimum of $V(x)$. We take an initial condition $\psi(0,x) = \psi_0(x)$ spread over both wells, which gurantees that the projection on the ground state is non-vanishing. Fig. 3 shows the solution $\psi(t)$ of Eq. (5) with the choice $\gamma = 0.12$. The probability mass "percolates" from the local to the global minimun (Fig. 3(b)), the energy is monotonically decreasing and the projection on the ground state of H_0 monotonically increasing(Fig. 3(b)).

4.2. *Discrete case*

As soon as discrete systems, such as spin lattices, are considered, a discrete version of the Kostin nonlinear term is required. In Ref. 12 we proposed the form:

$$K_D(\gamma, \psi(t,x)) = \gamma \sum_{y=2}^{x} \sin\left(S(t,y) - S(t,y-1)\right), \text{ with } \gamma > 0$$

which is a first order finite difference approximation of $K(\gamma, \psi)$ on a linear chain but cannot be extended to general n-dimensional lattices because of

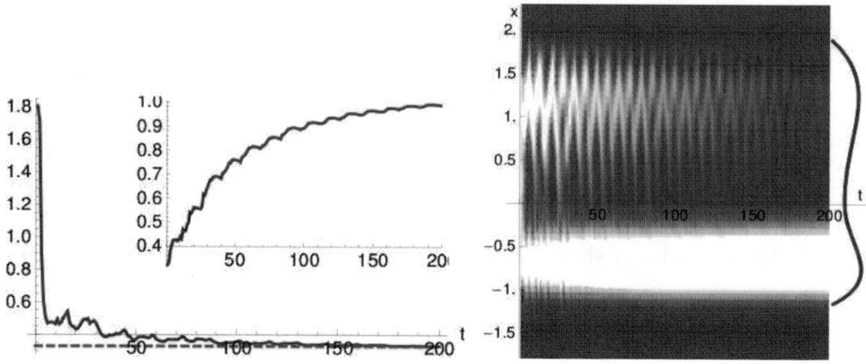

(a) (b)

Fig. 3. (Color online) $V(x)$ as in Fig. 2. $\psi(x,0) = I_{[-1,2]}/3$. (a) The probability density plot of $|\phi(t,x)|^2$: the probability mass percolates from the local to the global minimum, where the ground state of H_0 with $\frac{\nu^2}{2} = 0.025$ is located. The profile of the potential $V(x)$ is represented, out of scale, on the rightmost part of the graph. (b) The average energy $\langle\phi(t,x)|H_0|\phi(t,x)\rangle$ as a function of time; in the inset, the projection on the ground state, as a function of time. The oscillations of the energy (which is non "strictly" monotone decreasing) are due to numerical instability.

the different notion of neighbourhood of a site.

To generalize the discrete approximation of the frictional term, let us consider a generic undirected graph $G = (\mathcal{V}, \mathcal{E})$, $\mathcal{V} = \{v_1, v_2, \ldots, v_n\}$ being the set of vertices and $\mathcal{E} \subset \mathcal{V} \times \mathcal{V}$ is the set of edges. Given any $x \in \mathcal{V}$ we define

$$N(x) = \{v_k \in V : (v_k, x) \in \mathcal{E} \text{ or } (x, v_k) \in \mathcal{E}\}$$

that is the set of neighbors of the vertex x.

Here we propose the term:

$$K'_D(\gamma, \psi(t,x)) = \gamma \sum_{y \in N(x)} \sqrt{\rho(t,x)}(S(t,x) - S(t,y))$$

as the nonlinear term. It is easy to check that the term K'_D fulfills the requirements of Ref. 25. An example of the behaviour of the system in the simple case of a parabolic potential in a regular 2-dim lattice is provided in Fig. 4. Moreover, by taking the limit to the continuum, one sees that $K'_D(t,x)$ becomes proportional to $d\rho(t,x)/dt$, very close to the frictional term proposed by Davidson in Ref. 9.

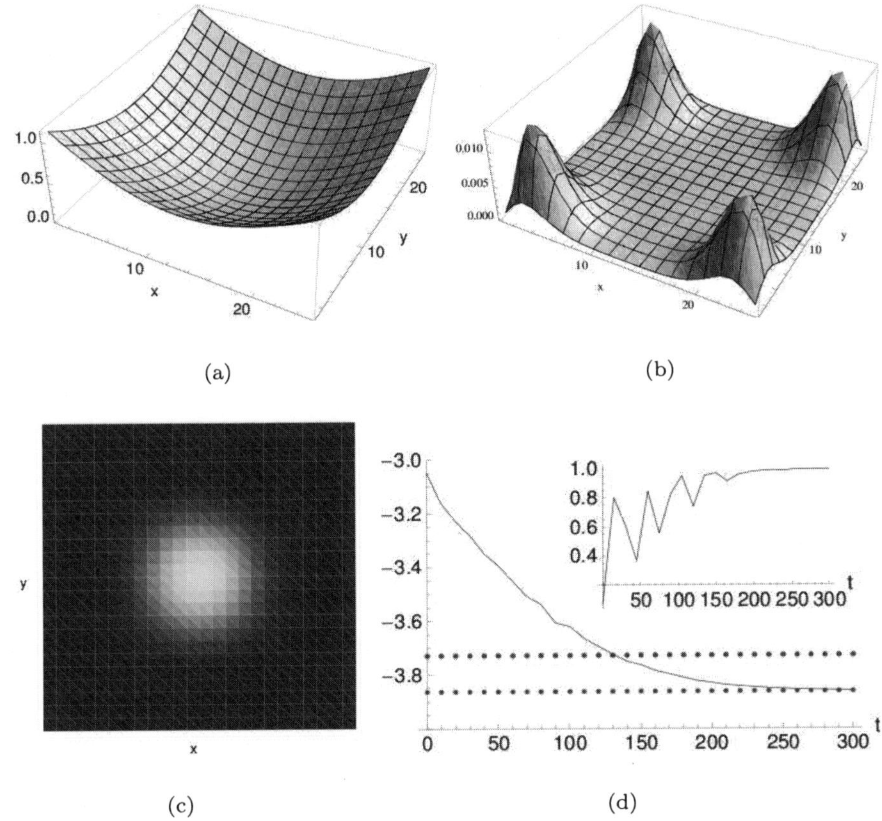

Fig. 4. (Color online) 2-dimensional square regular lattice; $\gamma = 2.0$. (a) The parabolic 2-dimensional potential. (b) Probability distribution $|\psi_0|^2$ of the initial condition. (c) Probability distribution associated to the $|\psi(t)|^2$, $t = 300$. (d) The average energy as a function of time; in the inset the projection, as a function of time, of $\psi(t)$ on the ground state of the Hamiltonian.

5. Conclusions and Outlook: Stabilized quantum annealing

Finding the ground state of an Hamiltonian is a computationally difficult task. The introduction of dissipation does not represent a solution to the problem but another approximating strategy which, in some case, works faster than adiabatic annealing. Moreover it represents an alternative to the unphysical step of *imaginary* time evolution, which, too, performs better than *real* time adiabatic dynamics. However there are critical points in the application of friction to quantum systems both from the physical and numerical points of view. For example, if we set to $\gamma = 0.2$ the friction constant in the asymmetric double well example of section 4.1, the dynamical convergence toward the ground state still happens, but the relaxation time increases exponentially. This would agree with Caldeira-Legget results[8] and represents a direction of future research. By setting $\gamma < 0.1$, instead, the numerical procedures used in this paper (essentially an adaptive fourth-order Runge-Kutta method) become wildly unstable. There is some hint addressing nonlinearity as the source of the problem but, for the time being, there is no analytical stability results.

References

1. Albeverio S., Hoeg-Krohn R. and Streit L., *Energy forms, Hamiltonians and distorted Brownian paths*, J. Math. Phys. 18 (1977), 907–917.
2. Amara P., Hsu D. and Straub J., *Global minimum searches using an approximate solution of the imaginary time Schrödinger equation*, J. Chem. Phys. 97 (1993), 6715–6721.
3. Anderson J., *A random-walk simulation of the Schrödinger equation: H_3^+* J. Chem. Phys. 63 (1975), 1499–1503.
4. Anderson P. W., *Absence of diffusion in certain random lattices*, Phys. Rev. 109 (1958), 1492–1505.
5. Apolloni B., Carvalho C. and de Falco D., *Quantum stochastic optimization*, Stoc. Proc. and Appl. 33 (1989), 223–244.
6. Apolloni B., Cesa-Bianchi N. and de Falco D., *A numerical implementation of Quantum Annealing* In: Albeverio et al. (Eds.), Stochastic Processes, Physics and Geometry, Proceedings of the Ascona-Locarno Conference, 4-9 July 1988, World Scientific, (1990), 97–111.
7. Bloch F., *Über die Quantenmechanik der Elektronen in Kristallgittern*, Z. Phys. 52 (1929), 555–600.
8. Caldeira A. O. and Leggett A. J., *Quantum tunnelling in a dissipative system*,

Ann. Phys. 149 (1983), 374-387.

9. Davidson A., *Damping in Schrödinger equation for macroscopic variables.* Phys. Rev. A 41 (1990), 3395–3398.
10. de Falco D. and Tamascelli D., *Dynamical kickback and non commuting impurities in a spin chain,* Int. J. Quant. Inf. 6 (2008), 807–813.
11. de Falco D. and Tamascelli D., *Quantum walks: a Markovian perspective,* In: V. Geffert et al. (Eds.) SOFSEM 08, LNCS 4910, Springer (2008), 519–530.
12. de Falco D. and Tamascelli D., *Quantum annealing and the Schrödinger-Langevin-Kostin equation,* Phys. Rev. A 79 (2009).
13. Farhi E. *et al., Quantum computation by adiabatic evolution,* arXiv:quant-ph/0001106v1, 2000.
14. Farhi E. *et al., A quantum adiabatic evolution algorithm applied to random instances of an NP-Complete problem,* Science 292 (2001).
15. Finnila A. et al., *Quantum annealing: a new method for minimizing multidimensional functions,* Chem. Phys. Lett. 219 (1994), 343–348.
16. Aharonov D. *et al., The power of quantum systems on a line,* Comm. Math. Phys. 287 (2009), 41–65.
17. Piela L. *et al., The multiple-minima problem of the conformational analysis of molecules. Deformation of the potential energy hypersurface by the diffusion equation method,* J. Chem, Phys. 93 (1989), 3339–3346.
18. Feynman R. P., *Simulating physics with computers,* Int. J. Theor. Phys. 21 (1982), 467–488.
19. Ford G. W., Kac M. and Mazur P., *Statistical mechanics of assemblies of coupled oscillators,* J. Math. Phys. 6 (1965), 504–515.
20. Gregor T. and Car R., *Minimization of the potential energy surface of Lennard-Jones clusters by quantum optimization,* Chem. Rev. Lett. 412 (2005), 125–130.

21. Grover L., *A fast quantum-mechanical algorithm for database search*, In: Proc. 28th Annual ACM Symposium on the Theory of Computing. New York: ACM, 1996.

22. Jona-Lasinio G., Martinelli F. and Scoppola E., *New approach to the semi-classical limit of quantum mechanics. Multiple tunnelling in one dimension*, Comm. Math. Phys. 80 (1981), 223–254.

23. Kirkpatrik S., Gelatt Jr G. D. and Vecchi M. P., *Optimization by simulated annealing*, Science 220 (1983), 671–680.

24. Kostin M. D., *On the Scrödinger-Langevin equation*, J. Chem. Phys. 57 (1972), 3589–3591.

25. Kostin M. D., *Friction and dissipative phenomena in quantum mechanics*, J. Stat. Phys. 12 (1975), 145–151.

26. Nelson E., *Quantum fluctuation.* Princeton Series in Physics. Princeton University Press, 1985.

27. Papadimitriou C. H. and Steiglitz K., *Combinatorial optimization: algorithms and complexity.* New York: Dover, 1998.

28. Eleuterio S. and Vilela Mendes S, *Stochastic ground-state processes*, Phys. Rev. B 50 (1994), 5035–5040.

29. Santoro G. E. and Tosatti E., *Optimization using quantum mechanics: quantum annealing through adiabatic evolution*, J. Phys. A: Math. Gen. 39 (2006), 393–431.

30. Santoro G. E. and Tosatti E., *Optimization using quantum mechanics: quantum annealing through adiabatic evolution.* J. Phys. A: Math. Theor. 41 (2001), 209801.

31. Schuch D., *Effective description of the dissipative interaction between simple model systems and their behaviour*, Int.J. Quant. Chem. 72 (1999), 537–547.

32. Shor P. W., *Polynomial-time algorithms for prime factorization and discrete*

logarithms on a quantum computer, SIAM J. Sci. Statist. Comput. 26 (1997), 14-84.

33. Stella L.,Santoro G. and Tosatti E., *Optimization by quantum annealing: Lessons from simple cases*, Phys. Rev. B 72 (2005).
34. Zener C., *A theory of the electrical breakdown of solid dielectrics*, Proc. R. Soc. Lond. A 145 (1934),523–529.

AUTHOR INDEX